Geometric Modeling: Techniques, Applications, Systems and Tools

Geometric Modeling: Techniques, Applications, Systems and Tools

edited by

Muhammad Sarfraz

*King Fahd University of Petroleum & Minerals,
Saudi Arabia*

KLUWER ACADEMIC PUBLISHERS
DORDRECHT / BOSTON / LONDON

Library of Congress Cataloging-in-Publication Data

ISBN 978-90-481-6518-6

Published by Kluwer Academic Publishers,
P.O. Box 17, 3300 AA Dordrecht, The Netherlands.

Sold and distributed in North, Central and South America
by Kluwer Academic Publishers,
101 Philip Drive, Norwell, MA 02061, U.S.A.

In all other countries, sold and distributed
by Kluwer Academic Publishers,
P.O. Box 322, 3300 AH Dordrecht, The Netherlands.

Printed on acid-free paper

Preface

Computer Aided techniques, Applications, Systems and tools for Geometric Modeling are extremely useful in a number of academic and industrial settings. Specifically, Computer Aided Geometric Modeling (CAGM) plays a significant role in the construction of designing and manufacturing of various objects. In addition to its critical importance in the traditional fields of automobile and aircraft manufacturing, shipbuilding, and general product design, more recently, the CAGM methods have also proven to be indispensable in a variety of modern industries, including computer vision, robotics, medical imaging, visualization, and even media.

This book aims to provide a valuable source, which focuses on interdisciplinary methods and affiliate research in the area. It aims to provide the user community with a variety of Geometric Modeling techniques, Applications, systems and tools necessary for various real life problems in the areas such as:

- *Font Design*
- *Medical Visualization*
- *Scientific Data Visualization*
- *Archaeology*
- *Toon Rendering*
- *Virtual Reality*
- *Body Simulation*

It also aims to collect and disseminate information in various disciplines including:

- *Curve and Surface Fitting*
- *Geometric Algorithms*
- *Scientific Visualization*
- *Shape Abstraction and Modeling*
- *Intelligent CAD Systems*
- *Computational Geometry*
- *Solid Modeling*

v

- *Shape Analysis and Description*
- *Industrial Applications*

The major goal of this book is to stimulate views and provide a source where researchers and practitioners can find the latest developments in the field of Geometric Modeling. Due to speedy scientific developments, there is a great deal of thirst among the scientific community worldwide to be equipped with state of the art theory and practice to get their problems solved in diverse areas of various disciplines. Although, a good amount of work has been done by researchers, yet a tremendous interest is increasing everyday due to complicated problems being faced in academia and industry.

This book is planned to have twenty 22 chapters distributed in three sections. These sections are meant for Geometric Modeling issues including:

- *Techniques*
- *Applications*
- *Systems and Tools*

First eight chapters, in Section I, are devoted towards new Geometric Modeling techniques for designing of objects using curves and surfaces. Section II is consisted of seven chapters on Applications related to Geometric Modeling. Rest of the seven chapters, in Section III, relate to Systems and Tools in the area.

The book is useful for researchers, practicing engineers, computer scientists, and many others who seek state of the art techniques, applications, systems and tools for Geometric Modeling. It is also useful for undergraduate senior students as well as graduate students in the areas of Computer Science, Engineering, and Mathematics. I hope this book will provide a useful resource of ideas and techniques for further research in the development and applications of CAGM.

The editor is thankful to the contributors for their valuable efforts towards the completion of this book. A lot of credit is also due to the various experts who reviewed the chapters and provided helpful feed back. The editor is happy to acknowledge the support of King Fahd University of Petroleum and Minerals towards the compilation of this book. This book editing project has been funded by King Fahd

University of Petroleum and Minerals under Project # ICS/GEOMETRIC/255.

M. Sarfraz

Contents

Section I: Techniques

Section III: Systems & Tools

Section I

Techniques

Chapter 01

Modeling and Visualization of Complex Geometric Environments

Modeling and Visualization of Complex Geometric Environments

Türker Yılmaz, Uğur Güdükbay, and Varol Akman
Department of Computer Engineering, Bilkent University, 06800 Bilkent, Ankara, TURKEY

Visualization of large geometric environments has always been an exciting project for computer graphics practitioners. Modern graphics workstations allow rendering of millions of polygons per second. Although these systems are impressive, they cannot catch up with the quality demanded by graphics systems used for visualizing complex geometric environments. After all, in such systems the amount of data that need to be processed increases dramatically as well. No matter how much graphics hardware evolves, it looks like practitioners are going to crave for what is impracticable for such hardware to render at interactive frame rates. In this chapter, we present some modeling techniques to overcome the problem of graphics hardware bottleneck in a particular context, viz. visualization of terrains and urban environments.

1 Introduction

In this chapter, we first present approaches towards modeling complex geometric environments comprising terrain height fields and urban scenery. Then, we discuss techniques to reduce the amount of workload in the graphics pipeline, thereby overcoming the graphics hardware bottleneck to some extent. These techniques include back-face culling, which eliminates the polygons that are back-facing from the viewer, and view-frustum culling, which eliminates the primitives outside the view-frustum with respect to a view position (so that they are not processed further in the graphics pipeline), as well as view-dependent refinement to selectively refine different parts of a scene using different simplification criteria. We also briefly mention the refinement criteria used in view-dependent visualizations of terrains. We study issues related with urban visualization and especially concentrate on the occlusion culling process by giving a taxonomy of methods in this area. Finally, we discuss techniques to speed-up stereoscopic visualization where the

4

second eye image is generated from the first eye image (in contrast to generating them separately). As we have already noted, our discussion takes place in the general context of terrain and urban visualization.

2 Modeling

The word *modeling* usually refers to the way data are represented in the computer memory and the way in which they are visualized. In the memory, data are kept in suitable data structures that are easy to access for the visualization algorithm. Here, we classify the types of structures into two, namely, *terrains* and *urban scenery*.

2.1 Terrain Representations

Terrains represent one of the most complex data sets in computer graphics because of their nonstandard nature. There is no simple mathematical characterization of terrains, and hence procedural methods for their representation cannot be applied. The data acquired can be stored and used as height fields, triangulated irregular network models, or quadtrees.

However, during the visualization of the terrains, it is cumbersome (and also unnecessary) to display triangles having all elevation points as their vertices. Therefore, the surface has to be approximated (while introducing some, preferably small, amount of error).

2.2 Height Fields

Terrain data are usually obtained by national imagery institutions or geo-science centers. One of the most common ways of acquiring terrain data is aerial photography and satellite imagery. Such data are usually in the form of grids collected at standard intervals, which we call the Digital Elevation Model (DEM). In DEM, the terrain data are not processed and the elevation values are acquired at regular intervals. The collection of elevation samples corresponds to *height fields*.

Height fields are elevation data sampled at regular intervals. The data acquired for terrains are stored in the DEM as height field representations. Naturally, this type of approach requires large amounts of storage because all elevation information are preserved regardless of the characteristics of the terrain surface. For example, the Digital Terrain Elevation Data (DTED) format, developed by NIMA (National Imagery and Mapping Agency of the U.S.), has two standard levels of data resolution (Fig. 1). One is DTED Level-1 in which there are three arc-seconds between two elevation points and the other is DTED Level-2 in which there is one arc-second between two elevation points (meaning higher resolution). Other types of data storage methods for terrain height fields such as gray-scale or vector format are also possible.

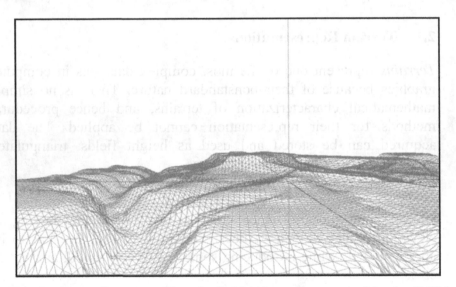

Fig. 1. *A sample view of height field information stored in the DEM, such as the DTED.*

2.2.1 Triangulated Irregular Networks

For representing the terrain, an efficient alternative to dense grids is the Triangulated Irregular Network (TIN). This stores a surface as a set of non-overlapping contiguous triangular facets of irregular size and shape [18]. The source of digital terrain data is dense raster models produced by automated orthophoto machines or direct sensors such as synthetic aperture radar.

A terrain surface can be characterized by a set of surface-specific points (peaks, pits, and passes), and a set of lines (ridges and channels) which connect them. The sample points in the TIN are chosen so that these features are contained as subgraphs of the model [18].

The surface is modeled as a set of contiguous non-overlapping triangles whose vertices are located adaptively on the terrain. The height field data are simplified using simplification algorithms and the resulting model is triangulated using the Delaunay criterion, finally yielding a TIN.

The TIN model is especially attractive because of its simplicity and memory efficiency. It is a significant alternative to the regular raster of the GRID model (Fig. 2). TINs can describe a surface at different resolutions.

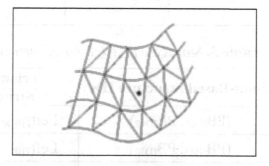

Fig. 2. *TIN generation.*

There are three ways of storing a triangulated network [34]:

- triangle-based structure (Table 1),
- point-based structure (Table 2), and
- edge-based structure (Table 3).

The first method is better for storing attributes (e.g., slope) for each triangle, but uses more storage space. The second one is better for generating contours and uses less storage, but attributes such as slope must be calculated and stored separately. The third one is an additional structure that must be maintained. Based on edge definitions, it provides neighboring information. (Here, it is necessary to store the previous two structures.) If an application needs this information, this structure is suitable.

Table 1. *Storage with triangle-based structure.*

ID	Triangle Vertex Coordinates	Neighbors
0	$(x_{01}, y_{01}, z_{01}), (x_{02}, y_{02}, z_{02}), (x_{03}, y_{03}, z_{03})$	8, 9, 1
1	$(x_{11}, y_{11}, z_{11}), (x_{12}, y_{12}, z_{12}), (x_{13}, y_{13}, z_{13})$	0, 12, 15
⋮	⋮	⋮

Table 2. *Storage with point-based structure.*

ID	Vertex Coordinate List of All Terrain	Triangle-Based Structure IDs
0	(x_0, y_0, z_0)	8, 9, 1, 7, 12
1	(x_1, y_1, z_1)	3, 4, 6, 13
⋮	⋮	⋮

Table 3. *Storage with edge-based structure.*

ID	Point-Based Structure IDs	Triangle-Based Structure IDs
0	(PB_{ID01}, PB_{ID02})	$Left_{TBID01}, Right_{TBID02}$
1	(PB_{ID11}, PB_{ID12})	$Left_{TBID11}, Right_{TBID12}$
⋮	⋮	⋮

Contour lines are one of the terrain features, representing the relief of the terrain with the same height. TINs can also be generated using elevation points along these contour lines and the interpolation is straightforward. However, TINs in island-like situations, as shown in Fig. 3, are special cases that warrant attention.

In the following section, we will describe another method for creating terrain surfaces. This method is also adaptive but at the expense of more storage. In return, it supports faster queries on the terrain structure.

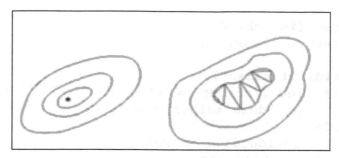

Fig. 3. *TINs in an island-like situations should be handled with care.*

2.2.2 Quadtree Representation

A quadtree is a rooted tree with internal nodes having four children. Each node is represented by at least four grid elevations, which correspond to squares. Constructing a quadtree for terrain data stored in a DTED file with grid elevations produces a dense representation of the terrain, i.e., all interval elevations are stored. The root node represents the whole terrain data with four corners. As we go down to the deeper levels in the quadtree hierarchy, the distance between the corners is halved and at the deepest level, there remains no other elevation point that is to be represented by the nodes of the quadtree (Fig. 4). The data structure obtained after the quadtree scheme applied is passed to a simplification algorithm. The model is simplified to eliminate the nodes having the same elevations within all children (Fig. 5). Further details of quadtrees and related hierarchical structures can be found in [39].

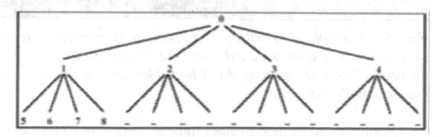

Fig. 4. *Quadtree structure.*

The quadtree structure can be represented as a 1D array (Fig. 4). In this tree, each level represents a different level-of-detail on the terrain. In order to traverse the nodes of the quadtree, represented as a 1D array, the following algorithms can be used:

```
parent(int child)
  return floor((child-1)/4);

child(int parent)
  if (level(parent)==MAXLEVEL)
      childnode=sibling(parent);
  else
      childnode=(4*parent)+1;
  return childnode;

sibling(int node)
  if (level(node)<level(node+1))
      return NIL;
  else
      if (parent(node)!=parent(node+1))
          return sibling(parent(node));
      else
          return (node+1);
```

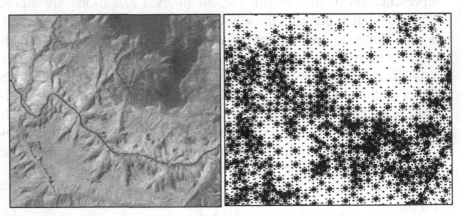

Fig. 5. *A simplifed quadtree representing a particular terrain (Grand Canyon data obtained from the United States Geological Survey (USGS). Processing by Chad McCabe of the Microsoft Geography Product Unit.).*

Quadtrees can be used to store indices, minimum and maximum elevations, and activation distances for the vertices, which are valuable for view-dependent refinement in terrain visualization. It is clear that an array-based representation can be used to eliminate the need for pointer manipulation. The numbering scheme used by the quadtree structure when it is stored in a one-dimensional array is illustrated in Fig. 6. The root is labeled as 0 and the other nodes are numbered recursively in the counterclockwise direction.

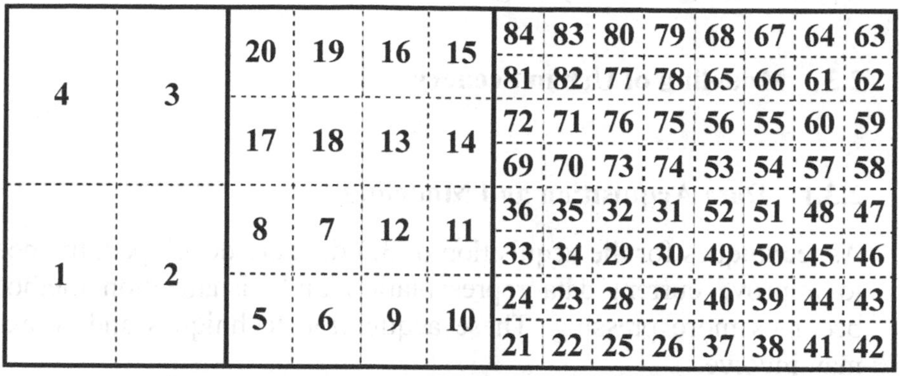

		20	19	16	15	84	83	80	79	68	67	64	63
4	3					81	82	77	78	65	66	61	62
		17	18	13	14	72	71	76	75	56	55	60	59
						69	70	73	74	53	54	57	58
		8	7	12	11	36	35	32	31	52	51	48	47
						33	34	29	30	49	50	45	46
1	2	5	6	9	10	24	23	28	27	40	39	44	43
						21	22	25	26	37	38	41	42

Fig. 6. *Numbering scheme for quad blocks in a quadtree when it is stored in a 1D array with levels 2, 3, and 4.*

2.2.3 Multi-resolution Representation using Quadtrees

Multi-resolution representation of data refers to those data structures, which provide a way to visualize data in different resolutions, depending on some criterion. In this section, we describe multi-resolution representation using quadtrees. For other methods of multi-resolution representation such as progressive meshes, the reader is referred to [26].

In order to visualize complex scenes (e.g., terrain height fields) at interactive frame rates, efficient data structures need to be used. The quadtree representation perfectly fits into grid elevation data. Generally, triangles are used as modeling primitives for complex scenes. The triangulation must be adaptive in order to reduce the number of polygons that should be processed and make efficient use of the limited memory sources. This means that high frequency elevation changes should be triangulated with more triangles than low frequency regions. While doing this, artifacts that can emerge on the terrain should be minimized as much as possible.

For multi-resolution representations, under the assumption that the structure in Fig. 4 is used, each level represents a different resolution on the terrain. The algorithms developed for multi-resolution representation generally make use of *view-dependent visualization*. This means the usage of some simplification criterion and traversing the nodes of the multi-resolution terrain representation hierarchy (Fig. 4) until the criterion is met.

2.3 Modeling of Urban Scenery

2.3.1 Data Acquisition and Modeling

As techniques for the acquisition of 3D data are developed, the need to improve current data representation and visualization methods becomes more pressing. These acquisition techniques and sources may involve:

- aerial (satellite) imagery of earth in high resolution,
- urban data automatically gathered by devices using laser range finding methods, and
- 3D views constructed with the help of radar systems.

Actually, data acquisition for terrains is less expensive than data acquisition for urban areas. For urban scenery creation, there are mainly three sources of information.

1. Constructing models from building footprints:

One of the main sources of information is building footprints, mostly available at official institutions of cartography. The buildings may easily be extruded from these footprints using the additional information stored, i.e., the number of floors or the construction type. However, the resultant geometry is a crude version, omitting many details such as balconies, pillars, etc. To make visualization more realistic, texturing may be applied. Moreover, there are techniques to automatically add selected features to the buildings, like balconies or windows [47]; see Fig. 7 for an illustration.

2. Constructing building models individually, and later populating them in a virtual environment:

In this technique, each building model is highly detailed and the resulting appearance is highly realistic. A sample application of this method is shown in Fig 8, where we generated a virtual city composed of nearly 300 buildings with more than 400K polygons.

Fig. 7. *A scene from [47], where buildings are extruded and additional features are added to the building geometry (© Peter Wonka. Reprinted with permission.).*

Fig. 8. *A virtual city that is a combination of individually modeled buildings.*

3. *(Semi)automatic reconstruction of buildings from aerial photographs:*

There are many ways of collecting urban information from aerial photographs. These images are procedurally corrected and methods are applied to obtain 3D data of an urban area. These data are later geo-corrected and other information is appended to them. To give an example [36], from various image maps given as input (e.g., land-water boundaries and population density), a system generates a network of highways and streets, divides the land into lots, and creates the appropriate geometry for buildings on the respective allotments (Fig. 9).

Fig. 9. *A (procedurally modeled) virtual city from [36]. (© Association for Computing Machinery. Reprinted with permission.)*

2.3.2 Building Representations

Naturally, buildings and terrain data are stored in spatial databases. The Database Management Systems (DBMSs) for spatial databases should manage access and retrieval of data to be sent to the graphics pipeline. There are mainly three types of building representations: *primitive-geometry*, *component-based*, and *space-based* [42].

Primitive-geometry representation: Most architectural drawings can be regarded as primitive-geometry representations. This kind of representation is based on geometric primitives, which give no explicit indication of what building entities they stand for. For this representation, storage space that can be used for the spatial database is crucial; the representation detail and the features that could be incorporated into the database require excessive amount of storage.

Component-based representation: We can cite CAD systems in this category. Most commercial design systems support explicit definitions for 3D building entities such as walls, windows, doors, floor slabs, and roofs. Representing buildings using this method allows designers to create and modify a single model rather than several (computationally unrelated) floor plans. In the representation scheme, the spatial locations of the components are defined. More importantly, entities can be accessed based on their respective locations in the scene. In component-based systems, the buildings may be assigned additional information, such as floor count, occupied area, type of construction, energy type used for heating, purpose of use, etc. The components of the building are stored and relations among them are defined. When there is a need to retrieve a component, the relationships among the currently displayed components can be used.

Space-based representation: This is used to define the spaces used by the components of buildings. Instead of obtaining spaces used by a component by the spatial locations of other components, the space is enclosed by another polygon and this polygon is used for space determination. The designer can achieve this by defining the polygons that enclose a space. In the former case, when the polygon is closed, floor and ceiling elements can be automatically generated. Thus, a space is always ensured to be a polyhedron.

3   Visualization

Visualization is defined as the transformation of the symbolic data into a geometric form to enable researchers to observe their simulations and computations. It can be used both for interpreting image data fed into a computer (=image understanding) and for generating images from complex multidimensional data sets (=image synthesis) [46]. Here, we use the word "visualization" to mean the generation of images of complex 3D scenes for different camera positions while moving the camera interactively (i.e., rendering complex scenes). If interaction is required, then the frame rate of the drawn scene should be more than 17 frames per second. If the frame rates are below this, then it means the system has a bottleneck either in the graphics pipeline, or in the display process. If the graphics load is not adjusted, then the user might experience problems during visualization such as jaggy motion. Taking the graphics hardware as a constant, the only approach that can overcome this bottleneck and achieve interactive frame rates is to adjust the load of the graphics pipeline by using suitable algorithms (i.e., decreasing the number of triangles processed for each frame using a predefined simplification criterion). These algorithms work according to the viewer position.

3.1 Culling Techniques

Before applying a view-dependent refinement process to reduce the number of triangles to be rendered, the portions of the scene that will not be displayed for a frame should be culled. In order to send only the related portions of the scene to the display processor, there are mainly three types of culling methods to get rid of the irrelevant portions of the geometry. The first one is *back-face culling*, discarding those polygons whose surface normals are facing away from the viewer. (This works only for convex objects.) The second one is *view-frustum culling*, discarding those objects that are out of the field of view. The last one is *occlusion culling*; this eliminates parts that are occluded by front objects and it is especially important for visualization of urban scenery where the buildings are occluding each other for different views. Back-face culling is explained below. View-frustum culling for terrain data and occlusion culling for urban scenery are discussed later in related subsections. In the case of *occlusion culling* for terrains, it is known that this culling method does not increase the performance significantly [28].

3.1.1 Back-Face Culling

Back-face culling is the process of discarding back-facing polygons. It is not possible to see them because they are not facing the viewer. However, back-face culling works only if:

- there are no holes or transparent passages in the objects, and
- the objects are convex.

Back-face culling is performed by evaluating the equations of the planes that form the object surface, i.e., triangles with respect to the viewpoint and viewing direction. Back-facing polygons are eliminated if the dot product of the viewing direction and polygon normal is greater than zero. Back-face culling is implemented in hardware in many graphics boards. In [30], some improvements are proposed and are compared to the hardware implementation. A sub-linear algorithm for back-face computation is presented. The polygonal model is partitioned into a hierarchy of clusters based on the similarity of orientation and physical proximity of polygons. The space is partitioned into back, front, and mixed regions with respect to each cluster. At run time, the algorithm uses the pre-computed cluster descriptions to locate the viewpoint in the corresponding region of each cluster.

3.2 Visualization of Terrain Data

Without decreasing the amount of geometry sent to the graphics pipeline, the quality and amount of the graphics primitives that can be viewed at interactive frame rates will be limited and insufficient. Therefore, the surface of a terrain has to be approximated up to a certain threshold, in order to decrease the number of triangles sent to the graphics pipeline without significant loss of image quality. While carrying out this process, the simplification part should cost significantly less than sending all graphics to the hardware and making the graphics pipeline do all the work.

3.2.1 View-Frustum Culling for Terrain Data

The view-frustum is a pyramid (generally chopped off using front and rear planes perpendicular to the viewing direction) and is based on the viewing parameters of the application. View-frustum culling process culls the parts of the view-frustum according to the six planes of the frustum.

A sample view-frustum over a terrain can be seen in Fig. 10. View-frustum culling for terrain data can be done as follows. The quadtree is traced from top to bottom and it is determined whether the nodes are viewable from the current viewing direction. Nodes of the quadtree are visited in preorder so that if a higher-level node is not in the frustum, then the children of that node are not further checked.

An efficient view-frustum culling (VFC) process is crucial to achieve interactive frame rates. The data structure is checked against the viewing frustum. To speed-up frustum culling process, frustum tests can be done using bounding spheres enclosing the nodes. In view-frustum culling, several optimizations are possible [24]:

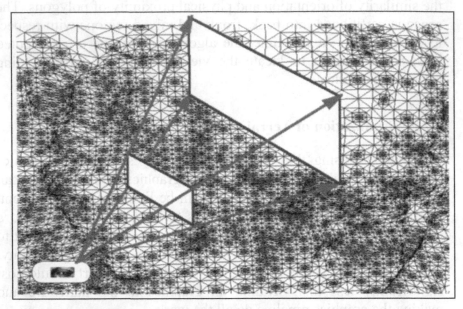

Fig. 10. *A sample view-frustum on terrain data.*

- One important optimization is to utilize the coherence between two frames when the user navigates through the terrain. If the user moves forward, then there is no need to cull the whole data again since the data have already been culled in the previous frame. So, previously culled blocks can be used for the current frame. This method is applicable if the frustum is not culled according to the far plane [2].

- Another method is *deferred VFC*, implying that VFC is not done for every frame but at predefined intervals. In this way, the overhead brought by the VFC step can be decreased. One problem with this approach is the navigation speed. If the user moves very fast involving rotation and backward motion, then the screen may not refresh on time. Accordingly, this approach is suitable for slow motion walkthroughs.

- As another approach, VFC depending on the deviation of the viewer location may be used. Deviation based culling is suitable for walkthroughs in which the viewer navigates very fast. The VFC algorithm can be run only if the user moves a prespecified distance from the previously culled position.

If the data are large, then we have to test for the far plane too. In this case, an altitude-based scheme can be used for far plane distance determination. If the altitude of the viewer is at lower levels in the data, the far plane is brought closer to the viewer, proportional to the altitude of the viewer because it may not be possible to see farther distances. This approach establishes a balance between the frustum distance and the data resolution, depending on the type of the data. For view-frustum culling in urban sceneries, the quadtree node explained above could be replaced with objects in the urban scene, where the application of it is straightforward.

3.2.2 View-Dependent Visualization using Quadtrees

During the visualization process, there is no reason to send the parts of the scene that cannot be seen from the viewer's position and viewing direction. The fundamental idea in view-dependent rendering is to perform culling of these unnecessary data and reduce the workload of graphics hardware. However, whenever view-dependent rendering is mentioned, usually only multi-resolution representations come to mind. A multi-resolution representation deals with simplifying parts of the scene, when the detailed view is not needed. View-dependent rendering covers all visibility algorithms, which try to speed up the rendering performance, because all culling algorithms are based on the viewer's position and viewing direction.

View-dependent rendering is mostly performed on-the-fly, in order to reduce the cost of secondary storage and provide a more realistic view of the scene. An alternative to dynamic view dependent visualization, where the scene is simplified on-the-fly based on the current view, using precomputed Level-of-Detail (LOD) samples during a fly-through is also commonplace. Simplification algorithms are applied to obtain a hierarchy of successively coarser approximations for the objects. Such multi-resolution hierarchies have been used in LOD based rendering schemes to achieve better frame rates per second [20, 33]. These hierarchies usually have a number of distinct levels of detail, usually five to 10 for a given object or a part of terrain [49]. During the visualization, if certain error criteria are satisfied, then one of the static detail levels is chosen and the object is displayed at that level.

The trend in surface simplification moved from statically defined levels of detail to dynamically created levels of detail after mid 1990's. Wavelet usage [16] and progressive meshes [26, 27, 28] have advanced simplification work a step further. These methods produce a continuous LOD through the entire scene, instead of a discrete number of levels of detail. Progressive meshes offer an elegant solution for continuous representation of polygonal meshes, which can be adapted to almost any kind of scene objects. In [49], the authors describe improvements on progressive meshes, by defining merge trees for performing edge collapses that permit adaptive refinement around any vertex. Progressive mesh usage is also adapted to regular grid approaches [32].

With quadtree representation, the grid structure of the elevation data is explored. The grid structure naturally lends itself to the quadtree representation. Each sequence of nine vertices comprehends to a quad block as in Fig. 11. These elevation points are selected in such a manner that no cracks occur during the simplification process and the elevation differences without the vertices do not disturb the quality of the resultant scene much – in any case, not more than a prespecified threshold value. The threshold values are defined in view of the simplification criterion employed by the simplification algorithm.

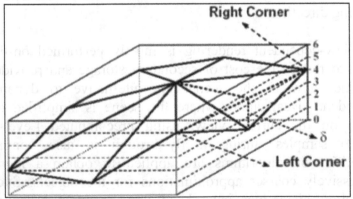

Fig. 11.   *The distance* δ *between original and removed positions of the tested vertex.*

There are several methods and criteria to simplify the data. The criterion for the evaluation of removal depends on several factors. One of these methods is *screen-space error criterion*. In this method, elevation differences are taken into account to evaluate a vertex for removal (Fig. 12). The number of projected pixels for the vertex is calculated for this purpose (Fig. 11). If the projected size of an object is below a prespecified threshold (in terms of pixels), then the vertices and associated triangles within this object are removed. If this size is greater than the pre-specified pixel tolerance, then the vertex is kept.

The problem here is that if the viewer is looking at the terrain from above, then the number of projected pixels to the camera plane will be very small, causing the vertex to be considered as unimportant and yielding to the elimination of it. This especially becomes a problem for stereoscopic visualizations where the preservation of the depth information is crucial. The problem can be illustrated via an example (Fig. 12). Suppose we are looking at a tower from above and we use the screen-space error tolerance. Since the projection of the elevation difference will be very small with respect to the position of the eye, the tested vertices will be removed, although they are important to the viewer (i.e., they will make the viewer see the height of the tower when viewed in stereo). Thus, while the screen-space error metric is suitable for the monoscopic view [32], it degrades the stereo effect and may result in incorrect stereoscopic vision.

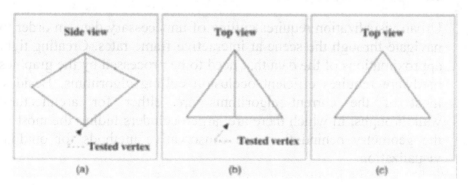

Fig. 12. *Screen-space error metric. (a) Side view of a quad block. (b) Top view of the same block. (c) Edge removal by screen-space error-based algorithm when the block is viewed from the top.*

Instead of using screen space error criterion, another one, called *distance-based angular error metric*, is defined [24]. This simplification criterion is suitable for both monoscopic and stereoscopic visualization, and yields faster algorithms for simplification. The elevation and distance of objects from the viewer are two important criteria that make us appreciate depth and differentiate between objects. Therefore, the threshold value must be specified adaptively so that it takes into account both of these parameters to reflect the correct depth information. Details of distance-based angular error threshold can be found in [24].

In order to prevent cracks over the terrain during simplification and provide a suitable heuristic for morphing to eliminate popping artifacts while switching between different resolutions of the terrain, a valid triangulation should be maintained. The vertex activation scheme using distance-based angular error threshold decides to activate or deactivate a vertex using the precomputed vertex activation values, and the distances between the viewer and the vertex location. Then, the area is triangulated accordingly. The vertex activation scheme is based on assigning correct activation distances to each vertex from the lowest level up to highest using a maximization operation and during visualization only measuring the distance between the viewer and the vertex. The details of vertex activation and triangulation can be found in [24].

3.3   Visualization of Urban Scenery

Urban visualization requires culling of unnecessary data, in order to navigate through the scene at interactive frame rates. Creating tight approximations of the data that need to be processed by the graphics hardware requires efficient occlusion culling algorithms. Besides, most of the current algorithms are either for architectural walkthroughs, in which there are large occluders hiding the most of the geometry behind them, or conservative methods for outdoor visualization.

For interactive walkthroughs of large building models or city-like scenes, a system must store in memory and render only a small portion of the model at each frame. The most important challenge is to identify the relevant portions of the model, swap them into memory by using a robust spatial database access, and render at interactive frame rates as the user changes position and viewing direction.

3.3.1 Occlusion Culling with Preprocessing

Occlusion culling algorithms can be classified based on targeted environments. They are generally effective in densely occluded scenes and do not offer much in terrain like scenes [28]. On the other hand, LOD control and impostors contribute mostly in wide-open sparsely occluded situations. The first classification is based on the suitability of the data for occlusion culling. According to this classification, in the first category there are algorithms that are suitable for scenes where much of the geometry is hidden behind potential occluders. In the second category, there are algorithms developed for general scenes. These are so complex that use of some special hardware might be more suitable. Most current occlusion culling algorithms work with occluders much smaller than they could actually handle and often need human intervention for finding effective occluders (or send hidden geometry to graphics pipeline unnecessarily). Therefore, tightness of the estimation of occluded parts is an important issue to be considered.

Most of the visibility-culling algorithms have computationally intensive preprocessing stages [15, 20, 21, 25, 31]. Preprocessing generally includes the computation of some kind of hierarchical data structure to store the scene and finding visible objects and majority of occluders for previously determined view cells. The clustering schemes are carefully chosen so that the algorithms make use of the data structures created during this step. For walkthroughs of outdoor environments, controlled subdivision using data structures like BSP trees or quadtrees is suitable [5, 19, 25, 35]. A 3D spatial partitioning data structure such as a quadtree, octree or kD-tree can be used for general fly-through application [12]. The visibility octree is an adaptive data structure that stores potentially visible sets at its terminal nodes [40, 50]. Unlike uniform grid structures, its size depends on the nature of the scene. Different schemes for occlusion culling are applied after this clustering step has been performed.

The goal of an efficient visibility-culling algorithm is to calculate a conservative and fast elimination of those parts of the scene that are definitely invisible. This means that if an object is visible it is certainly displayed. In the meantime, some unnecessary parts are sent to graphics pipeline. *Object space algorithms* are the ones that geometrically make computations on the scene and decide whether the objects are visible or not [8, 9, 10, 25, 29, 43]. The general approach of the previous work is to select some polygons to act as virtual occluders and check if they occlude any objects seen from the viewer. To reduce the cost of checking, occludees are usually approximated by bounding volumes. There is a conservative visibility preprocessing method for outdoor environments, which is described in [8]. A conservative superset of visible objects is computed for each cell by searching for a strong occluder for each object such that it cannot be seen from any point in the cell. The cellulization approach applied in this work is regular grid approach, which may not fit to real outdoor environments. The cellulization approach is best suited for architectural walkthroughs. In case of outdoor visualizations, a suitable cellulization is needed, which is not straightforward. Still, after suitable cellulization of the navigable area, most of the approaches used to visualize the indoor environments can also be used for outdoor environments.

Mostly, the target data for occlusion culling algorithms influence the way algorithms are designed. If an urban walkthrough restricts the viewer to move along the road paths, then cells should be associated with the roads only. For building interiors or ship-like scenes, most visibility algorithms decompose the model into cells [8, 20, 21, 43]. An occlusion region can be specified by an object space occlusion-culling algorithm using supporting planes [10]. These cells are connected by portals and the inter-cell visibility is computed. Since the walls of the buildings or doors of ships occlude a large amount of the geometry behind them, precomputing the potentially visible sets (PVSs) and later using this information to cull the invisible objects may be a promising approach [17, 20, 21, 43, 44]. This has the disadvantage of requiring large secondary storage for the PVS information. There are some algorithms to compress the data constituting the PVSs [3, 37, 38].

Conservative algorithms classify regions as invisible when they are completely occluded. Partially occluded objects are sent to the graphics pipeline as a whole. For urban environments that have less hidden geometry behind the objects, occlusion culling with a few large occluders is a popular approach. The navigable area is again subdivided into cells in many approaches and for each frame, a small set of (about five to 30) occluders that are likely to occlude a large part of the model is selected. The reason why these algorithms select only a small set of occluders is that the amount of data needed to store the potentially visible set for each cell is large. The selection schemes differ among the algorithms with respect to errors introduced into the resultant image, accurateness of the selection, tightness of the conservativeness, and the data that are needed to be stored with this potentially visible set [1, 14, 29].

Under the aforementioned classification, object space methods can be regarded as output sensitive algorithms. Output sensitive algorithms have their runtime depend only on the size of the output, and not the input. In the case of *image based algorithms*, the main idea for achieving the goal is to perform visibility computations for each frame by scan conversion of some potential occluders and checking if the projections of the other objects fall inside the image area of the projection of the occluders. Examples include the hierarchical z-buffer algorithm [22], hierarchical tiling algorithm [23], hierarchical occlusion maps [51], and others [4, 6, 11, 15, 45, 48]. Some of these classify the scene into both scene data structure and image replaceable parts, namely near and far fields. This sort of occlusion culling is very similar to radiosity calculations [7]. In image-based simplification methods, the whole scene parts are replaced with an *impostor* – a generated image of the scene [13, 41]. Unfortunately, one impostor is usually valid for a few frames and must be updated frequently. Other approaches use textured depth meshes that incorporate depth information for efficient impostor update. One of the advantages of image space algorithms is that the target data can be very complex (for which object space algorithms are at a disadvantage) and the occluded objects are within a very tight estimation range. A common deficiency of image space algorithms is that they are mostly hardware dependent and the screen resolution is fixed.

3.4   Stereoscopic Visualization

In stereoscopic visualization (Fig. 13), the two views must be generated fast enough to achieve interactive frame rates. Apparently, there will be limitations in terms of the features that could be incorporated to increase the realism (in comparison to monoscopic visualizations). Since the amount of data that can be processed decreases drastically, complex visualizations, such as visualization of urban scenery over a terrain, cannot be achieved easily.

There is some work done to decrease the time needed for generating the second eye image so that complex stereoscopic visualizations become possible. For this purpose, an algorithm has been proposed to speed up the generation of stereo pairs for stereoscopic view-dependent visualizations [24]. The algorithm, called Simultaneous Generation of Triangles (SGT), generates the triangles for the left and right eye images simultaneously using a single draw-list, thereby avoiding the need for performing the view frustum culling and the vertex activation operations needed for view-dependent refinement twice.

4 Summary

In this chapter, we presented some approaches toward terrain and urban scenery modeling. We discussed techniques to reduce the amount of workload in the graphics pipeline, thereby overcoming the graphics hardware bottleneck.

In the case of terrains, the most important problem is multi-resolution representation and simplification without significant loss of accuracy. In order to achieve this, an approach using quadtrees has been presented. Urban sceneries are somewhat different from terrains; they are more component based. An important problem is occlusion culling, because most geometry may be discarded without sending it to the graphics pipeline. In order to achieve these goals the algorithms must be adaptive: the algorithm should cost less compared to sending whole geometry to the graphics hardware and letting the machine do all the work and should keep accuracy of the geometry high as much as possible.

There may be other approaches to terrain and urban modeling, not mentioned in this chapter. However, we believe that the ones studied here constitute a useful, promising bunch.

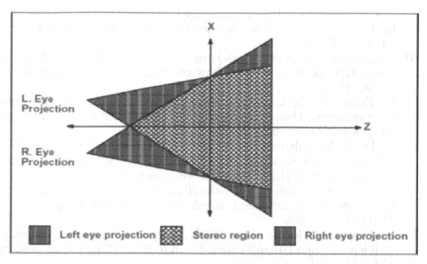

Fig. 13. *Stereoscopic projection with left and right eye frustums.*

References

1. C. Andùjar, C. Saona-Vàzquez, I. Navazo, and P. Brunet. Integrating Occlusion Culling and Levels of Detail Through Hardly-Visible Sets. Computer Graphics Forum, 19(3):499–506, 2000.
2. U. Assarsson and T. Möller, Optimized View Frustum Culling Algorithms for Bounding Boxes, Journal of Graphics Tools, 5(1): 9-22, 2000.
3. C. L. Bajaj, V. Pascucci, and G. Zhuang. Progressive Compression and Transmission of Arbitrary Triangular Meshes. In Proceedings of IEEE Visualization, pages 307–316, 1999.
4. D. Bartz, M. Meißner, and T. Hüttner. OpenGL-Assisted Occlusion Culling for Large Polygonal Models. Computers & Graphics, 23(5):667–679, 1999.
5. J. Bittner, V. Havran, and P. Slavik. Hierarchical Visibility Culling with Occlusion Trees. In Proceedings of Computer Graphics International, pages 207–219, 1998.
6. B. Chen, J. E. Swan, E. Kuo, and A. E. Kaufman. LOD-Sprite Technique for Accelerated Terrain Rendering. In Proceedings of IEEE Visualization, pages 291–298, 1999.
7. Y.-Y. Chuang and M. Ouhyoung. Clustering and Visibility Preprocessing of Hierarchical Radiosity for Object-Based Environment. In Proceedings of Computer Graphics Workshop'95, pages 102–111, 1995.
8. D. Cohen-Or, G. Fibich, D. Halperin, and E. Zadicario. Conservative Visibility and Strong Occlusion for Viewspace Partitioning of Densely Occluded Scenes. Computer Graphics Forum, 17(3):243–254, 1998.

9. S. Coorg and S. Teller. Temporally Coherent Conservative Visibility. In Proceedings of ACM Symposium on Computational Geometry, pages 78–87, 1996.
10. S. Coorg and S. Teller. Real-Time Occlusion Culling for Models with Large Occluders. In Symposium on Interactive 3D Graphics, pages 83–90, 1997.
11. L. Darsa, B. Costa, and A. Varshney. Walkthroughs of Complex Environments Using Image-Based Simplification. Computers & Graphics, 22(1):55–69, 1998.
12. D. Davis, W. Ribarsky, T. Y. Jiang, N. Faust, and S. Ho. Real-Time Visualization of Scalably Large Collections of Heterogeneous Objects. In Proceedings of IEEE Visualization, pages 437–440, 1999.
13. X. Decoret, F. Sillion, G. Schaufler, and J. Dorsey. Multi-Layered Impostors for Accelerated Rendering. Computer Graphics Forum, 18(3):61–73, 1999.
14. L. Downs, T. Möller, and C. H. Sèquin. Occlusion Horizons for Driving Through Urban Scenes. In SIGGRAPH'01 Proceedings, pages 121–124, 2001.
15. F. Durand, G. Drettakis, J. Thollot, and C. Puech. Conservative Visibility Preprocessing Using Extended Projections. In SIGGRAPH'00 Proceedings, pages 239–248, 2000.
16. M. Eck, T. DeRose, T. Duchamp, H. Hoppe, M. Lounsbery, and W. Stuetzle. Multiresolution Analysis of Arbitrary Meshes. In SIGGRAPH'95 Proceedings, pages 173–182, 1995.
17. C. Erikson, D. Manocha, and W. V. Baxter. HLODs For Faster Display of Large Static and Dynamic Environments. In SIGGRAPH'01 Proceedings, pages 111–120, 2001.
18. R. J. Fowler and J. J. Little. Automatic Extraction of Irregular Network Digital Terrain Models. Computer Graphics, 1979.
19. H. Fuchs. On Visible Surface Generation by a Priori Tree Structures. In SIGGRAPH'80 Proceedings, pages 124–133, 1980.
20. T. A. Funkhouser, C. H. Sèquin, and S. J. Teller. Management of Large Amounts of Data in Interactive Building Walkthroughs. In Proceedings of Symposium on Interactive 3D Graphics, pages 11–20, 1992.
21. C. Gotsman, O. Sudarsky, and J. A. Fayman. Optimized Occlusion Culling Using Five-Dimensional Subdivision. Computers & Graphics, 23(5):645–654, 1999.
22. N. Greene. Hierarchical Z-buffer Visibility. In SIGGRAPH'93 Proceedings, pages 231–238, 1993.
23. N. Greene. Efficient Occlusion Culling for Z-Buffer Systems. In SIGGRAPH'99 Proceedings, pages 78–79, 1999.
24. U. Güdükbay and T. Yılmaz. Stereoscopic View-dependent Visualization of Terrain Height Fields. IEEE Transactions on Visualization and Computer Graphics, 8(4):330–345, 2002.
25. J. Heo, J. Kim, and K. Wohn. Conservative Visibility Preprocessing for Walkthroughs of Complex Urban Scenes. In Proceedings of ACM Symposium on Virtual Reality Software and Technology, pages 115–128, 2000.
26. H. Hoppe. Progressive Meshes. In SIGGRAPH'96 Proceedings, pages 99–108, 1996.

27. H. Hoppe. Efficient Implementation of Progressive Meshes. Computers & Graphics, 22(1):27–36, 1998.

28. H. Hoppe. Smooth View-Dependent Level-of-Detail Control and its Application to Terrain Rendering. In Proceedings of IEEE Visualization, pages 35–42, 1998.

29. J. T. Klosowski and C. T. Silva. Efficient Conservative Visibility Culling Using the Prioritized-Layered Projection Algorithm. IEEE Transactions on Visualization and Computer Graphics, 7(4):365–379, 2001.

30. S. Kumar, D. Manocha, W. Garrett, and M. Lin. Hierarchical Back-Face Computation. Computers & Graphics, 23(5):681–692, 1999.

31. F. A. Law and T. S. Tan. Preprocessing Occlusion for Real-Time Selective Refinement. In Proceedings of Symposium on Interactive 3D Graphics, pages 47–54, 1999.

32. P. Lindstrom, D. Koller, W. Ribarsky, L. F. Hodges, N. Faust, and G. Turner. Real-Time Continuous Level of Detail Rendering of Height Fields. In SIGGRAPH'96 Proceedings, pages 109–118, 1996.

33. P. W. C. Maciel and P. Shirley. Visual Navigation of Large Environments Using Textured Clusters. In Proceedings of Symposium on Interactive 3D Graphics, pages 95–102, 1995.

34. S. Murai. GIS Workbook CD-ROM Version 1.0, Japan Association of Remote Sensing. 1999.

35. B. Naylor. Partitioning Tree Image Representation and Generation from 3D Geometric Models. In Proceedings of Graphics Interface, pages 201–212, 1992.

36. Y. I. H. Parish and P. Müller. Procedural Modeling of Cities. In SIGGRAPH'01 Proceedings, pages 301–308, 2001.

37. J. Popovic and H. Hoppe. Progressive Simplicial Complexes. In SIGGRAPH'97 Proceedings, pages 217–224, 1997.

38. J. Rossignac. Geometric Simplification and Compression in Multiresolution Surface Modeling. SIGGRAPH'97 Course Notes #25, 1997.

39. H. Samet. The Quadtree and Related Data Structures. ACM Computing Surveys, 16(2):187–260, 1984.

40. C. Saona-Vàzquez, I. Navazo, and P. Brunet. The Visibility Octree: A Data Structure for 3D Navigation. Computers & Graphics, 23(5):635–643, 1999.

41. F. Sillion, G. Drettakis, and B. Bodelet. Efficient Impostor Manipulation for Real-Time Visualization of Urban Scenery. In Proceedings of Eurographics'97, pages 207–218, 1997.

42. G. Suter and A. Mahdavi. Performance-Inspired Building Representations for Computational Design. In Proceedings of Building Simulation, Vol.3, pages 1203–1210, 1999.

43. S. J. Teller and C. H. Sequin. Visibility Preprocessing for Interactive Walkthroughs. In SIGGRAPH'91 Proceedings, pages 61–69, 1991.

44. S. Teller. Visibility Computations in Densely Occluded Environments. Ph.D. thesis, University of California, Berkeley, 1992.

45. M. Wand, M. Fischer, I. Peter, F. M. auf der Heide, and W. Straßer. The Randomized Z–Buffer Algorithm: Interactive Rendering of Highly Complex Scenes. In SIGGRAPH'01 Proceedings, pages 361–370, 2001.

46. A.Watt and M. Watt, Advanced Animation and Rendering Techniques: Theory and Practice, Addison-Wesley, 1997.
47. P. Wonka. Occlusion Culling for Real-Time Rendering of Urban Environments. Ph.D. Thesis, University of Vienna, 2001.
48. P. Wonka, M. Wimmer, and F. X. Sillion. Instant Visibility. In Proceedings of Eurographics, 20(3):411–421, 2001.
49. J. C. Xia, J. El-Sana, and A. Varshney. Adaptive Real-Time Level-of-Detail Based Rendering for Polygonal Models. IEEE Transactions on Visualization and Computer Graphics, 3(2):171–183, 1997.
50. K. Yamaguchi, T. L. Kunii, K. Fujimura, and H. Toriya. Octree-Related Data Structures and Algorithms. IEEE Computer Graphics and Applications, 4(1):53–59, 1984.
51. H. Zhang, D. Manocha, T. Hudson, and K. E. Hoff III. Visibility Culling Using Hierarchical Occlusion Maps. In SIGGRAPH'97 Proceedings, pages 77–88, 1997.

Chapter 02

Adaptive Surfaces Fairing by Geometric Diffusion

Adaptive Surfaces Fairing by Geometric Diffusion

Chandrajit L. Bajaj
Department of Computer Science, University of Texas, Austin, TX 78712, USA

Guoliang Xu
The Institute of Computational Mathematics, Chinese Academy of Sciences, Beijing, China

In triangulated surface meshes, there are often very noticeable size variances (the vertices are distributed unevenly). The presented noise of such surface meshes is therefore composite of vast frequencies. In this chapter, we solve a diffusion partial differential equation numerically for noise removal of arbitrary triangular manifolds using an adaptive time discretization. The proposed approach is simple and is easy to incorporate into any uniform timestep diffusion implementation with significant improvements over evolution results with the uniform timesteps. As an additional alternative to the adaptive discretization in the time direction, we also provide an approach for the choice of an adaptive diffusion tensor in the diffusion equation.

1 Introduction

The solution for triangular surface mesh denoising (fairing) is achieved by solving a partial differential equation (PDE), which is a generalization of the heat equation customized to surfaces. The heat equation has been successfully used in the image processing for about two decades. The literature on this PDE based approach to image processing is large (see [1, 10, 11, 17]). It is well known that the solution of heat equation $\partial_t \rho - \Delta \rho = 0$, based on the Laplacian Δ, at time T for a given initial image ρ_0 is the same as taking a convolution of the Gauss filter $G_\sigma(x) = \dfrac{1}{2\pi\sigma} \exp\left(-\dfrac{|x|^2}{2\sigma^2}\right)$ with standard deviation $\sigma = \sqrt{2T}$ and image ρ_0. Taking the convolution of $G_\sigma(x)$ and image ρ_0 is performing a weighted averaging process to

ρ_0. When the standard deviation σ become larger, the averaging is taken over a larger area. This explains the filtering effect of the heat equation to noisy images. The generalization of the heat equation for a surface formulation has recently been proposed in [4, 5] and shown to be very effective even for higher-order methods [3]. The counterpart of the Laplacian Δ is the Laplace-Beltrami operator Δ_M (see [7]) for a surface M. However, unlike the 2D images, where the grids are often structured, the discretized triangular surfaces are often un-structured. Certain regions of the surface meshes are often very dense, with a wide spectrum of noise distribution. Applying a single Gauss-like filter to such surface meshes would have the following side-effects:

(1). The lower frequency noise is not filtered (*under-fairing*) if the evolution period of time is suitable for removing high frequency noise,
(2). Detailed features are removed unfortunately, as higher frequency noise (*over-fairing*) if the evolution period of time is suitable for removing low frequency noisy components.

The bottom row of Fig 1 illustrates this under-fairing and over-fairing effects. For the input mesh on the top-left, three fairing results are presented at three time scales. The first figure exhibits the under-fairing for the head. The last figure exhibits the over-fairing for the ears, eyes, lips and nose. Hence, a phenomena that often appears for the triangular surface mesh denoising is that whenever the desirable smoothing results are achieved for larger features, the smaller features are lost. Prior work has attempted to solve the over-fairing problem by using an anisotropic diffusion tensor in the diffusion equation [3, 4]. However, this is far from satisfactory. The aim of this chapter is to overcome the under-fairing and over-fairing dilemma in solving the diffusion equation.

There are several situations where the produced triangular surface meshes have varying triangle density. One typical case is geometric modeling, where the detailed structures are captured by several small triangles while the simpler shapes are represented by fewer large ones. We may call such triangular meshes as *feature-adaptive*. Another case is the results of physical simulation, in which the researcher is interested in certain regions of the mesh. In these regions, accurate solutions are desired, and quite often, finer meshes are used. For example, in the acoustic pressure simulation [15], the interesting region is the ear canal for human hearing. Hence, more accurate and

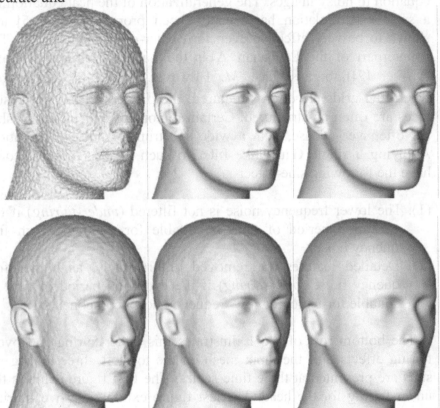

Fig. 1. *The left figure in the top row shows the initial geometry mesh. The top-middle and top-right figures illustrate the results of the adaptive timestep smoothing after 2 and 4 fairing steps with $\tau = 0.016$. The three figures in the bottom row are the results of the timestep $t = 0.001$ smoothing after 1, 2 and 4 fairing steps, respectively.*

finer meshes are used there. We call such meshes as *error-adaptive*. One additional case arises from the multiresolution representation of surfaces, for example using wavelet transforms or direct mesh simplifications. Each resolution of the representation is a surface mesh that approximates the highest resolution surface. The approximation error is usually adaptive and can vary over the entire surface. For instance, the mesh simplification scheme in [2], which is driven by the surface normal variation, results in meshes that are both feature-adaptive and error-adaptive.

Previous Work. For PDE based surface fairing or smoothing, there are several methods that have been proposed (see [3, 4, 5, 6]) recently. Desbrun et al in [5, 6] also use Laplacian, which is discretized as the umbrella operator in the spatial direction. In the time direction discretization, they propose to use the semi-implicit Euler method to obtain a stable numerical scheme. Clarenz et al in [4] generalize the Laplacian to the Laplace-Beltrami operator Δ_M, and use linear finite elements to discretize the equation. In [3], the problem is reformulated for 2-dimensional Riemannian manifold embedded in R^k aiming at smooth geometric surfaces and functions on surfaces simultaneously. The C^1 higher-order finite element space used is defined by the Loop's subdivision (box spline). One of the shortcomings of all these proposed methods that we address here is their non-adaptivity. All of them use uniform timesteps. Hence they quite often suffer from under-fairing or over-fairing problems.

Our Approach. For a feature-adaptive or error-adaptive mesh, the ideal evolution strategy would be to correlate the evolution speed relative to the mesh density. In short, we desire the lower frequency errors use a faster evolution rate and the higher frequency errors succumbs to a slower evolution rate. To achieve this goal, we present a discretization in the time direction, which is mesh adaptive. We use a timestep $T(x)$, which depends on the position x of the surface. The part of the surface, that is coarse, uses larger $T(x)$. The idea is simple and it is easy to incorporate it into any uniform timestep diffusion implementation. The improvements achieved over the evolution results with uniform timestep are significant. The top row of Fig 1 shows this adaptive time evolution improvement over the uniform timestep evolution results, shown in the bottom row. The middle and right figures in the top row are the smoothing results of the mesh on the left top after 2 and 4 fairing steps, respectively. As an alternative to the adaptive discretization in time direction, we also provide an approach for the adaptive choice of the diffusion tensor in the diffusion PDE equation.

The remaining of the chapter is organized as follows: Section 2 summarizes the diffusion PDE model used, followed by the discretization section 3. In the spatial direction, the discretization is realized using the C^1 smooth finite element space defined by the limit function of Loop's subdivision (box spline), while the discretization in the time direction is adaptive. The conclusion section 4 provides examples showing the superiority of the adaptive scheme.

2 Geometric Diffusion Equation

We shall solve the following nonlinear system of parabolic differential equations (see [3, 4]):

$$\partial_t x(t) - \Delta_{M(t)} x(t) = 0, \tag{1}$$

where $\Delta_{M(t)} = div_{M(t)} \circ \nabla_{M(t)}$ is the Laplace-Beltrami operator on $M(t)$, $M(t)$ is the solution surface at time t and $x(t)$ is a point on the surface. $\nabla_{M(t)}$ is the gradient operator on the surface. This equation is a generalization of heat equation $\partial_t \rho - \Delta \rho = 0$ to surfaces, where Δ is the Laplacian. To enhance sharp features, a *diffusion tensor* D, acting on the gradient, has been introduced (see [3, 4]). Then (1) becomes

$$\partial_t x(t) - div_{M(t)}(D\nabla_{M(t)} x(t)) = 0. \tag{2}$$

The diffusion tensor $D := D(x)$ is a symmetric and positive definite operator from $T_x M$ to $T_x M$. Here $T_x M$ is the tangent space of M at x. The detailed discussion for choosing the diffusion tensor can be found in [3, 4] for enhancing sharp features. In this chapter, we do not address the problem of enhancing sharp features. However, we shall use a scalar diffusion tensor for achieving an adaptive diffusion effect. The divergence $div_M \psi$ for a vector field $\psi \in TM$ is defined as the dual operator of the gradient (see [12]):

$$(div_M v, \phi)_M = -(v, \nabla_M \phi)_{TM}, \quad \forall \phi \in C_0^\infty(M), \tag{3}$$

where $C_0^\infty(M)$ is a subspace of $C^\infty(M)$, whose elements have compact support. TM is the tangent bundle, which is a collection of all the tangent spaces. The inner product $(\phi, \psi)_M$ and $(u, v)_{TM}$ are defined by the integration of $\phi\psi$ and $u^T v$ over M, respectively. The gradient of a smooth function f on M is given by

$$\nabla_M f = [t_1, t_2] G^{-1} \left[\frac{\partial(f \circ x)}{\partial \xi_1}, \frac{\partial(f \circ x)}{\partial \xi_2} \right]^T, \tag{4}$$

where

$$G^{-1} = \frac{1}{\det G}\begin{bmatrix} g_{22} & -g_{12} \\ -g_{21} & g_{22} \end{bmatrix}, \quad G = \begin{bmatrix} g_{22} & g_{12} \\ g_{21} & g_{22} \end{bmatrix}, \quad g_{ij} = \left(\frac{\partial}{\partial \xi_1}\right)^T \frac{\partial}{\partial \xi_2},$$

$x(\xi_1, \xi_2)$ is a local parameterization of the surface. G is known as the first fundamental form. For a vector field $X = \sum\limits_{i=1}^{2} X^i \frac{\partial}{\partial \xi_i} \in TM$, an explicit expression for the divergence is given by (see [8], page 84)

$$div_M X = \frac{1}{\sqrt{\det G}} \sum_{i=1}^{2} \frac{\partial}{\partial \xi_i}(\sqrt{\det G}\, X^i).$$

Then it is easy to derive that

$$div_M(h\nabla_M f) = (\nabla_M f)^T \nabla_M h + h\Delta_M f, \tag{5}$$

where f, h are smooth functions on M. From (4), (5) and the fact that $\Delta_M x = 2H(x)n(x)$, we could rewrite (2) as

$$\partial_t x(x) = \nabla_M D(x) + 2D(x)H(x)n(x), \tag{6}$$

where $H(x)$ and $n(x)$ are the mean curvature and the unit normal of M, respectively. Equation (6) implies that the motion of the sur-face $M(t)$ can be decomposed into two parts: One is the tangential displacement caused by $\nabla_M D(x)$, and the other is the normal dis-placement (mean curvature motion) caused by $2D(x)H(x)n(x)$.

Using (3), the diffusion problem (2) could be reformulated into the following variational form

$$\begin{cases} \text{Find a smooth } x(t) \text{ such that} \\ (\partial_t x(t), \theta)_{M(t)} + (D\nabla_{M(t)} x(t), \nabla_{M(t)}\theta)_{TM(t)} = 0, \\ M(0) = M, \end{cases} \tag{7}$$

for any $\theta \in C_0^\infty(M(t))$. This variational form is the starting point for the discretization.

We already know that equation (1) describes the mean curvature motion. Its regularization effect could be seen from the following equation (see [4, 13])

$$\frac{d}{dt}Area(M(t)) = -\int_{M(t)} H^2 dx, \quad \frac{d}{dt}Volume(M(t)) = -\int_{M(t)} H\, dx, \tag{8}$$

where $Area(M(t))$ and $Volume(M(t))$ are the area of $M(t)$ and volume enclosed by $M(t)$, respectively, H is the mean curvature. From these equations, we see that the evolution speed depends on the mean curvature of the surface but not on the density of the mesh. Hence if the mesh is spatially adaptive, the dense parts that have detailed structures, have larger curvatures, which very possibly be over-faired.

3 Discretization

We discretize equation (7) in the time direction first and then in the spatial direction. Given an initial value $x(0)$, we wish to have a solution $x(t)$ of (7) at $t = T(x(0))$. Using a semi-implicit Euler scheme, we have the following time direction discretization:

$$
\begin{cases}
Find \quad a \quad smooth \quad x(t) \quad such \quad that \\
\left(\dfrac{x(T) - x(0)}{T}, \theta \right)_{M(0)} + (D\nabla_{M(0)} x(T), \nabla_{M(0)} \theta)_{TM(0)} = 0,
\end{cases}
\tag{9}
$$

for any $\theta \in C_0^\infty(M(t))$. If we want to go further along the time direction, we could treat the solution at $t = T(x)$ as the initial value and repeat the same process. Hence, we consider only one time step in our analysis.

3.1 Spatial Discretization

The function in our finite element space is locally parameterized as the image of the unit triangle

$$
T = \left\{ (\xi_1, \xi_2) \in R^2 : \xi_1 \geq 0, \xi_2 \geq 0, \xi_1 + \xi_2 \leq 1 \right\}.
$$

That is, $(1 - \xi_1 - \xi_2, \xi_1, \xi_2)$ are the barycentric coordinate of the triangle. Using this parameterization, our discretized representation of M is $M = \bigcup_{\alpha=1}^k T_\alpha$, $\mathring{T}_\alpha \cap \mathring{T}_\beta = \phi$ for $\alpha \neq \beta$, where $\mathring{T}_\alpha$ is the interior of T_α. Each triangular patch is assumed to be parameterized locally as $x^\alpha : T \to T_\alpha$; $(\xi_1, \xi_2) \mapsto x^\alpha(\xi_1, \xi_2)$. Under this parameterization, tangents and gradients can be computed directly. The integration on surface M is given by

$$
\int_M f dx := \sum_\alpha \int_T f(x^\alpha(\xi_1, \xi_2)) \sqrt{\det(g_{ij})} d\xi_1 d\xi_2.
$$

The integration on triangle T is computed adaptively by numerical methods.

Let M_d be the given initial triangular mesh, $x_i, i = 1, \cdots, m$, be its vertices. We shall use C^1 smooth quartic Box spline basis functions to span our finite element space. The piecewise quartic basis function at vertex x_i, denoted by ϕ_i, is defined by the limit of Loop's subdivision for the zero control values everywhere except at x_i where it is one (see [3] for detailed description of this). For simplicity, we call it the *Loop's basis*.

Loop's Subdivision and Finite Element Function Space

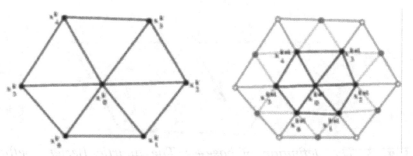

Fig. 2. *Refinement of triangular mesh around a vertex*

In Loop's subdivision scheme, the initial control mesh and the subsequent refined meshes consist of triangles only. In the refinement, each triangle is subdivided linearly into 4 sub-triangles. Then the vertex position of the refined mesh is computed as the weighted average of the vertex position of the unrefined mesh. Consider a vertex x_0^k at level k with neighbor vertices x_i^k for $i = 1, \cdots, n$ (see Fig 2), where n is the valence of vertex x_0^k. The coordinates of the newly generated vertices x_i^{k+1} on the edges of the previous mesh are computed as

$$x_i^{k+1} = \frac{3x_0^k + 3x_i^k + x_{i-1}^k + x_{i+1}^k}{8}, \quad i = 1, \cdots, n,$$

where index i is to be understood modulo n. The old vertices get new positions according to

$$x_0^{k+1} = (1 - na)x_0^k + a(x_1^k + x_2^k + \cdots + x_n^k),$$

where

$$a = \frac{1}{n}\left[\frac{5}{8} - \left(\frac{3}{8} + \frac{1}{4}\cos\frac{2\pi}{n}\right)^2\right].$$

Note that all newly generated vertices have a valence of 6, while the vertices inherited from the original mesh at level zero may have a valence other than 6. We will refer to the former case as *ordinary* and to the later case as *extraordinary*.

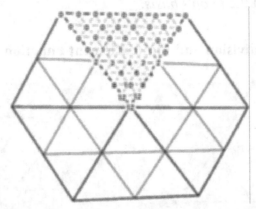

Fig. 3. *The definition of base ϕ_i: The quartic Bezier coefficients (each has a factor 1/24). The Bezier coefficients on the five macro-triangles are obtained by rotating the top macro-triangle around the center to the other five positions.*

Let e_j, $j = 1,\cdots,m_i$ be the 2-ring neighborhood elements. Then if e_j is regular (meaning its three vertices have valence 6), explicit Box-spline expressions exist (see [14, 16]) for ϕ_i on e_j. Using these explicit Box-spline expressions, we derive the BB-form expression for the basis functions ϕ_i (see Fig 3). These expressions could be used to evaluate ϕ_i in forming the linear system (11). If e_i is irregular, local subdivision is needed around e_i until the parameter values of interest are interior to a regular patch. An efficient evaluation method, that we have implemented, is the one proposed by Stam [14].

Compared with the linear finite element space, using the higher-order C^1 smooth finite element space spanned by Loop's basis does have advantages. The basis functions of this space have compact support (within 2-rings of the vertices). This support is bigger than the support (within 1-ring of the vertices) of hat basis functions that are used for the linear discrete surface model. Such a difference in the size of support of basis functions makes our evolution more efficient than those previously reported, due to the increased bandwidth of the affected frequencies. The reduction speed of high frequency noise in our approach is not that drastic, but still fast, while the reduction speed of lower frequency noise is not slow. Hence, the bandwidth of affected frequencies is wider. A comparative result showing the superiority of the Loop's basis function is given in [3].

Let $V_{M(0)}$ be the finite dimensional space spanned by the Loop's basis functions $\{\phi_i\}_{i=1}^m$. Then $V_{M(0)} \subset C^1(M(0))$. Let

$$x(0) = \sum_{i=1}^m x_i(0)\phi_i \in M(0), \quad x(T) = \sum_{i=1}^m x_i(T_i)\phi_i, \quad \theta = \phi_j.$$

Then equation (9) is discretized in $V_{M(0)}^3$ as

$$\sum_{i=1}^m (x_i(T_i) - x_i(0))(T_i^{-1}\phi_i, \phi_j)_{M(0)} +$$

$$\sum_{i=1}^m x_i(T_i)(D\nabla_{M(0)}\phi_i, \nabla_{M(0)}\phi_j)_{TM(0)} = 0 \quad (10)$$

for $j = 1, \cdots, m$, where $x_i(0) := x_i$ is the i-th vertex of the input mesh M_d, $T_i = T(y_i(0))$ and $y_i(0)$ is a surface point corresponding to vertex $x_i(0)$. Equation (10) is a linear system for unknowns $x_i(T_i)$.

3.2 Adaptive Timestep

We first use adaptive timesteps to achieve the adaptive evolution effect. In this case the diffusion tensor D is chosen to be identity, but $T(x)$ is not a constant function. Now (10) can be written in the following matrix form:

$$(S + L)X(T) = MX(0), \quad (11)$$

where $X(T) = [x_1(T_1), \cdots, x_m(T_m)]$, $X(0) = [x_1(0), \cdots, x_m(0)]$ and

$$S = \left((T_i^{-1}\phi_i, \phi_j)_{M(0)}\right)_{i,j=1}^m,$$

$$L = \left((D\nabla_{M(0)}\phi_i, \nabla_{M(0)}\phi_j)_{TM(0)}\right)_{i,j=1}^m.$$

$$(12)$$

Note that both S and L are symmetric. Since ϕ_1, ϕ_2, ... , ϕ_m are linearly independent and have compact support, S is sparse and positive definite. Similarly, L is symmetric and nonnegative definite. Hence, $S + L$ is symmetric and positive definite.

The coefficient matrix of system (11) is highly sparse. An iterative method for solving such a system is desirable. We solve it by the conjugate gradient method with a diagonal preconditioning.

Defining adaptive timesteps. Now we illustrate how $T(x)$ is defined. At each vertex x_i of the mesh M_d, we first compute a value $d_i > 0$, which measures the density of the mesh around x_i. We propose two approaches for computing it:

1. d_i is defined as the average of the distance from x_i to its neighbor vertices.

2. d_i is defined as the sum of the areas of the triangles surrounding x_i.

To make the $d_i's$ relative to the density of the mesh but not the geometric size, we always resize the mesh into the box $[-3,3]^3$. The experiments show that both approaches work well, and the evolution results have no significant difference. This value d_i is used as control value for defining timestep that is the same as defining the surface point:

$$T(x) = \tau D(x), \quad D(x) = \sum_{i=1}^{m} d_i \phi_i, \tag{13}$$

where $\tau > 0$ is a user specified constant. Hence, T is a function in the finite element space $V_{M(0)}$. Note that since T is not a constant any more, it is involved in the integration in computing the stiffness matrix S. Since $T(x) \in V_{M(0)}$, it is C^2, except at the extraordinary vertices, where it is C^1. However, $T(x)$ may also be noisy, since it is computed from the noisy data. To obtain a smoother $T(x)$, we smooth repeatedly the control value d_i at the vertex x_i by the following rule:

$$d_i^{(k+1)} = (1 - n_i l_i) d_i^{(k)} + l_i \sum_{j=1}^{n_i} d_j^{(k)}, \tag{14}$$

where $d_i^{(0)} = d_i$ for $i = 1, \cdots, m$, $d_j^{(k)}$ in the sum are the control values at the one-ring neighbor vertices of x_i, n_i is the valence of x_i, l_i and $a(n_i)$ are given as follows:

$$l_i = \frac{1}{n_i + 3/8a(n_i)}, \quad a(n_i) = \frac{1}{n_i}\left[\frac{5}{8} - \left(\frac{3}{8} + \frac{1}{4}\cos\frac{2\pi}{n_i}\right)^2\right].$$

The smoothing rule (14) is in fact for computing the limit value of Loop's subdivision (see [9], pp 41-42) applying to the control values $d_i^{(k)}$ at the vertices. In our examples, we apply this rule three times. Experiments show that even more times of smoothing of d_i are not harmful, but the influence to the evolution results are minor. The smoothing effect of (14) could be seen by rewriting it in the following form

$$\frac{d_i^{(k+1)} - d_i^{(k)}}{n_i l_i} = \frac{1}{n_i}\sum_{j=1}^{n_i}\left(d_j^{(k)} - d_i^{(k)}\right).$$

The left-handed side could be regarded as the result of applying the *forward Euler method* to the function $d_i(t)$, the right-handed side is the umbrella operator (see [5]). Hence, (14) is a discretization of the equation $\frac{\partial D}{\partial t} = \Delta D$. Since $n_i l_i < 1$, the stability criterion for (14) is satisfied.

A different view of adaptive timestep approach
Consider the following diffusion PDE

$$\partial_t x(t) - D(x)\Delta_{M(t)}x(t) = 0, \tag{15}$$

where $D(x)$ is a function defined by (13). Again, equation (15) describes the mean curvature motion with a compression factor $D(x)$. If we use a semi-implicit Euler scheme to discretize the equation with constant timestep τ, we could arrive at the same linear system as (11). Hence, solving the equation (1) with an adaptive timestep $\tau D(x)$ is equivalent to solving equation (15) with a uniform timestep τ. But (15) may be easier to handle in the theoretical analysis.

3.3 Uniform Timestep and Adaptive Diffusion Tensor

Now we use uniform timestep τ but a non-identity diffusion tensor $D(x)$. This $D(x)$ is the same as the one defined in (13), but we should regard it as $D(x)I$, where I is the identity diffusion tensor. The discretized equation (10) then becomes

$$\sum_{i=1}^{m}(x_i(\tau) - x_i(0))(\phi_i,\phi_j)_{M(0)} +$$

$$\sum_{i=1}^{m}x_i(\tau)(\tau D\nabla_{M(0)}\phi_i, \nabla_{M(0)}\phi_j)_{TM(0)} = 0. \tag{16}$$

From this, a similar linear system as (11) is obtained with

$$S = \left((\phi_i,\phi_j)_{M(0)}\right)_{i,j=1}^{m},$$

$$L= \left((\tau D\nabla_{M(0)}\phi_i, \nabla_{M(0)}\phi_j)_{TM(0)}\right)_{i,j=1}^{m}. \tag{17}$$

We know that $D(x)$ is a smooth positive function that characterizes the density of the surface mesh. The effect of this diffusion tensor is suppressing the gradient where the mesh is dense, and hence slows down the evolution speed. Comparing equation (11) with equation (16), we find that they are similar (since $\tau D = T$), though not equivalent. Indeed, if $D(x)$ is a constant on each triangle of M, then they are equivalent. In general, $D(x)$ is not a constant, but approximately a constant on each triangle, hence the observed behavior of (11) and (16) are often similar. The last two figures in Fig 4 exhibit this similarity, where the third and fourth figures are the evolution results using an adaptive timestep and an adaptive diffusion tensor, respectively. Since the results of the two adaptive approaches are very close, in the other examples provided in this chapter, we use only the adaptive timestep approach.

Homogenization Effect of D

It follows from (6) that the non-constant diffusion tensor $D(x)$ causes tangential displacement of the vertices. For the diffusion tensor $D(x)$ defined in the last sub-section, we know that it is adaptive to the density of the mesh in the sense that it takes smaller values at denser regions of the mesh. Consider a case where a small triangle is surrounded by large triangles. In such a case, function $D(x)$ is small on the triangle and larger elsewhere. This implies that the gradient of

$D(x)$ on the small triangle points to the outside direction, and the tangential displacement makes the small triangle become enlarged. If the density of the mesh is even, then $D(x)$ is near a constant. Then the tangential displacement is minor. Hence, the adaptive diffusion tensor we use has homogenizing effect. Such an effect is nice and important, as it avoids producing collapsed or tiny triangles in the faired meshes.

3.4 Algorithm Summary

For a given initial mesh, stopping control threshold values $\varepsilon_i > 0$, $i = 1,2$ and $\tau > 0$, the adaptive timestep evolution algorithm could be summarized via the following pseudo-code:

Compute function and derivative values of ϕ_i on the integration points;
   do {
      Compute d_i;
      Smooth d_i by (14);
      Compute matrices S and L by (12);
      Solve linear system (11);
      Compute $H(t)$;
   } while (none of (18)-(19) is satisfied);

Note that the evolution process does not change the topology of the mesh. Hence the basis functions could be computed before the multiple iterations.

We use two of the three stopping criteria proposed in [3] for terminating the evolution process: Let

$$H(t) = \int_{M(t)} \| H(t,x) \|^2 \, dx \Big/ \int_{M(0)} \| H(0,x) \|^2 \, dx,$$

where $H(t,x)$ is the mean curvature vector at the point x and time $tD(x)$. The stopping criteria are

$$|H'(t)| \le \varepsilon_1, \text{ or} \tag{18}$$

$$H(t) \le \varepsilon_2. \tag{19}$$

where ε_i are user specified control constants, $H'(t)$ is computed by divided differences.

4 Summary

We have proposed two simple adaptive approaches in solving the diffusion PDE by the finite element discretization in the spatial direction and the semi-implicit discretization in the time direction, aiming to solving the under-fairing/over-smooth problems that beset the uniform diffusion schemes. The implementation shows that the proposed adaptive schemes work very well.

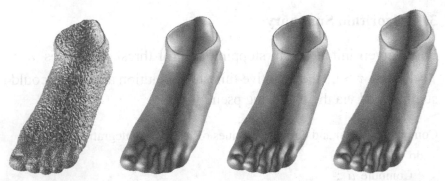

Fig. 4. *The left figure is the initial geometry mesh. The second figure is the faired mesh after 3 fairing iterations with uniform timestep t = 0.0011. The third and fourth figures are the faired meshes after 3 fairing iterations with adaptive timestep and alternatively adaptive diffusion tensor with uniform timestep τ = 0.025, respectively.*

Fig. 5. *The top figure is the initial geometry mesh. The second and the third figures are the faired meshes after 2 and 4 fairing iterations with uniform timestep t = 0.0001. The last two are the faired meshes after 2 and 4 fairing iterations with adaptive timestep and τ = 0.0016.*

Figure 4 and Fig 5 are used to illustrate the difference between the uniform timestep evolution and the adaptive timestep evolution. Since the adaptive timestep is not uniform, we cannot compare the evolution results for the same time. The comparing criterion we adopted here is we evolve the surface, starting from the same input, to arrive at similar smoothness for the rough/detailed features and compare the detailed/rough features. In Fig 4, the left figure is the input mesh, the second figure uses uniform timestep, the third and fourth figures use a adaptive timestep and a adaptive diffusion tensor with a uniform timestep, respectively. Comparing the three smoothing results, we can see that the large features look similar, but the toes of the foot are very different. The evolution results of the adaptive timestep and the adaptive diffusion tensor are much more desirable. Figure 5 exhibits the same effect. The top figure shows the input mesh, the next two are the results of the uniform timestep evolution. Comparing these to the bottom two figures, which are the results of the adaptive timestep evolution, many detailed features on the back and the snout of the crocodile are preserved by the adaptive approach. Furthermore, the large features of the uniform timestep evolution (compare the tails of the crocodiles) are less fairer than that of the adaptive timestep evolution, even though the detailed features are already over-faired.

Acknowledgments

This work has been supported in part by NSF grants ACI-998297 and KDIDMS-9873326 and in part by NSF of China and National Innovation Fund 1770900, Chinese Academy of Sciences.

References

1. Ed B. Harr Romeny. Geometry Driven Diffusion in Computer Vision. Boston, MA: Kluwer, 1994.
2. C. Bajaj and G. Xu. Smooth Adaptive Reconstruction and Deformation of Free-Form Fat Surfaces. TICAM Report 99-08, March, 1999, Texas Institute for Computational and Applied Mathematics, The University of Texas at Austin, 1999.
3. C. Bajaj and G. Xu. Anisotropic Diffusion of Noisy Surfaces and Noisy Functions on Surfaces. TICAM Report 01-07, February 2001, Texas Institute for Computational and Applied Mathematics, The University of Texas at Austin, 2001,.
4. U. Clarenz, U. Diewald, and M. Rumpf. Anisotropic Geometric Diffusion in Surface Processing. In Proceedings of Viz2000, IEEE Visualization , pages 397-505, Salt Lake City, Utah, 2000.
5. M. Desbrun, M. Meyer, P. Schröder, and A. H. Barr. Implicit Fairing of Irregular Meshes using Diffusion and Curvature Flow. SIGGRAPH99, pages 317-324, 1999.
6. M. Desbrun, M. Meyer, P. Schröder, and A. H. Barr. Discrete Differential-Geometry Operators in nD,
7. http://www.multires.caltech.edu/pubs/, 2000.
8. M. do Carmo. Riemannian Geometry. Boston, 1992.
9. J. Jost. Riemannian Geometry and Geometric Analysis, Second Edition. Springer, 1998.
10. C. T. Loop. Smooth subdivision surfaces based on triangles. Master's thesis. Technical report, Department of Mathematices, University of Utah, 1978.
11. P. Perona and J. Malik. Scale space and edge detection using anisotropic diffusion. In IEEE Computer Society Workshop on Computer Vision, 1987.

12. T. Preußer and M. Rumpf. An adaptive finite element method for large scale image processing. In Scale-Space Theories in Computer Vision, pages 232-234, 1999.
13. S. Rosenberg. The Laplacian on a Riemannian Manifold. Cambridge, Uviversity Press, 1997.
14. G. Sapiro. Geometric Partial Differential Equations and Image Analysis. Cambridge, University Press, 2001.
15. J. Stam. Fast Evaluation of Loop Triangular Subdivision Surfaces at Arbitrary Parameter Values. In SIGGRAPH '98 Proceedings, CD-ROM supplement, 1998.
16. T. F. Walsh. P Boundary Element Modeling of the Acoustical Transfer Properties of the Human Head/Ear. PhD thesis, Ticam, The University of Texas at Austin, 2000.
17. J. Warren. Subdivision method for geometric design, 1995.
18. J. Weickert. nisotropic Diffusion in Image Processing. B. G. Teubner Stuttgart, 1998.

Chapter 03

Finite Difference Surface Representation Considering Effect of Boundary Curvature

Finite Difference Surface Representation Considering Effect of Boundary Curvature

Lihua You and Jian J. Zhang
National Centre for Computer Animation, Bournemouth University, Poole, Dorset
BH12 5BB, United Kingdom

Surface representation as a boundary-valued problem of partial differential equations (PDE) is an important topic of computer graphics and computer aided design. In existing references, various free-from surfaces were created with a fourth order PDE which is only able to meet the tangential conditions at the surface boundaries. The need for a sixth order PDE in surface modeling arises in two situations: one is to generate surfaces with curvature continuity; and the other is to use curvature values as a user handle for surface shape manipulation. In this chapter, we introduce such a sixth order PDE for free-form surface generation and develop a finite difference method to solve this PDE. We also investigate the effects of boundary curvature and the vector-valued shape parameters on the surface shape.

1   Introduction

Traditional parametric surface representations, such as Bézier, B-spline and NURBS [7, 18-29] are common methods of surface modeling. Surfaces and curves are manipulated through the use of control points.

In the real world, all objects have certain physical properties and will change shape when they are subject to internal and external actions. But the traditional surface modeling methods do not attempt to describe the effect of physics laws.

Physically based surface modeling methods have attracted more and more attention in the communities of computer graphics and computer-aided design to complement the traditional geometrically based modeling methods. Terzopolous et al. were among the first researchers to study physically based surface modeling approaches for computer graphics. In their initial work, an elastic deformation model was proposed by Terzopolous et al. [30]. Based on the theories of elasticity and the finite element technique, deformable curve and surface finite element formulae were implemented by Celniker and Gossard [4]. The elastic deformation model was later extended to include viscoelasticity, plasticity and fracture by Terzopolous, et al. [19, 20]. In order to describe dynamic surface deformation, various physically based dynamic surface modeling methods were developed by Terzopolous, Qin, Mandal and Vemuri [13, 16, 17, 33]. These static and dynamic surface modeling methods were developed from the theories of elasticity, plasticity and fracture mechanics and implemented using the finite element method and finite difference method. Applying other mechanics approaches to generate surfaces has also been studied by some researchers. For example, Léon and Verson [9], and Guillet and Léon [8] applied a bar network mechanics method to deform free-form surfaces.

Partial differential equation based surface modeling method should also be classified as physically based modeling. For cloth and other flexible fabrics, their governing equations that represent the deformed surfaces can be derived from the physical laws of elastic cloth deformation and motion which are the fourth order dynamic PDEs involving the mechanical properties [34, 36]. PDE based surface modeling was first proposed by Bloor et al. [2]. In order to extend the PDE based methods to solve more complicated surface modeling problems, some numerical methods, such as the finite element method [10, 11, 12], finite difference method [5, 6] and collocation method [3] have been developed. Zhang and You [25] also discussed the effectiveness and efficiency of surface modeling using the second, fourth and mixed order PDEs.

In the above work, only the second and fourth order PDEs were applied. For a fourth order PDE, tangent continuity at the boundary curves of the surfaces to be created can be taken into account. Therefore, such an equation is effective in generating surfaces requiring tangent continuity. However, in many engineering design tasks, curvature continuity must be considered to satisfy the functional or visual requirements. For example, the streamlined surfaces of an aircraft, ship and submarine with curvature continuity do not only look pleasant, but more importantly reduce the risk of flow separation and turbulence. In contrast, a cam designed without curvature continuity between two connected surfaces will cause abrupt changes in acceleration resulting in harmful impact.

Pegna [14] proposed a method with which curvature continuous fairing surfaces can be interactively generated. For designing curvature continuous blending surfaces, Pegna and Wolter [15] developed a linkage curve theorem. Curvature continuity between two rectangular or triangular patches was examined by Zheng et al. [38] with the two patches being represented with rational Bézier surfaces. Aumann [1] proposed the so-called normal ringed surfaces to form curvature continuous connections of cones and/or cylinders.

In this chapter, we will use a sixth order PDE for free-from surface representation to account for the effect of boundary curvature. Since finding the closed form solution of a sixth order PDE is more difficult than that of a fourth order PDE, we also develop a finite difference technique to solve the proposed sixth order partial differential equation. How boundary curvature and the shape parameters affect the surface shape will be investigated. Surface representation examples will be given to demonstrate the applications of the developed method.

2   Governing Equations and Boundary Conditions

Boundary curvature continuity can be taken into account with a sixth order PDE. Since the vector-valued shape parameters in the equation exert a strong effect on surface shape, they act as user handles for surface manipulation. Therefore, instead of using one shape parameter proposed by Bloor et al. [2], we introduce four shape parameters and propose to use the following sixth order PDE to represent free-from surfaces considering the curvature boundary conditions

$$(\mathbf{a}\frac{\partial^6}{\partial u^6} + \mathbf{b}\frac{\partial^6}{\partial u^4 \partial v^2} + \mathbf{c}\frac{\partial^6}{\partial u^2 \partial v^4} + \mathbf{d}\frac{\partial^6}{\partial v^6})\mathbf{x} = 0 \tag{1}$$

where $\mathbf{a} = \begin{bmatrix} a_x\ a_y\ a_z \end{bmatrix}^T$, $\mathbf{b} = \begin{bmatrix} b_x\ b_y\ b_z \end{bmatrix}^T$, $\mathbf{c} = \begin{bmatrix} c_x\ c_y\ c_z \end{bmatrix}^T$, $\mathbf{d} = \begin{bmatrix} d_x\ d_y\ d_z \end{bmatrix}^T$ are   vector-valued   shape   parameters, $\mathbf{x} = \begin{bmatrix} x(u,v),\ y(u,v),\ z(u,v) \end{bmatrix}^T$ represents a vector-valued function, and u, v are the parametric variables.

Depending on the boundary conditions and the shape parameters, PDE (1) can represent various free-form surfaces.

In order to solve a PDE for the generation of a surface, we must define proper boundary conditions first. Normally, the curvature of a surface $\mathbf{x} = \mathbf{x}(u,v)$ can be described with its second partial derivatives. In this chapter, we only consider the curvature crossing the boundary curves, i. e., $\dfrac{\partial^2 \mathbf{x}}{\partial u^2}$. In this way, the boundary conditions which include the effects of boundary tangent and curvature can be written as

$$u = 0, \quad \mathbf{x} = \mathbf{G}_1(v) \quad \frac{\partial \mathbf{x}}{\partial u} = \mathbf{G}_2(v) \quad \frac{\partial^2 \mathbf{x}}{\partial u^2} = \mathbf{G}_3(v)$$

$$u = 1, \quad \mathbf{x} = \mathbf{G}_4(v) \quad \frac{\partial \mathbf{x}}{\partial u} = \mathbf{G}_5(v) \quad \frac{\partial^2 \mathbf{x}}{\partial u^2} = \mathbf{G}_6(v)$$

(2)

The solution of the sixth order partial differential equation under boundary conditions (2) will result in the representation of a surface.

3 Finite Difference Technique

The analytical solution of PDE (1) subject to boundary conditions (2) is usually very difficult to obtain. Only for some simple boundary conditions and special combinations of the shape parameters, does the closed form solution of Eq. (1) exist. Generally, Eq. (1) can only be solved with numerical methods such as the above-mentioned finite element method, finite difference method and weighted residual method [35].

Among them, the finite difference method transforms a PDE to a system of algebraic equations by replacing all the partial derivatives in the differential equation with their discretized approximations. In the following, we will develop such a finite difference technique to solve PDE (1) subject to boundary conditions (2).

To facilitate the description, we define a new vector product operator whose operands are two vectors of the same dimension and each element of the resultant vector is the product of the corresponding elements of the two vectors, i. e.,

$$\mathbf{pq} = \begin{bmatrix} p_x q_x & p_y q_y & p_z q_z \end{bmatrix}^T$$

(3)

where $\mathbf{p} = \begin{bmatrix} p_x & p_y & p_z \end{bmatrix}^T$ and $\mathbf{q} = \begin{bmatrix} q_x & q_y & q_z \end{bmatrix}^T$ are two column vectors.

Using the Taylor series expansion of a continuous function $f(u,v)$, we can derive its central finite difference approximations of the first and second partial derivatives at a typical node 0 shown in

Figure 1. Then using the basic finite difference approximations of the first and second derivatives to formulate the fourth and sixth partial derivatives, we can obtain the finite difference approximations of all the partial derivatives required. Substituting these finite difference approximations into PDE (1) and using the notation of Eq. (3), we obtain its finite difference formula at the typical node 0 as follows

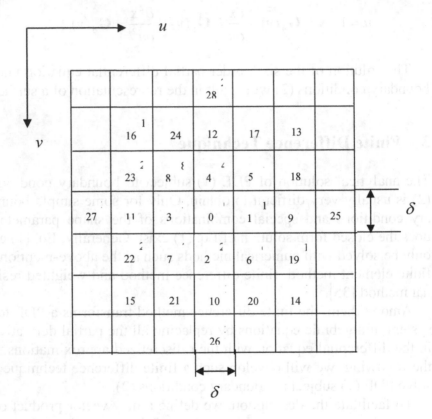

Fig. 1. *Typical finite difference nodes*

$$
\begin{aligned}
&-4(5\mathbf{a}+3\mathbf{b}+3\mathbf{c}+4\mathbf{d})\mathbf{x}_0 + (15\mathbf{a}+8\mathbf{b}+6\mathbf{c})\mathbf{x}_1 + (6\mathbf{b} \\
&+8\mathbf{c}+15\mathbf{d})\mathbf{x}_2 + (15\mathbf{a}+8\mathbf{b}+6\mathbf{c})\mathbf{x}_3 + (6\mathbf{a}+8\mathbf{c}+15\mathbf{d}) \\
&\mathbf{x}_4 - 4(\mathbf{b}+\mathbf{c})(\mathbf{x}_5+\mathbf{x}_6+\mathbf{x}_7+\mathbf{x}_8) - 2(3\mathbf{a}+\mathbf{b})(\mathbf{x}_9 \\
&+\mathbf{x}_{11}) - 2(\mathbf{c}+3\mathbf{d})(\mathbf{x}_{10}+\mathbf{x}_{12}) + \mathbf{b}(\mathbf{x}_{18}+\mathbf{x}_{19}+\mathbf{x}_{22} \\
&+\mathbf{x}_{23}) + \mathbf{c}(\mathbf{x}_{17}+\mathbf{x}_{20}+\mathbf{x}_{21}+\mathbf{x}_{24}) + \mathbf{a}(\mathbf{x}_{25}+\mathbf{x}_{27}) \\
&+\mathbf{d}(\mathbf{x}_{26}+\mathbf{x}_{28}) = 0
\end{aligned}
\tag{4}
$$

where $\mathbf{x}_i$ $(i = 0 \sim 12, 17 \sim 28)$ represents the co-ordinate value of function $\mathbf{x}$ at node i.

Substituting the basic finite difference approximations of the first and second derivatives into boundary conditions (2), the finite difference formulae at the typical node 0 can be written as

$$
\begin{aligned}
u = 0 \qquad & \mathbf{x}_0 = \mathbf{G}_1(v_0) \\
& \mathbf{x}_1 - \mathbf{x}_3 = 2\delta\mathbf{G}_2(v_0) \\
& \mathbf{x}_1 + \mathbf{x}_3 - 2\mathbf{x}_0 = \delta^2\mathbf{G}_3(v_0) \\
u = 1 \qquad & \mathbf{x}_0 = \mathbf{G}_4(v_0) \\
& \mathbf{x}_1 - \mathbf{x}_3 = 2\delta\mathbf{G}_5(v_0) \\
& \mathbf{x}_1 + \mathbf{x}_3 - 2\mathbf{x}_0 = \delta^2\mathbf{G}_6(v_0)
\end{aligned}
\tag{5}
$$

where v_0 is the value of the parametric variable v at node 0.

Dividing the resolution domain into m by n discrete nodes, we can obtain the same finite difference formula (4) for all the nodes at the inner resolution domain and finite difference formula (5) for all the nodes at the boundaries. Writing these finite difference formulae into a matrix form, we obtain a set of linear algebraic equations which has the form of

$$
\mathbf{KX} = \mathbf{F}
\tag{6}
$$

The resolution of equations (6) will lead to the finite difference solution of partial differential equation (1) under boundary conditions (2).

4   Effect of Boundary Curvature

In our previous discussions, we have mentioned that the boundary curvature has a great influence on the shape of surfaces to be created. In order to explore this effect and develop it into a user handle for shape manipulation, in this section, we will investigate the effect of boundary curvature on the surface shape by a concrete problem of vase surface representation.

The vase surface to be generated is defined by two boundary curves represented with two concentric circles of different radii. The boundary conditions considering tangent and curvature effects can be written as

$$u = u_i \quad x = r_i \cos 2\pi v \quad \frac{\partial x}{\partial u} = r_i' \cos 2\pi v \quad \frac{\partial^2 x}{\partial u^2} = r_i'' \cos 2\pi v$$

$$y = r_i \sin 2\pi v \quad \frac{\partial y}{\partial u} = r_i' \sin 2\pi v \quad \frac{\partial^2 y}{\partial u^2} = r_i'' \sin 2\pi v \tag{7}$$

$$z = h_i \quad \frac{\partial z}{\partial u} = h_i' \quad \frac{\partial^2 z}{\partial u^2} = h_i''$$

$(i = 0, \ 1)$

where $u_0 = 0$, $u_1 = 1$, $h_0 = 0$, and $h_1 = h$.

For the surface to be generated, the resolution domain is a unit square in a two-dimensional space of parametric variables u and v. Uniformly dividing this square resolution domain into 20×20 nodes and solving the above-developed finite difference equation (6), we obtain the co-ordinate values of the surface at these nodes. In this section, the basic geometric parameters are taken to be: $r_0 = 0.3$, $r_0' = r_0'' = r_1 = 0.8$, $h_0' = h_0'' = h_1'' == r_1' = 1$, $r_1'' = 0$, $h = 3$ and $h_1' = 0.1$.

In order to investigate the effect of boundary curvature on surface shape, we fix the shape parameters in a vector-valued form $\mathbf{a} = \mathbf{c} = \mathbf{d} = \begin{bmatrix} 1 & 1 & 1 \end{bmatrix}^T$ and $\mathbf{b} = \begin{bmatrix} -10 & -10 & -10 \end{bmatrix}^T$ for all the case studies in this section and only change the boundary curvature. Let us first consider the effect of the boundary curvature whose value is determined by r_1'' on the vase surface shape. The initial value of r_1'' given above produces the vase surface in Figure 2a. Then it is set to $r_1'' = -20$ and the image in Figure 2a is changed to that in Figure 2b. Further setting it to $r_1'' = 50$ gives the surface shape in the Figure 2c. Next, we investigate the effect of the boundary curvature whose value is determined by r_0'' on vase surface shape. With the above basic geometric parameters but changing r_0'' to 30, the vase shape is changed to that in Figure 2d. The image in Figure 2e is created with $r_0'' = -10$.

Similarly, the boundary curvature determined by h_0'' and h_1'' also have an obvious effect on surface shape. Firstly, taking $h_0'' = 80$, we obtained the vase shape in Figure 2f. It is changed to that in Figure 2g when $h_0'' = -50$. For the boundary curvature determined by h_1'', a value of 30 gives the shape in Figure 2h and the value of -90 creates that in Figure 2i.

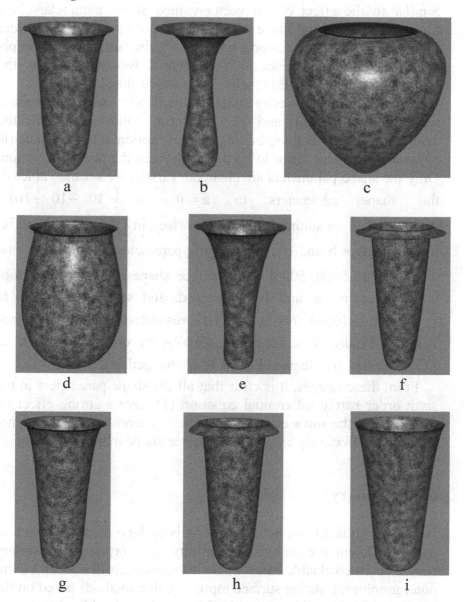

Fig. 2. *Effect of boundary curvature on surface shape*

These images clearly show that the boundary curvature can effectively affect the shape of the surfaces. By adjusting the values of the coefficients of the boundary curvature functions in Eq. (7), we can obtain different surface shapes.

5 Effect of Vector-valued Shape Parameters

Similar to the effect of the vector-valued shape parameters in a fourth order PDE, the shape parameters in our proposed sixth order PDE should also have a strong influence on the surface shape. In order to study this influence, in this section, we examine how the shape parameters in PDE (1) affect the surface shape.

For simplicity, we demonstrate this with the same vase surface representation example and the same node collocation within the resolution domain as those in the previous section. All the geometric parameters are also taken to be those given in the previous section. Only the shape parameters are changed. Firstly, we set the values of the shape parameters to $\mathbf{a} = \mathbf{0}$, $\mathbf{b} = \begin{bmatrix} -10 & -10 & -10 \end{bmatrix}^T$, $\mathbf{c} = \mathbf{d} = \begin{bmatrix} 1 & 1 & 1 \end{bmatrix}^T$ resulting in the vase surface in Figure 3a. Then fixing parameters $\mathbf{b}$ and $\mathbf{d}$, and changing parameters $\mathbf{a}$ to $\begin{bmatrix} 1 & 1 & 1 \end{bmatrix}^T$ and $\mathbf{c}$ to $\begin{bmatrix} 5000 & 5000 & 5000 \end{bmatrix}^T$, the surface shape in Figure 3b is obtained. Keeping $\mathbf{a}$ and $\mathbf{b}$ unchanged, and setting $\mathbf{c}$ and $\mathbf{d}$ to $\begin{bmatrix} 1 & 1 & 1 \end{bmatrix}^T$ and $\begin{bmatrix} 70000 & 70000 & 70000 \end{bmatrix}^T$, respectively, the vase surface shape in Figure 3c is generated. Finally, by only changing $\mathbf{b}$ to $\begin{bmatrix} 30 & 30 & 30 \end{bmatrix}^T$, the shape in Figure 3c is changed to that in 3d.

From these images, it is clear that all the shape parameters in the sixth order partial differential equation (1) have a strong effect on the shape of the surfaces to be generated. Therefore, these parameters can be effectively applied as a surface shape manipulation tool.

6 Summary

Free-form surfaces can be treated as a boundary-valued problem of partial differential equations. The ability of incorporating boundary curvature is a valuable merit in both engineering design and computer graphics. Existing surface representation methods based on the solution of PDEs did not tackle this problem. In this chapter, we have proposed a sixth order PDE which is capable of coping with such boundary conditions for surface generation. In addition, the proposed PDE introduces four vector-valued shape parameters, which can be effectively used as surface shaping tools.

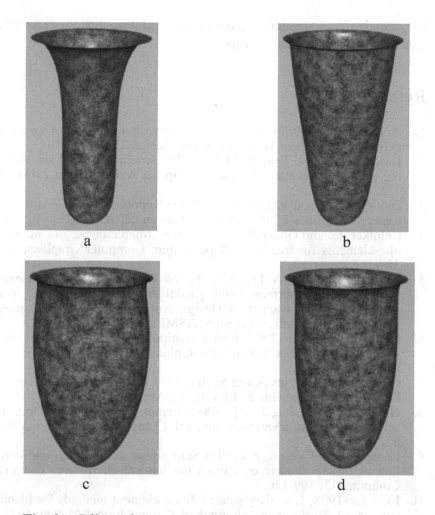

a   b

c   d

Fig. 3. *Effect of vector-valued parameters on surface shape*

Since the analytical solution of a sixth order PDE is far more difficult to achieve than that of a fourth or lower order PDE, a numerical resolution method is usually sought as an alternative. In order to solve the proposed sixth order PDE effectively, we have derived the basic finite difference approximation formulae taking up to the sixth partial derivatives. We have also transformed the proposed sixth order PDE and the boundary conditions into a resultant finite difference equation represented in a matrix form.

We have made some investigations to understand the effects of the boundary curvature and the shape parameters in shape manipulation. It was found that all boundary curvature conditions and the shape parameters have a great influence on the surface. Therefore, they can be implemented into a surface modeling system as an effective tool for surface deformation.

References

1. Aumann, G. (1995) Curvature continuous connections of cones and cylinders. Computer Aided Geometric Design 27(4), 293-301.
2. Bloor, M.I.G. and Wilson, M.J. (1989) Generating blend surfaces using partial differential equations. Computer Aided Design 21(3), 165-171.
3. Bloor, M.I.G. and Wilson, M.J. (1990) Representing PDE surfaces in terms of B-splines. Computer Aided Design 22(6), 324-331.
4. Celniker, G. and Gossard, D., 1991, Deformable curve and surface finite-elements for free-form shape design, Computer Graphics, 25(4), 257-266.
5. Cheng, S.Y., Bloor, M.I.G., Saia A. and Wilson, M.J. (1990) Blending between quadric surfaces using partial differential equations. in Ravani, B. (Ed.), Advances in Design Automation, Vol. 1, Computer Aided and Computational Design, ASME, 257-263.
6. Du, H. and Qin, H. (2000) Direct manipulation and interactive sculpting of PDE surfaces. Computer Graphics Forum (Eurographics'2000 Proc.) 19(3), 261-270.
7. Farin, G. (1997) Curves and Surfaces for Computer Aided Geometric Design: A Practical Guide. 4th Edition. Academic Press.
8. Guillet, S. and Léon, J.C. (1998) Parametrically deformed free-form surfaces as part of a variational model. Computer-Aided Design 30(8), 621-630.
9. Léon, J.C. and Veron, P. (1997) Semiglobal deformation and correction of free-form surfaces using a mechanical alternative. The Visual Computer 13, 109-126.
10. Li, Z.C. (1998) Boundary penalty finite element methods for blending surfaces, I. Basic theory. Journal of Computational Mathematics 16, 457-480.
11. Li, Z.C. (1999) Boundary penalty finite element methods for blending surfaces, II. Biharmonic equations. Journal of Computational and Applied Mathematics 110, 155-176.
12. Li, Z.C. and Chang, C.-S. (1999) Boundary penalty finite element methods for blending surfaces, III, Superconvergence and stability and examples. Journal of Computational and Applied Mathematics 110, 241-270.
13. Mandal, C., Qin, H. and Vemuri, B. (2000) Dynamic modeling of butterfly subdivision surfaces. IEEE Transactions on Visualization and Computer Graphics 6(3), 265-287.
14. Pegna, J. (1989) Interactive design of curvature continuous fairing surfaces. Proceedings of the Eighth International Conference on Offshore Mechanics and Arctic Engineering, ASME, New York, 191-198.
15. Pegna, J. and Wolter, F.-E. (1992) Geometrical criteria to guarantee curvature continuity of blend surfaces. Journal of Mechanical Design, Transactions of the ASME 114, 201-210.

16. Qin, H. and Terzopoulos, D. (1995), Dynamic NURBS Swung Surfaces for Physical-Based Shape Design, Computer-Aided Design 27(2), 111-127.
17. Qin, H. and Terzopoulos, D. (1997) Triangular NURBS and their dynamic generations. Computer Aided Geometric Design 14, 325-347.
18. Sarfraz, M, (2003), Weighted Nu Splines with Local Support Basis Functions, Advances in Geometric Modeling, Ed.: M. Sarfraz, John Wiley, 101 - 118.
19. Sarfraz, M. (2003), Optimal Curve Fitting to Digital Data, International Journal of WSCG, Vol 11(1).
20. Sarfraz, M. (2003), Curve Fitting for Large Data using Rational Cubic Splines, International Journal of Computers and Their Applications, Vol 10(3).
21. Sarfraz, M., and Razzak, M. F. A., (2003), A Web Based System to Capture Outlines of Arabic Fonts, International Journal of Information Sciences, Elsevier Science Inc., Vol. 150(3-4), 177-193.
22. Sarfraz, M., and Razzak, M. F. A., (2002), An Algorithm for Automatic Capturing of Font Outlines, International Journal of Computers & Graphics, Elsevier Science, Vol. 26(5), 795-804.
23. Sarfraz M. (2002) Fitting Curves to Planar Digital Data, Proceedings of IEEE International Conference on Information Visualization IV'02-UK: IEEE Computer Society Press, USA, 633-638.
24. Sarfraz, M, and Raza, A, (2002), Visualization of Data using Genetic Algorithm, *Soft Computing and Industry: Recent Applications*, Eds.: R. Roy, M. Koppen, S. Ovaska, T. Furuhashi, and F. Hoffmann, ISBN: 1-85233-539-4, Springer, 535 - 544.
25. Sarfraz, M. (1995) Curves and Surfaces for CAD using C^2 Rational Cubic Splines, Engineering with Computers, 11(2), 94-102.
26. Sarfraz, M. (1994) Cubic Spline Curves with Shape Control, Computers and Graphics 18(4), 707-713.
27. Sarfraz, M. (1993) Designing of Curves and Surfaces using Rational Cubics. Computers and Graphics 17(5), 529-538.
28. Sarfraz, M. (1992) A C^2 Rational Cubic Alternative to the NURBS, Computers and Graphics 16(1), 69-77.
29. Sarfraz, M. (1992) Interpolatory Rational Cubic Spline with Biased, Point and Interval Tension, Computers and Graphics 16(4), 427-430.
30. Terzopoulos, D., Platt, J., Barr, A. and Fleischer, K. (1987) Elastically deformable models. Computer Graphics 21(4), 205-214.
31. Terzopoulos, D. and Fleischer, K. (1988) Modeling inelastic deformation: viscoelasticity, plasticity, fracture. Computer Graphics 22(4), 269-278.
32. Terzopoulos, D. and Fleischer, K. (1988) Deformable models. The Visual Computer (4), 306-331.
33. Terzopoulos, D. and Qin, H. (1994) Dynamic NURBS with geometric constraints for interactive sculpting. ACM Transactions on Graphics 13(2), 103-136.
34. You, L.H., Zhang, J.J. and Comninos P. (1999) Cloth deformation modeling using a plate bending model. The 7th International Confer-

ence in Central Europe on Computer Graphics, Visualisation and Interactive Digital Media 485-491.

35. You, L.H., Zhang, J.J. and Comninos, P. (2000) A solid volumetric model of muscle deformation for computer animation using the weighted residual method. Computer Methods in Applied Mechanics and Engineering 190, 853-863.

36. Zhang, J.J, You, L.H. and Comninos P. (1999) Computer simulation of flexible fabrics. The 17th Eurographics UK Conference 27-35.

37. Zhang, J.J. and You, L.H. (2001) Surface representation using second, fourth and mixed order partial differential equations. International Conference on Shape Modeling and Applications, Genova, Italy, 7-11, May.

38. Zheng, J.M., Wang, G.Z., and Liang, Y.D. (1992) Curvature continuity between adjacent rational Bézier patches. Computer Aided Geometric Design 9(5), 321-335.

Chapter 04

A New and Direct Approach for Loop Subdivision Surface Fitting

A New and Direct Approach for Loop Subdivision Surface Fitting

Weiyin Ma, Xiaohu Ma and Shiu-Kit Tso
City University of Hong Kong, Department of Manufacturing Engineering and
Engineering Management, 83 Tat Chee Avenue, Kowloon, Hong Kong SAR,
China

This chapter presents a new and direct approach for Loop subdivision surface fitting from a dense triangular mesh with arbitrary topology. The initial mesh model is first simplified with a topology- and feature-preserving mesh decimation algorithm. The simplified mesh is further used as the topological model of a Loop subdivision surface. The control vertices of the subdivision surface are finally fitted from a subset of vertices of the original dense mesh. During the fitting process, both the subdivision rules and position masks are used for setting up the observation equations. The emphasis of this chapter is on fitting issues. While only the Loop subdivision scheme is discussed in this chapter, the approach is applicable to any stationery subdivision scheme.

1 Introduction

In practical applications, one often needs to reconstruct surfaces from laser scanning or other range data. This chapter presents a direct approach for automatic reconstruction of a Loop subdivision surface from a dense triangular mesh with arbitrary topology. In literature, one could find several approaches on subdivision surface fitting.

Hoppe et al [6] presented a method for automatically fitting Loop subdivision surfaces from a dense triangular mesh. This method makes use of all the input data. The fitting criterion is defined such that the final fitted control mesh after two subdivisions will be most close to the initial dense mesh, which provides sufficient approximation for practical applications. Both smooth and sharp models can be handled. This approach produces high quality models. However, the computing time is very long due to an optimization procedure involved.

66

The method of Suzuki et al [13] for subdivision surface fitting starts from an interactively defined initial control mesh. For each of the control vertices, the corresponding limit position is obtained from the initial dense mesh. The positions of the control vertices are then inversely solved following the relationship between the control vertices and the corresponding limit positions. The fitted subdivision surface is checked against a pre-defined tolerance. In case of need, the topological structure of the control mesh is further subdivided and a refined mesh of limit positions is obtained. A new subdivision surface is then fitted through the refined limit positions. The resulting surface uses the same number of control vertices as that of the selected limit positions and is actually an interpolation surface going through the selected subset of vertices of the initial mesh. Since only a subset of vertices of the initial dense mesh is involved in the fitting procedure, the computing speed is extremely fast. Sharp features are however not handled.

One can also find a parametrization-based approach for fitting Catmull-Clark subdivision surfaces by Ma and Zhao [9]. A network of boundary curves is first interactively defined for topological modelling. A set of base surfaces is then defined from the topological model for sample data parametrization. A set of observation equations is defined using the exact evaluation scheme of Stam [12]. Final Catmull-Clark surface fitting is achieved through linear least squares fitting. This approach uses all input data and the final criterion ensures best fit between the input data and the final fitted subdivision surface. Sharp features are not handled with this approach.

All the above fitting approaches are built upon schemes for limit surface query, either at discrete positions [6, 13, 14] or at arbitrary parameters [9]. In literature, one can also find schemes interpolating a set of surface limit positions with or without surface normal vectors instead of fitting [5, 10]. Theoretically, all interpolation schemes can be extended as a fitting scheme if the number of known surface conditions, such as limit positions, normal vectors and other constraints, are more than that of the unknown control vertices of the subdivision surface. One can also find an approach for adaptively fitting a Catmull-Clark subdivision surface to a given shape through a fast local adaptation procedure [7]. The fitting process starts from an initial approximate generic model defined as a subdivision surface in the same type.

In this chapter, the input data is an existing dense triangular mesh. All sharp features of the model are first identified and marked for later processing. The marked triangular mesh is further simplified to the require resolution while preserving the same topology and sharp features as those of the original mesh. A Loop subdivision surface is finally fitted from a subset of vertices of the original dense mesh. For the fitting method of this chapter, the number of vertices from the original dense mesh is about four times the number of the vertices of the final control mesh, which ensures good approximation to the initial dense mesh. Different from the approaches of Hoppe et al [6] and Suzuki et al [13], both subdivision rules and position masks are used for setting up the fitting equation. Shape features are also preserved during the fitting process. For solving the fitting equation, a conjugate gradient method without any constraints is used and the computing time is pretty fast.

In the following sections, we first summarize the Loop subdivision scheme [6, 8, 13] in section 2 for later development. Schemes for limit position evaluation are discussed in section 3. Details on surface fitting are presented in section 4. Some examples are also provided in section 4 for demonstrating the proposed method. A conclusion is finally summarized in section 5.

2   The Loop Subdivision Scheme

In literature, one can find many subdivision schemes [1-3,15]. The Loop subdivision scheme is one of the subdivision methods based on triangular control meshes and an one-to-four splitting scheme as shown in Fig. 1 (the shaded area) [6, 13, 15]. During the subdivision procedure, a new vertex is inserted into each of the edges and each of the old vertices is also updated based on given subdivision rules. Each of the triangular face is thus split into four smaller triangles and the number of the triangular faces after one subdivision will be four times of the number before subdivision.

In this chapter, we use the following conventions. Let M be a triangular mesh and p_i is the i-th vertex of M. Each of the vertex is connected to k neighboring vertices through edges. For an interior vertex, k is also the number of neighboring triangles of the i-th vertex and is called the *valence* of the corresponding vertex. To represent models with sharp features, we use the following classifications proposed in [6]:

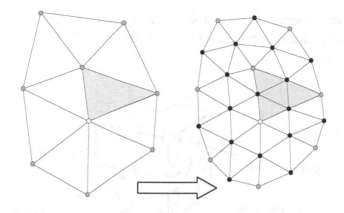

O   ordinary vertex with valence six
◐   extraordinary vertex with valence other than six,
     or a boundary vertex
●   newly inseted vertex

Fig.1. *One-to-four subdivision [13].*

- *Edge classification*: The model under discussion may have one or more *crease edges* over the entire surface. Boundary edges of an open surface are also regarded as sharp edges during the sub-division process. All other edges are smooth edges. Sharp edges are initially tagged through a procedure based on the variation of the angles between neighboring faces.

- *Vertex classification*: Let the number of sharp edges meeting at a vertex be s. We can then classify the vertices as follows:

  ◊ A *smooth vertex* is one where the number of meeting sharp edges s is zero.
  ◊ A *dart vertex* is one where a crease edge terminates with $s=1$.
  ◊ A *crease vertex* has $s=2$ and an *interior corner vertex* has $s>=3$. A boundary vertex with $s=2$ can also be defined as a *corner vertex* if the boundary curve is C^0 and the surface goes through that vertex.

Crease vertices can be further defined as *regular* and *non-regular* depending on the arrangement of the smooth edges. An interior crease vertex is regular if it has valence 6 with exactly two smooth edges on each side of the crease. A boundary crease vertex is regular if it has valence 4. All other crease vertices, whether interior or boundary, are non-regular. Figure 1 and Fig. 2 show some types of the vertices.

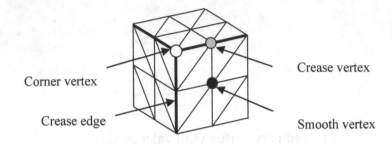

Fig. 2. *Edge and vertex classification.*

After edge and vertex classification, one can apply the Loop subdivision rules for producing a refined mesh [6, 8]. For a smooth vertex p, the updated vertex p' is defined by the following equation

$$p' = (1 - k\beta)p + \beta(p_1 + p_2 + \ldots + p_k) \tag{1}$$

where k is the valance of p and $\{p_i\}_{i=1}^{k}$ are the immediate neighboring vertices of p before subdivision. The term β is defined by the following formula

$$\beta = \begin{cases} \frac{3}{16} & (k = 3) \\ \frac{1}{k}\left(\frac{5}{8} - \left(\frac{3}{8} + \frac{1}{4}\cos\frac{2\pi}{k}\right)^2\right) & (k > 3) \end{cases} \tag{2}$$

For each edge, the newly inserted vertex is defined as follows

$$p' = \tfrac{1}{8}(3p_1 + 3p_2 + p_3 + p_4) \tag{3}$$

where p_1 and p_2 are the two end vertices of the corresponding split edge, and p_3 and p_4 are those of the other two vertices of the triangles sharing this edge.

In the above formulation, both the updated and the newly inserted vertices are thus defined by averaging incident vertices with appropriate weights. Diagrams called masks or stencils are often used to graphically represent the weighted-averaging formula. The subdivision masks for the smooth vertices and smooth edge vertices are shown in Fig. 3.

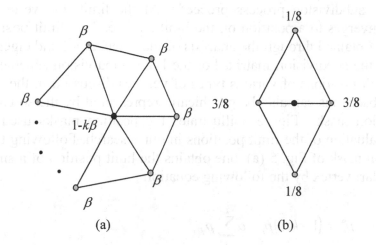

Fig. 3. *Subdivision masks for (a) a smooth or dart vertex of valence k and (b) new vertex on a smooth edge.*

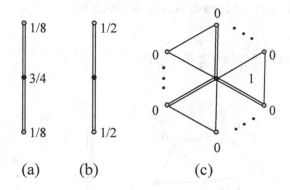

Fig. 4. *Subdivision masks for (a) a crease or boundary vertex, (b) an inserted vertex on a crease or boundary edge, and (c) a vertex at a corner point.*

For vertices at sharp features or on boundary edges, similar masks can be defined. Figure 4 illustrates the subdivision masks for vertices on a crease edge, a boundary curve and a corner point, respectively. A corner point in the control mesh remains unchanged after subdivision.

3 Limit Position Evaluation

For later surface fitting, we also need to evaluate the corresponding limit position of a control point [6,13]. For a given vertex p_i of the control mesh, the position of the vertex can be repeatedly updated as the subdivision process proceeds. At the limit, the vertex finally converges to a position on the limit surface. This limit position can be obtained through the analysis of the eigenvalues and eigenvectors of the subdivision matrix. For the Loop subdivision scheme, all the limit positions of various types of vertices discussed in the previous subsections can also be graphically represented by the so-called position masks. Figure 5 illustrates the position masks used for the evaluation of the limit positions in our research. Following the position mask of Fig. 5 (a), one obtains the limit position of a smooth or a dart vertex by the following equation

$$p_i^\infty = (1 - k\alpha)p_i + \alpha \sum_{j=1}^{k} p_j \tag{4}$$

where α is defined by $\alpha = (\frac{3}{8\beta} + k)^{-1}$. Similarly, the limit position of a boundary or crease vertex is given by

$$p_i^\infty = \frac{1}{6}p_1 + \frac{2}{3}p_i + \frac{1}{6}p_2. \tag{5}$$

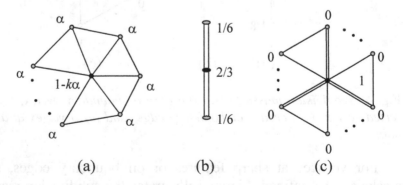

(a)　　　　　　(b)　　　　　　(c)

Fig. 5. *Limit position masks for (a) a smooth or dart vertex of valence k, (b) a crease or a boundary vertex, and (c) a dart vertex.*

Following Fig. 5(c), the limit position of a corner remains the same as that of the initial control vertex.

4 Least Squares Fitting

Starting from the initial dense mesh, we first mark the feature edges and feature vertices. The marked model is further simplified into a coarse mesh based on a given tolerance. Techniques for feature-preserving model simplification can be found in [1] and for general meshes in [4, 11, 14]. With our fitting approach, the control mesh has exactly the same topology and sharp features as those of the simplified mesh model.

To further proceed, let $Q_n = \{q_i \mid \text{for } i = 1,2,\cdots,n\}$ be the set of vertices of the simplified triangular mesh of the original dense mesh. We can produce a refined triangular mesh by subdividing the simplified mesh in one step through mid-point linear subdivision. Each of the newly inserted vertices is further replaced by the nearest vertex on the original mesh surface. We denote the set of vertices of this refined triangular mesh by $Q_m = \{q_j \mid j = 1,2,\cdots,m\}$, which will be used as the limit positions for surface fitting.

Now let $X_n = \{x_i \mid \text{for } i = 1,2,\cdots n\}$ be a set of vertices of the initial control mesh of a Loop subdivision surface with exactly the same topology as the simplified control mesh $\{Q_n\}$. Let $X_m = \{x_j \mid \text{for } j = 1,2,\cdots,m\}$ be the set of vertices of a refined control mesh after one subdivision of $\{X_n\}$. It should be noticed that this refined control mesh $\{X_m\}$ also has the same topology as that of the refined triangular mesh $\{Q_m\}$. Following the Loop subdivision scheme, we have

$$X_m = B_{m \times n} \cdot X_n \tag{6}$$

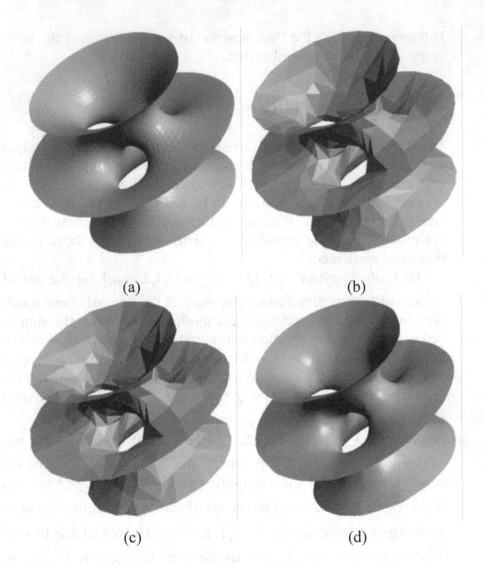

Fig. 6. *Smooth subdivision surface fitting: (a) the original dense mesh with 3571 vertices & 6928 triangles; (b) the simplified mesh with 282 vertices & 496 triangles; (c) the initial contro mesh with 282 vertices & 496 triangles; and (d) the fitted subdivision surface model illustrated by a refined control mesh after three subdivisions.*

where $B_{m \times n}$ is the Loop subdivision matrix with m rows and n columns that can be defined using the subdivision masks discussed in Section 2. Similarily, following the position masks discussed in section 3, we can also obtain a square matrix $A_{m \times m}$ such that

$$A_{m \times m} \cdot X_m = Q_m \tag{7}$$

Base on Equations (6) and (7), we thus have:

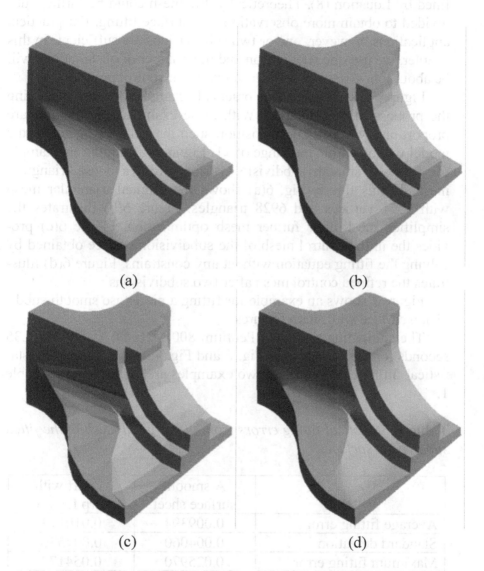

(a) (b)

(c) (d)

Fig. 7. *Piecewise smooth subdivision surface fitting: (a) the original dense mesh with 1698 vertices and 3392 triangles; (b) the simplified mesh with 173 vertices and 342 triangles; (c) the initial control mesh with 173 vertices and 342 triangles; and (d) the fitted subdivision surface model illustrated by a refined control mesh.*

$$C_{m \times n} \cdot X_n = Q_m \tag{8}$$

where $C_{m \times n} = A_{m \times m} \cdot B_{m \times n}$ is the coefficient matrix with m rows and n columns. We can thus obtain X_n by solving the linear system defined by Equation (8). Theoretically, the mesh could be further subdivided to obtain more observations for surface fitting. For practical applications, however, one or two subdivisions are sufficient. In this chapter, we use one subdivision and the number of observations will be about four times the number of unknown control points.

Figures 6-7 illustrate two practical examples for demonstrating the proposed fitting method. With these examples, the models are proportionally scaled and reposition such that all coordinates of the model will be within the range of ± 1. Figure 6 illustrates an example for fitting a smooth subdivision surface from a dense triangular mesh. In this figure, Fig. 6(a) shows the original triangular mesh with 3571 vertices and 6928 triangles. Figure 6(b) illustrates the simplified mesh after further mesh optimization. Figure 6(c) provides the initial control mesh of the subdivision surface obtained by solving the fitting equation without any constrains. Figure 6(d) illustrates the refined control mesh after two subdivisions.

Figure 7 shows an example for fitting a piecewise smooth subdivision surface with sharp features.

The computing time on a Pentium 800 MHz PC is 5.8 and 1.35 seconds for the examples of Fig. 6 and Fig. 7, respectively. The statistical fitting errors of these two examples are summarized in Table 1.

Table 1. *Statistical fitting errors from the original mesh to the fitted subdivision surface.*

	A smooth surface sheet	A part with sharp features
Average fitting error	0.009394	0.010113
Standard deviation	0.004060	0.007530
Maximum fitting error	0.025970	0.054177
Minimum fitting error	0.000637	0.000000

5 Summary

This chapter presents a new approach for reconstructing a smooth or a piecewise smooth subdivision surface from a dense triangular mesh. The initial mesh is first simplified into a coarse triangular mesh for capturing the topological structure. A subset of vertices of the initial dense mesh, including those of the simplified coarse mesh, is used as the limit positions of the subdivision surface to be identified. A set of fitting equations is further set up. Different from existing approaches, both subdivision rules and limit position masks are used when setting up the fitting equation. The control mesh of the subdivision surface is finally obtained by solving the fitting equation using a fast conjugate gradient method without any constraints. The reported approach preserves all feature edges and corners. The computing time is pretty fast. Although this chapter only demonstrates the case for Loop subdivision surfaces, it can also be applied to most of the stationary subdivision schemes.

Acknowledgements

The work described in this chapter is supported by a SRG grant 7001241 and an AoE grant 9360030 - CIDAM, both from CityU.

References

1. Catmull, E. and J. Clark (1978), Recursively Generated B-Spline Surfaces on Arbitrary Topological Meshes. Computer Aided Design **10**, pp.350-355.
2. DeRose, T., M. Kass and T. Tien (1998), Subdivision Surface in Character Animation, Compter Graphics Proceedings (SIGGRAPH'98), pp. 85-94.
3. Doo, D. and M. Sabin (1978), Analysis of the Behaviour of Recursive Division Surfaces near Extraordinary Points. Computer Aided Design **10**, pp.356-360.
4. Garland, M. and P. S. Heckbert (1992), Surface Simplification Using Quadric Error Metrics. Computer Graphics (Proc. SIGGRAPH'92), pp209-216.
5. Halstead, M., M. Kass and T. DeRose (1993), Efficient, Fair Interpolation using Catmull-Clark Surfaces, Computer Graphics Proceedings (SIGGRAPH'93), pp. 35-44.

6. Hoppe, H., T. DeRose, T. Duchamp, M. Halstead, H. Jin, J. McDonald, J. Cchweitzer and W. Stueltzle (1994), Piecewise Smooth Surface Reconstruction. Computer Graphics Proceedings (SIGGRAPH'94), pp. 295-302.

7. Litke, N., A. Levin and P. Schröder (2001), Fitting Subdivision Surfaces. *Proceedings of Scientific Visualization,* (http://www.multires.caltech.edu/pubs/).

8. Loop, C. (1987), Smooth Subdivision Surface Based on Triangles. Master's thesis, Department of Mathematics, University of Utah, USA.

9. Ma, W. and N. Zhao (2000), Catmull-Clark surface fitting for reverse engineering applications. In Proc. Geometric Modeling and Processing 2000, IEEE Computer Society, pp. 274-283.

10. Nasri, A. (1987), Polyhedral Subdivision Methods For Free-Form Surfaces, Communication of ACM, Transactions on Graphics **6**, January 1987, pp. 29-73.

11. Schroeder, W., J. Zarge and W. Lorensen (1992), Decimation of triangle meshes. Computer Graphics (SIGGRAPH'92 Proceedings), pp. 65-70.

12. Stam, J. (1998), Exact evaluation of catmull-clark subdivision surfaces at arbitrary parameter values. In Computer Graphics (Proc. SIGGRAPH'98), pp. 395-404.

13. Suzuki, H., S. Takeuchi and T. Kanai (1999), Subdivision Surface Fitting to a Range of Points. In Proc. 7th Pacific Conference on Graphics and Applications (Pacific Graphics'99), pp. 158-167.

14. Takeuchi, S., T. Kanai, H. Suzuki, K. Shimada and F. Kimura (2000), Subdivision Surface Fitting with QEM-based Mesh Simplification and Reconstruction of Approximated B-spline Surfaces. In Proc. 8th Pacific Conference on Graphics and Applications (Pacific Graphics'99), IEEE Computer Society, pp. 202-212.

15. Zorin, D. and P. Schroder (2000), Subdivision for Modeling and Animation, ACM SIGGRAPH'2000, Course Notes.

Chapter 05

A New Algorithm for Computing Intersections of Two Surfaces of Revolution

- **Introduction**
- **Preprocessing**
- **Intersection Determination**
- **Intersection Computation**
- **Experimental Results**
- **Conclusion**
- **References**

A Novel Algorithm for Computing Intersections of Two Surfaces of Revolution

Jinyuan Jia
Department of Computer Science, The Hong Kong University of Science & Technology, Clear Water Bay, Kowloon, Hong Kong.

Kai Tang
Department of Mechanical Engineering, The Hong Kong University of Science & Technology, Clear Water Bay, Kowloon, Hong Kong.

Ki-wan Kwok
Department of Computing, The Hong Kong Polytechnic University, Hung Hom, Kowloon, Hong Kong.

This chapter presents a novel method for computing the intersection curves of two surfaces of revolution RSIC. In this method, each surface of revolution is decomposed into a collection of coaxial spherical stripes along the generatrix, by subdividing its generatrix into a collection of C^0 or C^1 coaxial circular arcs centered on the revolute axis. Thus, computing intersections of two surfaces of revolution RSIC is reduced to computing intersection curves of two spherical stripes SSIC. RSIC can be represented as a piecewise C^0 or C^1 circular approximation, which is quite convenient for various operations such as offsetting, blending and so on, To avoid the unnecessary intersection computations, cylindrical bounding shell CBS is devised and valid intersection intervals VII is introduced. Finally, a simple algorithm is designed to trace RSIC for classification.

1 Introduction

Surface intersection continues to be one of critical and fundamental tasks in computer graphics, computational geometry, geometric modeling, and CAD/CAM applications. It is essential to Boolean operations necessary in the creation of Boundary Representation in solid modeling; it is also needed for trimming off the region bounded by the self-intersection curves of offset surfaces when designing complex objects in CAD and NC machining. Much work has been done in this field and many algorithms have been suggested for the intersections of two surfaces. However, it is still an extremely challenging work for one to develop a surface intersection algorithm that is accurate, robust, efficient, and without user intervention, even in the case of simple surfaces such as quadratic surfaces, tori and cyclides. In this paper we present a novel algorithm for finding the intersection of two surfaces of revolution.

1.1 Previous work on intersections of surfaces of revolution

Basically, there are four major methodologies, *subdivision method*, *lattice method*, *algebraic method* and *tracing method* for solving surfaces intersection problems. Among them the subdivision method is the most robust; but traditional triangulation produces too many triangles during the subdivision procedure that result in slow computing and huge memory demand. To overcome these drawbacks, we shall attend to enhance the *subdivision method* by proposing a new subdivision scheme for surfaces of revolution in this paper.

Using the traditional *subdivision method*, surfaces of revolution can be decomposed into $m{\times}n$ quadrilateral meshes or $2{\times}m{\times}n$ triangular meshes. However, such a subdivision must be dense for good approximation quality, and its robustness is difficult to guarantee. So traditional quadrilateral or triangular subdivision method becomes inefficient when it is applied directly to surfaces of revolution. Baciu *et al.* [1] subdivided surfaces of revolution into a set of truncated cones for rendering purposes by linear approximation of the generatrix, so it produces densely truncated cones.

Based on *circle decomposition*, as a special case of ringed surfaces, Heo *et al.* [3] reformulated the intersection problem of two surfaces of revolution as the zero-set of a bi-variate function $\lambda(u,v) = 0$, $||f(u)+g(v)||^2 = |R_1(v)|^2$ in the parameter space, where $f(u)$ and $g(v)$ are space curves of degrees $2{\times}max(d_1,1)$ and $2{\times}max(d_2,1)$, and d_1 and d_2 are the degrees of the generatrix functions $R_1(u)$ and $R_2(v)$ of the two surfaces of revolution respectively. However, the zero-set searching problem must use numerical solutions which demand large computing power and encounter convergence problem at some special cases. Thus, Heo's method is better suited for rendering *RSIC*, not the parameterization and classification for *RSIC*.

1.2 Main idea and our contributions

We propose a new idea of decomposing a surface of revolution into a collection of coaxial spherical strips, based on the following considerations:

- It can be done easily and naturally along the revolute axis.
- It reduces the task of computing the *RSIC* of two surfaces of revolution to that of computing the intersection curve of two spherical strips (*SSIC* for short).
- *An SSIC* can always be represented by a collection of circular arcs and can be computed geometrically, easily, accurately and robustly.
- Compared with traditional quadrilateral/triangular subdivision and Heo's *uv*-subdivision in parameter plane, the subdivided spherical stripes are far fewer in number and larger in size than the subdivided triangles.

Our other contributions include (a) proposing *spherical decomposition* for approximating surfaces of revolution; (b) proposing *valid intersection intervals* (VII) for avoiding unnecessary intersection computation; and (c) tracing the *RSIC* arc by arc and organizing the loop (branch) by loop (branch).

Generally, there are four main steps for computing a *RSIC*: (1) preprocessing (performing spherical decomposition and building the cylindrical bounding shell tree); (2) intersection detection for computing VII; (3) computing the *RSIC*; and (4) representing and storing the *RSIC*.

2 Preprocessing

2.1 Spherical decomposition

Spherical decomposition of a surface of revolution is achieved by rolling an elastic ball step by step along its axis. The center and radius of the ball varies adaptively in each step to fit the shape of the surface of revolution based on the specified error bound. Equivalently, using the 2D profile, we can roll an elastic circle along the axis to fit its symmetric generatrix pair, step by step. This corresponds to solving two problems: (1) which fitting method should be used for best local approximation of the generatrix by a circle; and (2) what is an appropriate method for adaptively modifying the step-size for the rolling ball in each step. To the first question, we choose circular *least square fitting* (*LSF*) as the locally optimal approximation. To the second question, we employ a *gold ratio marching*.

(1) C^0 and C^1 least square circular arc fitting

A general circle can be represented by $x^2 + y^2 + Dx + Ey + F = 0$; a circle, symmetric about the *x*-axis, can be simplified to $x^2 + y^2 + Dx + F = 0$. Assume that a collection of fitting circular arcs C_i (for step T_i) ($i = 1, 2, ..., m$) are defined in the interval $[x_i, x_{i+1}]$ and their equations are g_i: $x^2 + y^2 + D_ix + F_i = 0$, $i=1,2,...,m$. Within the interval $[x_i, x_{i+1}]$, $k+1$ sampling points x_{ij} ($0 \leq j \leq k$, $x_{i0} = x_i$, $x_{ik} = x_{i+1}$) are used for circular *LSF*, the goal is to find the best D_i and F_i for C_i from those sampling points. By minimizing the sum of the difference of the square distance on all the $k+1$ sampling points between the original generatrix segment G_i and its fitting circular arc C_i, then, by adding two interpolating constraints on two endpoints x_i and x_{i+1}, a C^0 circular *LSF* f is given as:

$$f = \sum_{j=1}^{k-1}(x_{ij}^2 + y_{ij}^2 + D_ix_{ij} + F_i)^2 + \lambda_1(x_{i0}^2 + y_{i0}^2 + D_ix_{i0} + F_i) + \lambda_2(x_{ik}^2 + y_{ik}^2 + D_ix_{ik} + F_i).$$

Differentiate f with respect to $D_i, F_i, \lambda_1, \lambda_2$

$$f'_{D_i} = \sum_{j=1}^{k-1}2(x_{ij}^3 + x_{ij}y_{ij}^2) + D_i\sum_{j-1}^{k-1}2x_{ij}^2 + F_i\sum_{j-1}^{k-1}2x_{ij} + \lambda_1x_{i0} + \lambda_2x_{ik} = 0,$$

$$f'_{F_i} = \sum_{j=1}^{k-1}2(x_{ij}^2 + y_{ij}^2) + D_i\sum_{j-1}^{k-1}2x_{ij} + F_i\sum_{j-1}^{k-1}2 + \lambda_1 + \lambda_2 = 0,$$

$$f'_{\lambda_1} = x_{i0}^2 + y_{i0}^2 + D_ix_{i0} + F_i = 0,$$

$$f'_{\lambda_2} = x_{ik}^2 + y_{ik}^2 + D_ix_{ik} + F_i = 0,$$

We can obtain D_i and F_i by solving the following linear equation system:

$$\begin{bmatrix} \sum 2x_{ij}^2 & \sum 2x_{ij} & x_{i0} & x_{ik} \\ \sum 2x_{ij} & \sum 2 & 1 & 1 \\ x_{i0} & 1 & 0 & 0 \\ x_{ik} & 1 & 0 & 0 \end{bmatrix} \begin{bmatrix} D_i \\ F_i \\ \lambda_1 \\ \lambda_2 \end{bmatrix} = \begin{bmatrix} -\sum 2(x_{ij}^3 + x_{ij}y_{ij}^2) \\ -\sum 2(x_{ij}^2 + y_{ij}^2) \\ -(x_{i0}^2 + y_{i0}^2) \\ -(x_{ik}^2 + y_{ik}^2) \end{bmatrix}. \tag{1}$$

Similarly, we can derive the formulation of C^1 circular *LSF* to the generatrix by adding the constraint of a tangential interpolation at each of the two endpoints of g_i.

(2) Adaptive global coaxial circular arc fitting

The basic idea is to approximate the generatrix with the C^0 or C^1 circular *LSF* for an initial interval Δx in the first step. Then we compare the specified tolerance e and the distance difference d between the original and the fitting conic arcs. If $d = e$, we fix the interval **dt** and begin fitting the next circular arc; if $d < e$, increase the size of interval of the fitting arc **dt** with the *inverse golden ratio* to $1.618dt$; otherwise, if $d > e$, decrease the interval size of fitting arc **dt** with the *golden ratio* to $0.618dt$. This procedure is applied recursively, trying to stretch each fitting circular arc as long as possible. Since such circular *LSF* can only ensure the optimal fitting locally in each step, this method does not provide globally optimal fitting.

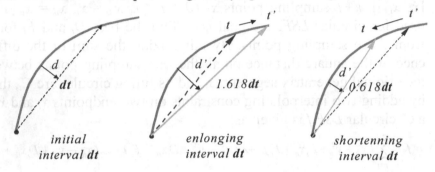

Fig. 1. *Adaptive fitting by golden ratio section*

2.2 Cylindrical bounding shell

Bounding boxes and bounding spheres are very frequently used for fast intersection tests in computer graphics. Considering the characteristics of surfaces of revolution, a *cylindrical bounding shell*, defined by its exterior and interior bounding cylinders, is used to enclose a surface of revolution both externally and internally as shown in Fig. 2 and 4. This shell encloses a surface of revolution more tightly than a bounding box, a bounding sphere or a single bounding cylinder. It is easy to determine the intersection between two cylinders geometrically by comparing distance d of the two axes and the sum of the two radii r_1+r_2. If $d > r_1 + r_2$, there is no intersection between the two cylinders; otherwise, they must intersect each other, if $d < r_1$ or $d < r_2$, then one cylinder traverses the other. A *binary cylindrical bounding shell tree* (BCBST) is organized by recursively bisecting the surface as shown in Fig. 2.

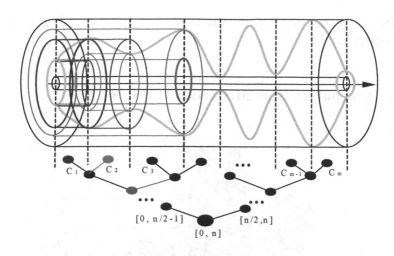

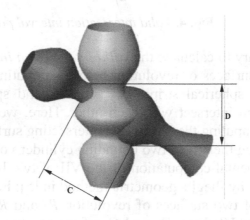

Fig. 2. *Binary cylindrical bounding shell tree*

3 Intersection determination

3.1 Valid intersection intervals VII

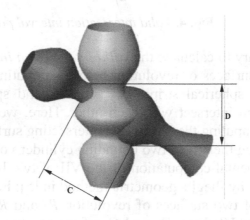

Fig. 3. *Valid intersection interval pair*

Two surfaces of revolution are decomposed into m and n spherical strips respectively. A brute-force algorithm would require the computation of intersection between $m \times n$ pairs of spherical strips. Usually, only few of them have real intersection with each other. These intersections occur within the overlapped intervals C and D (see Fig. 3), which are called *valid intersection intervals* (VII). Clearly, those spherical strips beyond the VII cannot intersect and hence need not be considered for intersection computation

3.2 Computing valid intersection intervals VII

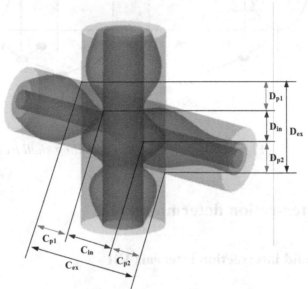

Fig. 4. *Valid intersection interval pair*

It is necessary to calculate the *valid intersection interval* VII pair (C, D) of two surfaces of revolution before computing the intersection. Only those spherical stripes S_i within C and spherical stripes S_j within D can intersect with each other. Here, we propose an algorithm of computing the VII of two intersecting surfaces of revolution by computing the VII of two bounding cylinder volumes recursively. The fundamental computation of the VII of two bounding cylinders can be done by Piegl's geometric method in [5]. For a pair of bounding shells of two surfaces of revolution R_1 and R_2 (see Fig. 4), we first compute VII pairs D_{in} and C_{in} of the two interior bounding cylinders, and D_{ex} and C_{ex} of the two exterior bounding cylinders respectively. D_{p1} and D_{p2} are two subintervals inside D_{ex} but outside D_{in}. C_{p1} and C_{p2} are other two subintervals inside C_{ex} but outside C_{in}. So, all the revolute quadrics of R_1 and R_2 can be categorized into three groups as follows:

- *Non-intersecting portion* - the revolute quadrics of R_1 outside D_{ex} and the revolute quadrics of R_2 outside C_{ex} cannot intersect; their intersection tests are not needed;
- *Definite intersecting portion* - the revolute quadrics of R_1 inside D_{ex} definitely intersect the revolute quadrics of R_2 inside C_{ex};
- *Potential intersecting portion* - the revolute quadrics of R_1 inside D_{p1} (D_{p2}) may intersect the revolute quadrics of R_2 inside C_{p1} (C_{p2}) or not, further intersection determination is expected.

Just by refining these interval pairs recursively, we can get the VII of the two surfaces of revolution R_1 and R_2. Algorithm1 below outlines the above idea of computing the VII of R_1 and R_2.

(1) ComputingVII_RR(R_1, R_2)
1. If two exterior bounding cylinders C_{01} and D_{01} do not intersect, return nil L_1 and L_2;
2. $k = 0$;
3. (C_{ex}, D_{ex}) = ComputingVII_CC(C_{01}, D_{01});
4. (C_{in}, D_{in}) = ComputingVII_CC(C_{02}, D_{02});
5. Compute the potential VIIs (C_{p1}, C_{p2}) from (C_{ex}, C_{in}) and (D_{p1}, D_{p2}) from (D_{ex}, D_{in});
6. Locate all spherical strips of R_1 within C_{in} by searching its BCBST;
7. Insert them into L_1 in order;
8. Locate all revolute quadrics of R_2 within D_{in} by searching its BCBST;
9. Insert them into L_2 in order;
10. Locate all revolute quadrics of R_1 within potential intervals C_{p1} and C_{p2} by searching BCBST and Form the new cylindrical bounding shells $C_1(C_{11}, C_{12})$ and $C_2(C_{21}, C_{22})$ respectively;
11. Locate all revolute quadrics of R_2 within potential intervals D_{p1} and D_{p2} by searching BCBST and Form the new cylindrical bounding shells $D_1(D_{11}, D_{12})$ and $D_2(D_{21}, D_{22})$ respectively;
12. $k=k+1$;
13. ComputingVII_SS(C_k, D_k);
14. $k=k+1$;
15. ComputingVII_SS(C_{k+1}, D_{k+1});
16. Return L_1 and L_2.

We terminate *Algorithm1* after a specified maximum number of iterations or when the number of revolute quadrics within VII cannot be diminished after two or three recursive iterations. Assume that L_1 and L_2 contains m_1 and n_1 overlapped spherical stripes respectively, then, the complexity of intersection computation is reduced from

$O(m \times n)$ to $O(m_1 \times n_1)$, which depends on the configuration of (relative orientations and positions) of two surfaces of revolution and the shape characteristics.

4 Intersection Computation

4.1 Computing Intersections of Two Spherical Stripes *SSIC*

Computing the intersections of two surfaces of revolution is reduced to computing the intersection curves of two spherical stripes (*SSIC*s). The intersection curve of two spheres is always a circle; a pair of parallel bounding planes trim it into two symmetric arcs in general, and these two arcs can be further classified according to their overlapping configurations; there are totally five cases of overlapping configurations (see Fig. 5): (1) no intersection (there is no overlapping between the two trimmed circular arcs; (2) one circular arc;(3) two circular arcs; (4) a complete circle; and (5) an isolated point, *singularity*, which happens when one bounding plane is tangential to the intersecting circle.

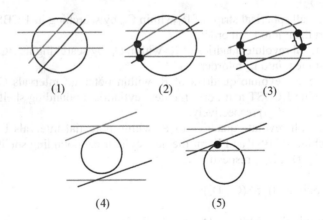

Fig. 5. *Five configurations of 4 bounding planes truncating an SSIC*

4.2 Tracing *RSIC*

Recognizing loops and singularity of the intersection curves is always important in geometric modeling and CAD. Because our subdivision scheme is based on one-dimensional subdivision along axes, tracing a *RSIC* becomes much easier than in a two-dimensional subdivision (e.g. a triangular subdivision) – a *RSIC* can be traced sequentially along the axis of the surface of revolution. We can generate the final *RSIC* by concatenating the associated *SSICs* one by one sequentially; the concatenation cases of two adjacent *SSICs* can be classified as shown in Fig. 6. Each *SSIC* has two possible cases (two-branches or one branch). There is no contact between the two adjacent *SSICs* at their endpoints in (1), so no concatenation is needed to do in this case; a two-branches *SSIC* touches a one-branch *SSIC* at one common endpoint in (2) and (3); a two-branches *SSIC* (doubly) touches another two-branches *SSIC* at their two common endpoints in (4); a one-branch *SSIC* touches another one-branch *SSIC* at a single point in (5), singularity occurs in this case; a one-branch *SSIC* touches a two-branches *SSIC* at their two common endpoints in (6); and finally, a one-branch *SSIC* (doubly) touches another one-branch *SSIC* at their two common endpoints in (7), they form a closed loop. With respects to these 7 conjunction cases, we give the following Algorithm2 to merge all individual *SSICs* together into the final *RSIC*.

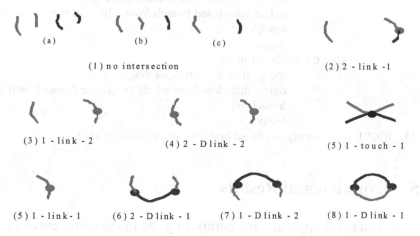

Fig. 6. *Seven conjunction cases of two SSIC arcs*

(2) Algorithm2. TracingRSIC(R_1,R_2)
1. Compute D_1 and D_2 (VII of R_1 and R_2);
2. branch = 0; k = 0;
3. Initialize the active branch list ABL into empty;
4. for each spherical stripe SS_i of R_1 in D_1
5.   for each spherical stripe SS_j of R_2 in D_2
6.   if $SS_i \cap SS_j = \phi$, continue;
7.   k = k+1;

8.   Compute $SSIC_k$;

9.   Match $SSIC_k$ with all the current $SSIC_m$ in ABL by checking the co-incidence type of their current open endpoints in Fig. 6;

10.   Switch (conjunction type of $SSIC_k$) {

          case 1 (no conjunction):

              branch++;

              insert new branch which begins with $SSIC_k$ into ABL;

              break;

          case 2 (one-touch-one):

              report singular point;

              merge it to the corresponding branch;

              refresh its active endpoints for next matching;

              break;

          case 3 (one-link-one, one-link-two, two-link-one or two-Dlink-two):

              merge it to the corresponding branch;

              refresh its active endpoints for next matching;

              break;

          case 4 (one-Dlink-two):

              merge it to the corresponding branch;

              refresh its active endpoints for next matching;

              break;

          case 5 (two-Dlink-one):

              merge it to the corresponding branch;

              delete this closed branch from ABL and store it into ILL;

              branch++;

              break;

          case 6 (one-Dlink-one):

              merge it to the corresponding branch;

              delete this closed branch from ABL and store it into ILL;

              branch++;

              break;

11. if ABL is not empty, store all branches one by one into ILL.

5   Experimental Results

The prescribed approach for computing the intersection curve of two surfaces of revolution (*RSIC*) was implemented; we present three examples here. The first one is a three-branches *RSIC* generated by two bowl-shaped revolve objects with perpendicular configurations as shown in Fig. 7. The second one is a two-branches *RSIC* generated by two dumbbells with a slight angle shown in Fig. 8. The last one is a single-branch *RSIC* generated by two dumbbells with perpendicular configurations shown in Fig. 9.

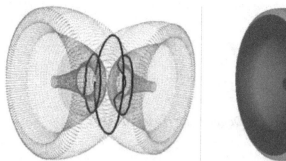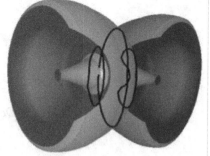

Fig. 7. *Two bowl-shaped objects with three branches of RSIC*

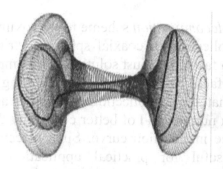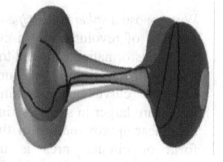

Fig. 8. *Two dump-bells with two branches RISC*

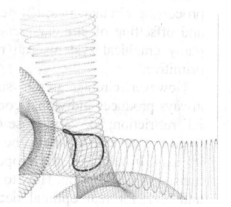

Fig. 9. *Two dumbbells with one loop RSIC*

Finally, we note two special degenerate cases:

- *RSIC* degenerates to a list of parallel circles or common spherical stripes if two surfaces of revolution become coaxial;
- *RSIC* is the surface itself if two identical surfaces of revolution coincide completely.

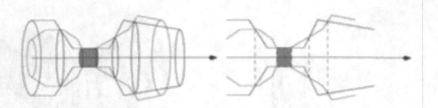

Fig. 10. *Degenerate Cases of RSIC*

6 Conclusion

We propose a *spherical stripe decomposition* scheme to approximate surfaces of revolution by a collection of coaxial spherical stripes. This transformation can lead to fast and robust solutions for computing the intersection of two surfaces of revolution. The resulting intersection curves are approximated as a collection of circular arcs, which are larger in size, less in number and of better continuity than the linear approximation of the intersection curve. Such piecewise form of circular arcs is useful for practical applications in CAD/CAM attributed to the following reasons: (1) many industrial computing devices have direct and special processing capability for processing circular arcs; (2) geometric computations such blending and offsetting of circular arcs are easy, simple and exact; and (3) many graphical systems can render a circular arc as a graphical primitive.

However, circular arcs in such *spherical decomposition* scheme always produces individual convex curves due to the strong 'coaxial' restriction; so, very dense coaxial spherical stripes are required to fit the generatrix where the generatrix becomes concave toward its axis. In addition, the proposed circular *LSF* method is only a quasi-optimal solution. How to better deal with a concave generatrix and seek for real optimal decomposition method for surfaces of revolution are open for future work.

References

1. Baciu G., Jia J. and Lam G. (2001) Ray Tracing Surface of Revolution: An Old Problem with A New Perspective", *Proceedings of CGI'01*, pp. 129-135.

2. Kim M. S. (2000) The Intersection of Two Simple Sweep Surfaces, *Proceedings of the Riken Symposium on Geometric Processing for Innovative Applications*, pp. 1-17.
3. Heo H. S., Hong S. J., Seong J. K. and Kim M. S. (2001) The Intersection of Two Ringed Surfaces and Some Related Problems, *Graphical Model*, 63(4).
4. Patrikalakis N. and Johan H. (2001) Intersection Problem, to appear in *Handbook of Computer Aided Geometric Design*, Farin G., Hoschek J. and Kim M. S. (Eds.), Elsevier, Amsterdam.
5. Kwok Ki-wan (2001) Efficient Computing Intersection Curves of Two Surfaces of Revolution, MS Thesis, Hong Kong University of Science and Technology.
6. Piegl L. (1989) Geometric Method of Intersecting Natural Quadrics Represented in Trimmed Surface Form", *CAD*, 21(4), pp. 201-212.
7. Les Piegl and Wayne Tiller (1997) *The NURBS Book*, Springer.
8. Shene C. K. and Johnstone J. K. (1994) On The Lower Degree Intersection of Two Natural Quadrics, *ACM Transactions on Graphics*, 13(4), pp. 400-424.
9. Sarfraz M. and Habib, Z. (2000) Rational cubic and conic representation: A practical approach", *IIUM Engineering Journal*, Malaysia, 1(2), pp.7-15.

Chapter 06

An Improved Scan-line Algorithm for Display of CSG Models

An Improved Scan-line Algorithm for Display of CSG Models

Kohe Tokoi
Faculty of Systems Engineering, Wakayama University, 930, Sakaedani, Waka-yama, 640-8510, Japan.

Tadahiro Kitahashi
 School of Science & Technology, Kwansei Gakuin University, 2-1, Gakuen, Sanda, 669-1337, Japan

A scan-line algorithm uses a one-dimensional set operation instead of a depth comparison to find a visible surface of the object shape defined by Constructive Solid Geometry (CSG). However, this method requires considerable computing time for retrieving the CSG data structure to find visible surface by the set operation.

In this chapter, we discuss 3 steps to improve the scan-line algorithm to display the CSG models on a screen. First, changing the order of application of the partial procedures in the algorithm can improve the performance in the interactive modelers. Secondly, substituting set operations for simple manipulations of the elements in two kinds of lists reduces computing time. Finally, using the connection information at the adjoining facets drastically reduces the number of set operation executions.

1   Introduction

Constructing object models is still one of the biggest problems in Computer Graphics (CG for short). A geometric modeler equipped with an interactive mechanism is convenient for CG model designers. However, it must be able to immediately generate half-tone image of the object in order to check whether the present operation in the design procedure is appropriate or not.

A variety of algorithms are already proposed for this purpose. However, changing the conceptual model of existing rendering schemes opens the way to improving the performance of some of them. In this chapter, we will show an example of this for scan-line-based rendering algorithms.

Constructive Solid Geometry (CSG for short) is one of the most popular ways to describe geometric models of objects for CG. In CSG, a three-dimensional (3D for short) model is generated by applying set operations to the selected geometric primitives located at the desired positions in the 3D space.

Even for solid modelers that employ a boundary representation to describe objects for CG, set operations are often useful to create complex shaped models.

In an interactive modeler, it is necessary to display the created model on a screen in order to check its shape, and if it is different from the expected one, the added modification should be cancelled to restore the previous model.

Noticing the actual effect of a rendering algorithm of the CSG model, we have succeeded in improving the rendering speed by only changing the order of the data processing of a created model and its checking procedure.

That is, the hidden-surface removal procedure for a CSG model consisting of two geometric primitives produces the same image as that of the data after being completed the set operations in case of casting the target model. This indicates that it is possible to get an image in order to check the currently obtained shape before changing its geometric data by applying the set operations. This is a desirable property for interactive solid modelers.

Accordingly, it is useful to generate an image directly from the description formula of the CSG model before applying the set operations to the data of the model especially in interactive modelers. Several types of algorithm can be used to do this: a ray casting algorithm[1], a depth-buffering algorithm[2], an extended depth buffer algorithm[3], a stencil buffer algorithm and a scan-line CSG algorithm[4]. All these algorithms find a visible surface by applying one-dimensional set operations to the parts of facets overlapping in the depth direction. The processing time of this task depends on the complexity of Boolean combinations, but not on the total number of the facets consisting of the models. Consequently, the processing time each algorithm takes is about $O(m+n)$ for the number of all facets $m+n$ that come from the two geometric primitives. This order is equivalent to a normal hidden surface elimination procedure.

Some algorithms can be accelerated using special hardware or parallel processing. However, for methods other than a scan-line-based algorithm, the acceleration rate will be small whenever the display resolution is raised and the description of the set operations becomes complex, since they execute the tasks pixel-by-pixel. In contrast, a scan-line-based algorithm is more efficient because the execution time does not depend on the number of pixels on a scan-line, but on the number of sample spans on it.

Even in this algorithm, the execution time of the set operations depends on the number of scan-lines on the screen and the total number of facets of all the objects. The computing time of these occupies a large part of the whole rendering process.

To overcome these difficulties, the performance of the scan-line-based algorithm can be improved by using a pattern matrix [6] that performs the one-dimensional set operations by means of simple incremental calculation. In addition, the number of executions of the set operations can be drastically reduced by knowledge about the connections of the adjoining facets obtained from the Winged Edge Data structure (WED for short). This results in a robust and high-speed CSG display algorithm relevant to interactive solid modelers.

2 The CSG Display Algorithm using the Spanning Scan-Line Algorithm

In a direct CSG display using a scan-line-based algorithm, the visible surface is found by applying one-dimensional set operations to the facets overlapped in depth direction in each sample span instead of the normal depth comparison used by the hidden surface elimina-

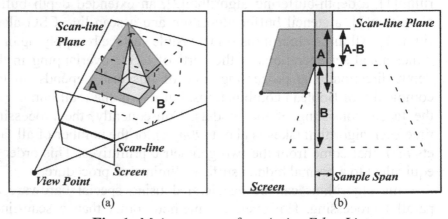

(a)   (b)

Fig. 1. *Maintenance of an Active Edge List*

tion algorithm. In Fig. 1 (a), the object shape is defined by the subtraction of primitive B from primitive A. The visible surface in this part is a back-face of the primitive B in the sample span as shown in Fig. 1 (b). The outline of the whole procedure is as follows.

2.1 Maintenance of an Active Edge List

In this section, we mention how to maintain an Active Edge List (AEL for short). An AEL stores the edges that cross the scan-line currently being processed. Fig. 2 shows the flow of the procedure that we employed.

(1) Sort each of the edges by the *y*-coordinates of their lower endpoints and store each of them in the individual *y*-Entry Lists.

(2) Repeat the following processes from bottom to top of the scan-lines on the screen.

(3) Consider a *y*-Entry List corresponding to the current scan-line. Sort the edges in the *y*-Entry List by the *x*-coordinates of the lower endpoints of the edges, and merge them into the AEL.

(4) If the current AEL is not empty, execute the procedure ex-

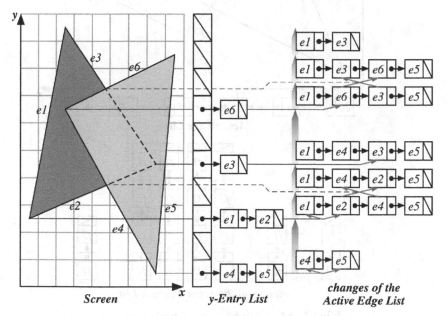

Fig. 2. *Maintenance of an Active Edge List*

99

plained below.

(5) Remove the edges whose higher endpoints are located at the current scan-line position from the AEL.

(6) With respect to the edges that remain in the AEL, add each of unit value of inclination of them to their x and z coordinates, and find the intersection position of them with next scan-line.

(7) Sort the edges of the current AEL by their x-coordinates.

2.2 Maintenance of an Active Polygon List

In this section, we mention how to maintain an Active Polygon List (APL for short). An APL stores the facets that overlap the sample span currently being processed. These facets are selected from all facets based on an AEL. Fig. 3 shows the flow of this procedure.

The visible surface is readily found by selecting the facet closest to the screen from those in the APL in a normal hidden surface elimi-

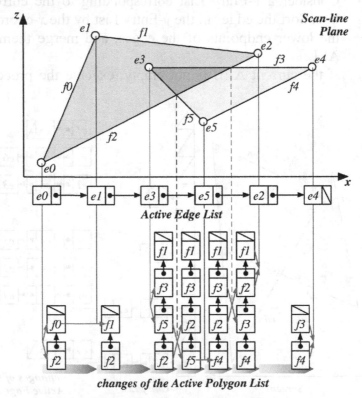

Fig. 3. *Maintenance of Active Polygon List*

nation procedure. However, in the CSG display procedure, the APL needs to keep all the facets overlapping the sample span under processing in depth order. Furthermore, in this algorithm, two facets share an edge because a primitive is defined by WED. Thus, the following procedure is somewhat more complicated than in normal hidden surface elimination.

(1) Set the initial left-end position of a sample span at the x-coordinate of the leading edge of the AEL.

(2) Extract all the edges that are located at the current left-end position of the sample span taken from the AEL.

(3) Check whether the two facets sharing an edge are already registered to the APL or not.

    (a) If both facets are not registered, like $f0$ and $f2$ at $e0$, register them in the APL. A simple insertion algorithm may be used for this registration to keep the facets in order.

    (b) If both facets are already registered, like $f1$ and $f2$ at $e2$, remove them from the APL.

    (c) If one facet is already registered and the other is not, like $f0$ and $f1$ at $e1$, replace the registered facet ($f0$) with the unregistered one ($f1$).

(4) Set the initial right-end position of the sample span at the x-coordinates of the leading edge remaining in the AEL.

(5) If the current APL is not empty, then execute the procedure explained next section.

(6) Set the left-end position of the next sample span at the right-end position of the sample span, and then repeat from (2) to (6) until the AEL becomes empty.

2.3 Subdivision of a sample span

In a normal hidden surface elimination, the sample span is divided recursively until no intersections of facets are found, before the visible surface is found. However, this method has certain problems, such as below, when applied CSG display.

- The APL keeps the facets in depth order by the procedure described in section 2.2. However, the normal method does not use this property when sorting the facets.
- When a sample span is divided, the depths of the registered facets in the APL must be recalculated every time.

- It tends to get an incorrect visible surface by the error when the length of a sample span is short and the angle of the two facets that intersects is small. This problem destroys the generated image when the span coherency is used to improve efficiency of finding a visible surface.

Therefore, the method we propose in this chapter uses the following procedure that is similar to a bubble sort algorithm to find a visible surface.

(1) Calculate the depth of the registered facets in the APL at the right-end of a sample span (Fig. 4 (a)).

(2) Examine whether the two neighboring facets are in depth order in the APL. If they are not in depth order, at least one facet intersects another facet within a sample span.

(3) If the pair of facets that intersect exists, record them and find the position of the intersection, and then move the right-end position of that sample span to the intersection position that is the closest to the left-end of the sample span (Fig. 4 (b)).

(4) Now, there is no intersection within a current sample span (Fig. 4 (c)). If its length is longer than zero, execute the one-dimensional set operations described below in section 3.2 and find a visible surface.

(5) This procedure stops if there is no recorded pair of facets.

(6) Swap the facets of the recorded pair in the APL. The facets of the APL are in depth order at the right-end position of a sample span by this operation (Fig. 4 (d)).

(7) Move the left-end position of the sample span to the right-end of the current sample span.

(8) Repeat from (2).

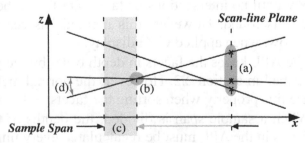

Fig. 4. *Division of a sample span*

3   Set Operation

By the procedure mentioned above in section 2.3, the APL keeps active facets in depth order, and no intersections exist within the current sample span. The one-dimensional set operation is executed using the APL in order to generate an image of Boolean-combined polyhedrons. This can be done by simple incremental calculation using a pattern matrix [6] describing the Boolean combinations of the primitives.

3.1   A pattern matrix

A pattern matrix is a shape description language introduced in a solid modeler TIPS-1 [6]. A pattern matrix describes a shape in terms of the segment S_j defined by the intersection of the several elements (E_{ij}). Each element is called a primitive.

$$S_j = \bigcap_{i=1}^{m} E_{ij} \tag{1}$$

A union of the segments defines the shape T.

$$T = \bigcup_{j=1}^{n} S_j \tag{2}$$

This can be also described by matrix notation. This is called a pattern matrix.

$$T = \begin{pmatrix} E_{11} & E_{21} & \cdots & E_{m1} \\ E_{12} & E_{22} & \cdots & E_{m2} \\ \vdots & \vdots & \ddots & \vdots \\ E_{1n} & E_{2n} & \cdots & E_{mn} \end{pmatrix} \tag{3}$$

A row of this matrix expresses a segment. The negative sign of the element implies the subtraction operation. Any CSG tree can be converted into this matrix by a normalization algorithm [7]. We express a pattern matrix in the form of an element list and segment list as shown in Fig. 5. A *counter* shown in this figure is used for the set operation.

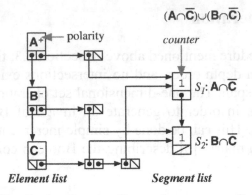

$(A \cap \overline{C}) \cup (B \cap \overline{C})$

Fig. 5. *An element list and a segment list*

3.2 Performing a one-dimensional set operation by incremental calculation

A segment exists among all positive elements. Therefore, following simple incremental calculation can perform the one-dimensional set operation.

(1) The number of positive elements that belong to each segment is stored in a *counter*. (Fig. 5)

(2) Then, each facet registered in the APL is examined in its order. If the front of a facet faces the screen, decrease by one the *counter* of the segment that the facet belongs to. If not, it increases by one. (Fig. 6)

(3) When a *counter* of any segment becomes zero after this operation, that facet is made a visible surface. The facets that belong to the negative element are processed inside out.

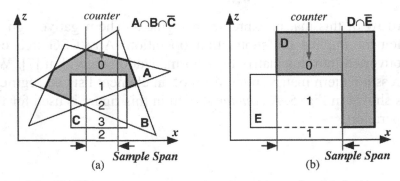

Fig. 6. *The set operation using incremental calculation*

4 Improving Efficiency of Finding the Visible Surface

A set operation using incremental calculation, still takes more time than normal hidden surface elimination. Therefore, it is necessary to reduce the number of execution times of the set operations to improve efficiency.

4.1 Using sample span coherency

Coherency of neighboring sample spans allows us to reduce the number of repetitions of set operations. A visible surface at the previous sample span is used at the current sample span if it does not satisfy all of the following conditions.

- Another facet is inserted in front of the visible surface.
- The visible surface is removed from the APL.
- The current sample span is divided at the intersection.

The visible surface, however, is valid when another facet is inserted in front of it by the operation of section 2.2. (c). The visible surface replaced by another facet in this operation makes that facet the visible surface. Consequently, the one-dimensional set operation is only executed on the silhouette line and on the intersection line. This is the same technique as the segment chain[8].

4.2 Omitting the intersection behind the visible surface

In the procedure described in section 2.3, the sample span under processing is divided by step (3) if an intersection exists. Then perform the one-dimensional set operation on the right part of the dividing point. However, when this intersection is behind the visible surface, this intersection does not influence the visible surface. Therefore, it is not necessary to divide the sample span in this case. This reduces the number of set operation repetitions to less than half of that when the sample span is divided.

5   Experiments

5.1   Instruments and conditions of the experiments

The computer used for the experiments was a SGI Indy R5000 (R5000PC, 150MHz, 96MB main memory, XZ graphics, IRIX 6.2). No optimization options were designated because the profiler was used for measuring the processing time.

5.2   5.2. The processing time of this algorithm

The processing time for generating the images shown in Fig. 7 is measured. Each shape consists of nine polyhedral cylinders. One is a union of all the cylinders (indicated by zero segments). Others are the images presenting eight types of the remaining constructions after removing parts of cylinders one by one.

Fig. 9 shows that the processing time of the one-dimensional set operation for this algorithm does not depend on the number of all facets of objects. This is why Fig. 8 shows that the order of the total processing time of this algorithm is equivalent to the normal scanline hidden surface removal procedure.

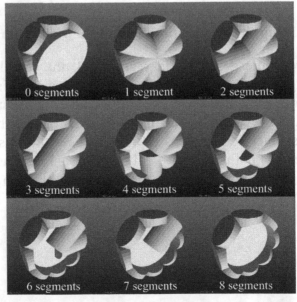

Fig. 7. *Sample data for the experiment in section 5.2.*

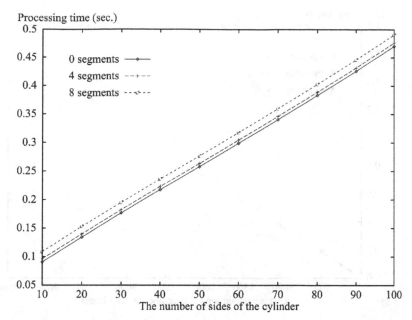

Fig. 8. *The relationship between the number of the sides of a cylinder and the processing time*

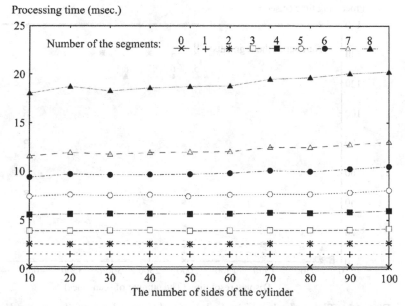

Fig. 9. *The relationship between the number of the sides of a cylinder and the processing time of the set operation (screen resolution is 512 x 512)*

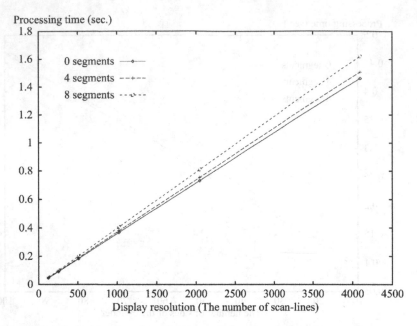

Fig. 10. *The relationship between the number of scan-lines on a screen and processing time (32 sides, total 306 facets)*

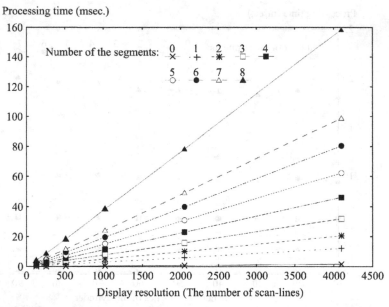

Fig. 11. *The relationship between the number of the scan-lines of the screen and the processing time of the set operations (32 sides, total 306 facets)*

Fig. 10 and Fig. 11 show that processing time of this algorithm is in proportion to the number of scan-lines of the screen, but not to the number of pixels. They also show that the property of this algorithm is equivalent to the normal scan-line hidden surface removal procedure.

5.3 The effect of improvement increasing efficiency

The effect of the improvement described in section 4 is examined. Fig. 12 shows each processing time of part of the program. (a), (b) and (c) means as follows.
  (a)   No improvement technique used.
  (b)   Using span coherency.
  (c)   Omitting the intersection behind the visible surface.
Fig. 12 shows that use of span coherency is very effective. In contrast, omitting the intersection behind the visible surface is not. However, the overhead of this technique is constant, so it will be effective when the Boolean combination become more complex.

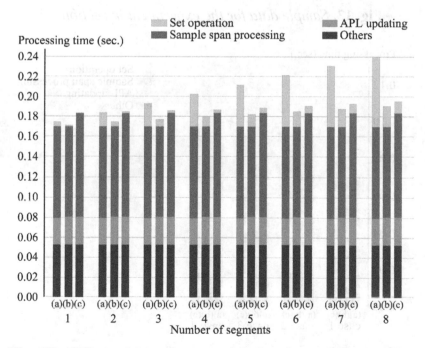

Fig. 12. *Effect of improvement for the shape shown in Fig. 7 (screen resolution is 512 x 512, 32 sides, total 306 facets)*

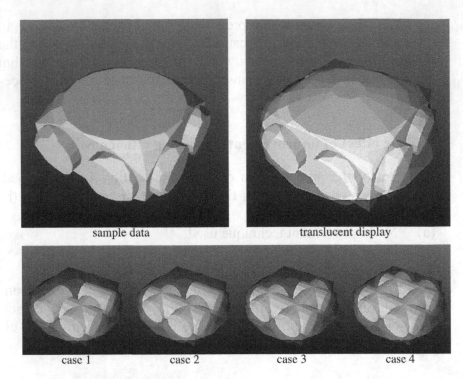

sample data   translucent display

case 1   case 2   case 3   case 4

Fig. 13. *Sample data for the experiment in section 5.3.*

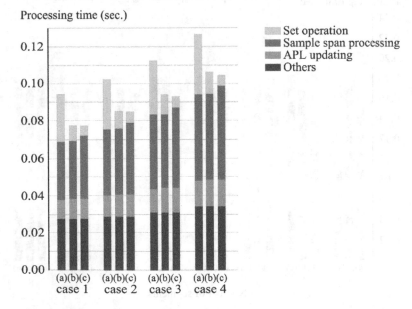

Fig. 14. *Effect of improvement for the shape shown in Fig. 13*

The same experiment as the above is done for the shape shown in the sample data of Fig. 13. The shape is a union of four polyhedral cylinders and a rectangular solid, which intersects surrounding polyhedral ellipsoid as shown in translucent display. There are no changes of the images generated from the shapes shown in cases 1 to 4, but they have different numbers of hidden intersections. Fig. 14 shows that the above method is more effective when the hidden intersections are increased.

6 Summary

The processing time of this CSG scan-line algorithm is proportional to the number of scan-lines on the screen and the number of all facets of the objects, but it does not depend on the number of the pixels on a scan-line. In addition, it has almost no dependence on the complexity of CSG description.

Besides this algorithm is able to realize translucent display shown in Fig. 13. So it becomes easy to confirm a section or position a primitive with a translucent display of negative primitive.

Consequently, we think this algorithm is suitable for interactive solid modeling environment.

References

1. Roth, S. D., "Ray Casting for Modeling Solids", Computer Graphics and Image Processing, Vol. 18, pp. 109-144 (1982)
2. Rossignac, J. R, and Requicha A. A. G., "Depth Buffering Display Techniques for Constructive Solid Geometry", IEEE Computer Graphics and Applications, Vol. 6, No. 9, pp. 29-39 (1986)
3. Okino, N., Kakazu, Y, and Morimoto, M., "Extended Depth-Buffer Algorithms for Hidden-Surface Visualization", IEEE Computer Graphics and Applications, Vol. 4, No. 5, pp. 79-88 (1984)
4. Atherton, P. R., "A Scan-line Hidden Surface Removal Procedure for Constructive Solid Geometry", Computer Graphics (Proc. SIGGRAPH83), Vol. 17, No. 3, pp. 73-82 (1983)
5. Tokoi, K. and Kitahashi, T., "Acceleration Techniques for a Scan-line Display Algorithm of Constructive Solid Geometry", Transactions of Information Processing Society of Japan, Vol. 34, No. 11, pp. 2412-2420 (1993) (Japanese)

6. Okino, N., Kakazu, Y. and Kubo, H., "TIPS-1, Technical Information Processing System for Computer Aided Design, Drawing and Manufacturing", Proceedings of PROLAMAT'73, (1973)
7. Goldfeather, J. and College, C., "Near Real-Time CSG Rendering Using Tree Normalization and Geometric Pruning", IEEE Computer Graphics and Applications, Vol. 9, No. 3, pp. 20-28 (1989)
8. Yamaguchi, F. and Tokieda, T., "Development of Hidden Line Elimination Algorithm and Hidden Surface Elimination Algorithm by Classification Method", Proceedings of Graphics and CAD Symposium, pp. 133-140 (1983) (Japanese).

Chapter 07

Rotating Squares to Boundary Cubes: Boundary Representation and Classification Techniques for Two- and Three-dimensional Objects

Rotating Squares to Boundary Cubes:
Boundary Representation and Classification Techniques for Two- and Three-dimensional Objects

Carsten Maple
Department of Computing and Information Systems, University of Luton, Luton, Bedfordshire, LU1 3JU. United Kingdom.

In this chapter we present algorithms for the representation and classification of the boundaries of two- and three-dimensional objects. The techniques covered are based on the Marching Cubes algorithm of Lorenson and Cline and differ from standard approximations such as those highlighted in Rosin as they do not involve any loss of data when using pixelised values.

The techniques also permit a method for the charcterisation of objects by their boundaries; all that is required is the application of standard graph comparison algorithms on these modified data structures.

1   Introduction

The Marching Cubes algorithm of Lorensen and Cline [5], is a surface rendering technique used to produce 3D images given point data from 3-space. The data is given on orthogonal grids over parallel planes, such as can be seen in Fig. 1. The distance between the planes is the same as the grid mesh size in each plane. Data is given at each of the vertices of every square in every slice. This pixel data is transformed into values of either 0 or 1; a 0 corresponding to a vertex lying outside the surface, a 1 corresponding to all other vertices.

The traditional data structure for this type of information is a standard three-dimensional array, where the value of the pixel, 0 or 1, at position (i, j, k) is stored in position $[i, j, k]$ of the array. There is a simple mapping from a pixel-wise representation to a marching cube representation. Two consecutive slices, k and $k-1$ say, are used to define a layer of marching cubes. Four neighbouring pixels that form a square from each of the two slices comprise a cube. Fig. 2 shows a series of cubes formed by two consecutive slices.

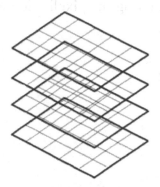 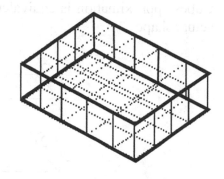

Fig. 1. *Pixelwise data is given over aligned grids on parallel planes.*

Fig. 2. *Two consecutive slices are used to construct a set of Marching Cubes.*

The marching cubes algorithm forms a piecewise-planar approximation to an object. The boundary of the object is assumed to pass exactly half way between a pixel that is exterior to the boundary of an object and a neighbouring pixel that is not. This determines the way in which the surface intersects a cube, if indeed it does. There are 8 vertices to each cube, each of which can be outside the surface of the object or not. There are, therefore, 256 different ways in which a cube can intersect the surface. In order to uniquely identify each of the 256 ways we assign a weighting value to each vertex of a cube as shown in Fig. 3.

If the pixel at the front bottom-left of the cube is outside the cube it is assigned a value of 0, and a 1 otherwise; if the pixel at the back top-right of the cube is outside the cube it is assigned a value of 0, and 64 otherwise, and so forth. A cube is given the value that is the sum of the value of its vertices. Two examples of cubes, their values and approximations is given in Fig. 4. Once an object is represented in this form, the normals for each of the 256 possible cubes can be found by simply consulting a table. The Marching Cubes algorithm then uses rendering techniques produce 3D images. In this chapter we are not concerned with the normals to the surface nor the rendering of the surface, but simply with the way of representing the data in voxels. Equally, the fact that the Marching Cubes algorithm only gives an approximation to an object is also not of consequence.

115

Since every unique shape of an object has a single Marching Cubes approximation (for a given pixelisation) and there exists a simple mapping between a shape and its approximation, storing a Marching Cubes approximation is equivalent to storing the pixelisation of the actual shape.

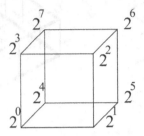

Fig. 3. *The weightings given to the vertices of each cube.*

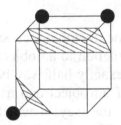

Fig. 4. *Two examples of cubes and their marching cubes approximations. The left cube has a value of 6 and the right a value of 193.*

2 Representation Techniques for Two-dimensional Objects

We will begin by defining the Marching Square. This can be used to give a piecewise-linear approximation to a two-dimensional. An image is, usually, electronically scanned and the bitmap pixel values then used to give function values over a grid. A shape may also be defined by a series of equations over intervals; these equations can then be used to determine the two-dimensional object they enclosed, which in turn can be used to define the pixel values over a uniform grid.

2.1 Marching Squares

In a manner analogous to that of marching cubes, we assign each vertex of each square a value of either 0 or 1; one corresponding to a vertex on the boundary or in the interior of the shape with which we are concerned, and zero corresponding to all other vertices. These values can then be used to assign a value to each square that denotes the manner in which the shape boundary passes through the square. If we allow the lower left vertex of a square to contribute 0 or 1 to the square's value, the lower right to contribute 0 or 2, and so on round the four vertices, then the binary numbers 0000 to 1111 can be used to uniquely identify this intersection. We store the decimal equivalent of these binary numbers.

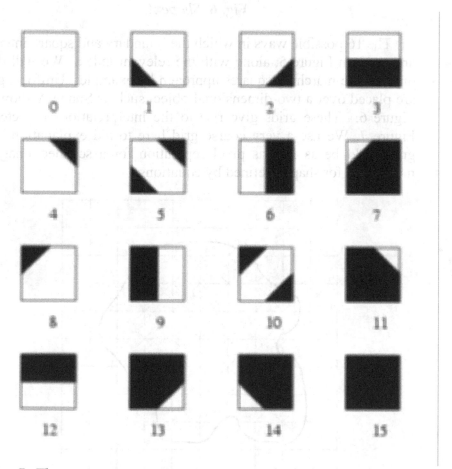

Fig. 5. *The approximations to the way in which the boundary intersects a grid square.*

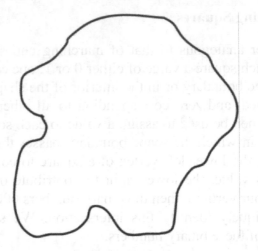

Fig. 6. *Shape A.*

The 16 possible ways in which the boundary and square intersect are given in Figure 5, along with the relevant index. We will demonstrate the marching squares approach by example. Uniform grids are placed over a two-dimensional object such as Shape A shown in Figure 6. These grids give rise to the interpretations depicted in Figure 7. We use a very coarse grid here to aid explanation. This grid could be as fine as pixel separation for a scanned image or much finer for shapes defined by equations.

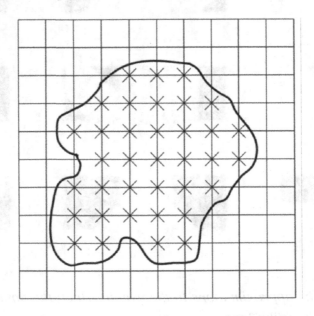

Fig. 7. *Coarse pixelisation of Shape A.*

Assigning the vertices the weightings given in Fig. 8, we can represent this object using the array 2D_Object as follows:

$$2D\_Object = \begin{pmatrix}
0 & 0 & 0 & 0 & 0 & 0 & 0 & 0 & 0 & 0 \\
0 & 0 & 0 & 2 & 3 & 3 & 1 & 0 & 0 & 0 \\
0 & 0 & 2 & 11 & 15 & 15 & 7 & 1 & 0 & 0 \\
0 & 2 & 11 & 15 & 15 & 15 & 15 & 7 & 1 & 0 \\
0 & 8 & 14 & 15 & 15 & 15 & 15 & 15 & 5 & 0 \\
0 & 2 & 11 & 15 & 15 & 15 & 15 & 13 & 4 & 0 \\
0 & 10 & 15 & 15 & 15 & 15 & 13 & 4 & 0 & 0 \\
0 & 10 & 15 & 13 & 14 & 15 & 5 & 0 & 0 & 0 \\
0 & 8 & 12 & 4 & 8 & 12 & 4 & 0 & 0 & 0 \\
0 & 0 & 0 & 0 & 0 & 0 & 0 & 0 & 0 & 0
\end{pmatrix}$$

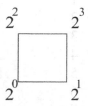

Fig. 8. *The weightings given to the vertices of each square.*

The correspondence between the pixels in Fig. 7 and the elements of 2D_Object should be obvious. However, we can see that a number of the elements in 2D_Object correspond to interior points of the object, namely those elements with a value of 15. Since the object is completely defined by its boundary, there is no need to store these interior points, if a suitable (meaningful) method can be used to store the boundary of the object. Such a method is the Boundary Squares algorithm.

2.2 Boundary Squares

We now describe an algorithm in which only significant functional values are stored. That is, those with a value greater than 0 and less than 15. Indeed once a single non-trivial function value has been found no other trivial values need to be computed nor stored. Avoidance of computation of non-trivial values relies upon realising that, given the function value of a square and the direction of entry of the boundary to that square, the direction of exit of the boundary from that square can be uniquely determined. In order to ensure the direction of entry to a square is known, we have to establish the direction of travel around the boundary. For example, when travelling around the boundary in a clockwise direction, a square with a function value of 11 indicates that the boundary edge of the shape continues at the adjacent square on the right. Given the 16 possibilities shown in Figure 5, and traversing the boundary in a clockwise direction, we can produce a table showing the next boundary square given the function value for square $A_{i,j}$. This is shown in Table 1.

Table 1. *Exit face given through square Ai,j, following the boundary in a clockwise manner .*

Square Value	Exit Face
1	$A_{i,j-1}$
2	$A_{i+1,j}$
3	$A_{i+1,j}$
4	$A_{i-1,j}$
5	$A_{i,j+1}$ or $A_{i,j-1}$
6	$A_{i,j+1}$
7	$A_{i,j+1}$
8	$A_{i-1,j}$
9	$A_{i,j-1}$
10	$A_{i+1,j}$ or $A_{i-1,j}$
11	$A_{i+1,j}$
12	$A_{i-1,j}$
13	$A_{i,j-1}$
14	$A_{i-1,j}$

The boundary squares representation of this image can be seen to be as follows

```
2D_Boundary = (2, 3, 3, 1, 7, 1, 7, 1, 5, 4,
               13, 4, 13, 5, 4, 12, 8, 14, 13,
```

$$4, \ 12, \ 8, \ 10, \ 10, \ 2, \ 11, \ 14, \ 8,$$
$$2, \ 11, \ 2, \ 11)$$

To obtain this representation, the first non-zero entry in the 2D_Object array is found, in this case at (2,4). This coordinate system is defined using the top-left square as (0,0) with the first coordinate increasing from top to bottom and the second from left to right. The boundary is then traced in a clockwise fashion. This is a straightforward operation, again refer to [1]. When the boundary arrives back at (2,4) the array is closed.

It is immediately obvious that this method of representation requires far less storage. We also note that the representation is now in the form of a single-dimensioned array. As such all information regarding the coordinates of the cubes is lost (earlier the entry in 2D_Object[i,j] corresponded to the pixels in positions $i, j, i+1$ and $j+1$.) However, the placing of the origin is relative to the coordinates used, and as such, it is only important to keep account of relative coordinates; any surface analysis should be invariant to translations.

If a similar method could be devised to represent the **surface** of a three-dimensional object with finite continuous surface, then this could lead to an appropriate reduction in the amount of storage space required to define the object.

When considering a two-dimensional object, tracking the boundary in a **clockwise** manner is straight forward. However, due to the increased degree of freedom, tracking a boundary in three dimensions is non-trivial. A method of using rotational tracking around each **slice** of the array (that is, fixing each k in the 2D_Object array and then travelling around the surface in a rotational manner) is not effective as the correct connections between slices are difficult to maintain.

2.3 Rotating Squares

Let us consider Shape A rotated by a factor of $\pi/2$. It would be appropriate for this to have the same representation as the original - though not according to the Boundary Squares algorithm. This is also evident if a shape is rotated by a factor of π and $3\pi/2$. One method to combat this is to rotate one of the images by a factor of $\lambda\pi/2$, where $\lambda= 1,...,3$, and then apply the Marching or Boundary

Squares algorithm again. This, however, requires four times as much work as the original experiment. A possibility would be to take the function values from one orientation, and apply three mappings, each corresponding to a rotation of the image. This again leads to increased computation time and an inefficient algorithm. It is this problem that is counteracted in the Rotating Squares algorithm.

The Rotating Squares algorithm is an adaptation of Boundary Squares. It addresses the issue that an image, when rotated through a multiple of $\pi/2$, should have the same representation. Thus, the Rotating Squares algorithm produces a representation that is invariant over rotations modulo $\pi/2$. The value that each vertex of a square contributes to that square's function value on a shape's boundary can be examined in terms of the direction that the grid square is approached. This then provides the vertex weighting depicted in Figure 9.

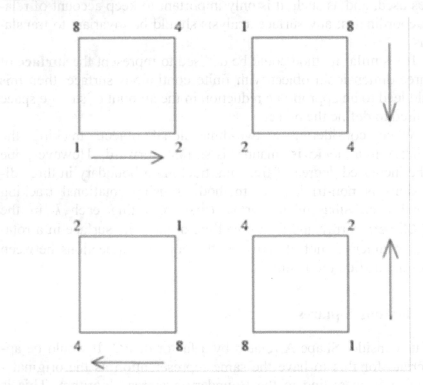

Fig. 9. *The weightings given to the vertices of each square dependent on direction of approach (indicated by arrow.)*

Using the Rotating Squares algorithm on Figure 6 we achieve the representation

```
2D_RS = (1, 3, 3, 1, 7, 1, 7, 1, 5, 1, 7,
         1, 7, 5, 1, 3, 1, 7, 7, 1, 3, 1,
         10, 10, 1, 7, 7, 1, 1, 7, 1, 7),
```

which is invariant to rotation modulo $\pi/2$.

3 Representation Techniques for Three-dimensional Objects

The key to determining a suitable method is to consider the representation for two-dimensional case. The boundary of the object is stored in a one-dimensional array. However it is clear that the tail and head of the array should, in fact, be as connected as the second and third elements in the array. Thus, though we store the boundary as an array, it should be thought of as more of a ring. Moreover, if we consider Fig. 7 we see that the boundary of the object goes from square (2,4) to square (2,5) **right** of square (2,4). Conversely, the boundary goes from the square at (2,5) to (2,4) through the **left** of square (2,5). Hence the structure can be represented graphically as in Fig. 10.

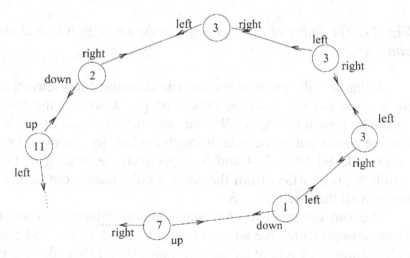

Fig. 10. *A graphical representation of the boundary of a two-dimensional object.*

Considering the structure in this way gives us a natural progression into higher dimensions. Here the boundary is represented in a type of surface map. The above representation as a directed graph with edge and vertex labels. Furthermore, any vertex with a label of 2, as in the above figure, will have two edges, a **down** edge and a **right** edge; similarly any vertex with a label of 1 will also have two edges, a **left** edge and a **down** edge. In much the same way, we can characterise the path of the boundary over a three-dimensional object. We do this in such a way as to determine the faces (as opposed to edges in the two-dimensional case) through which the boundary passes through a particular cube. We index the six faces through which the boundary can pass as follows:

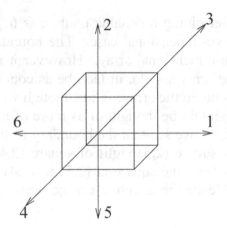

Fig. 11. *The index of the faces of a cube through which a surface can pass.*

Using this direction index, we can describe every direction that the surface passes out of the cube through. Consider the two cube values as shown in Fig. 4. We can see that from the first cube, the surface passes out of the cube through the left, top, front and bottom faces, indexed by 1, 2, 4 and 5 respectively, according to Fig. 11. Similarly, the surface from the second cube passes out of the cube through all the faces 1, ... ,6.

We can construct a table that contains information about the faces through which the surface passes for each of the 254 possible cube values. We ignore cubes with index 0 or 255 as these are trivial in the sense that the boundary of the object does not pass through such cubes. An excerpt from such a table is given in Table 2. It can be seen that cubes with index 1 and 254 have surfaces which pass through exactly the same faces; also cubes with index 2 and 253 have surfaces which pass through the same faces. In general, surfaces from cubes with values x and 255 - x pass through the same face due to the dual nature of the two cubes and so it is only required to store information for cube values up to and including 127. It can also be seen that the surfaces from cubes with value 5, 7 and 10 pass through the same faces. To minimise the storage required for the cube value-face list table, the information can be stored in a two-dimensional array as in

Table 2. *Table showing surface exit values for each possible non-trivial cube value.*

Cube Value	Corresponding Exit Faces
1	4, 5, 6
2	1, 4, 5
3	1, 4, 5, 6
4	1, 2, 4
5	1, 2, 4, 5, 6
6	1, 2, 4, 5
7	1, 2, 4, 5, 6
8	2, 4, 6
9	2, 4, 5, 6
10	1, 2, 4, 5, 6
...	...
253	1, 4, 5,
254	4, 5, 6

Table 3. *A complete, alternative list detailing the exit faces for every non-trivial cube value from 1-127.*

Cube Value	Corresponding Exit Faces
1	4, 5, 6
2	1, 4, 5
3	1, 4, 5, 6
4	1, 2, 4
5, 7, 10, 11, 13, 14	1, 2, 4, 5, 6

6	1, 2, 4, 5
8	2, 4, 6
9	2, 4, 5, 6
12	1, 2, 4, 6
15	1, 2, 5, 6
16	3, 5, 6
17	3, 4, 5, 6
18, 19, 33, 49, 50	1, 3, 4, 5, 6
20-23, 26-30, 37, 39, 40-46, 52-	1, 2, 3, 4, 5, 6
63, 65, 67, 69, 71, 73, 75, 77, 78,	
81-94, 97, 99, 101, 104-109,	
113,114, 116, 117, 120-122, 125	
24, 25, 103, 110, 118, 126	2, 3, 4, 5, 6
31, 35, 47, 79, 80, 95, 112	1, 2, 3, 5, 6
32	1, 3, 5
34	1, 3, 4, 5
36, 38, 66, 70, 98, 100	1, 2, 3, 4, 5
48	1, 3, 5, 6
51	1, 3, 4, 6
64	1, 2, 3
68	1, 2, 3, 4
72, 76, 115, 123	1, 2, 3, 4, 6
96	1, 2, 3, 5
102	2, 3, 4, 5
111	2, 3, 5, 6
127	2,3,6

Given the directions that a surface can exit a cube with a particular cube value, following the boundary of an object is now straightforward. From a given cube value, we can establish the exit faces by consulting a table such as Table 3, and track the boundary using a Case statement such as that given in below:

```
Case lowest_face of
        1:  x:=x+1
        2:  y:=y+1
        3:  z:=z+1
        4:  z:=z-1
        5:  y:=y-1
        6:  x:=x-1
```

4   An Algorithm to Obtain a Boundary Representation

Now we require a data structure that will allow a three-dimensional pixelwise representation to be transformed into a boundary method representation. The boundary method gives rise to a graphical representation similar to the two-dimensional version shown in Fig 10. A selection of an example of a graphical representation arising using the boundary method on a three-dimensional object is given in Fig. 12. Every cube that is constructed from pixels on the boundary of the object is represented by a node in the graph. It is typical that, for a computer representation of a standard graph, information about each node is stored in an element of a one-dimensional array. For a graph such as that in Fig. 12, it is required that the cube value and the connection list for each node is stored.

For purposes of constructing the boundary representation, the coordinates of each cube must be stored. Storing the coordinates of cubes allows a check to be made as to whether any information about the cube has already been stored. The graph is complete once every connection between cubes has been considered. For each node, we introduce a variable, `complete`, which is set to 1 if all the connections of that cube have been stored (checked by consulting a table such as Table 3 and 0 otherwise. Once every element in the boundary list has `complete` = 1 the boundary representation is complete. We can write the data structure to be used in the algorithm as:

```
Boundary_List (I) = (complete, coordinates,
                          connection_list)  (1)
```

Here the connections list is a list of elements (`exit_face`, `next_cube`), where `exit_face` is an integer in 1, ... , 6 and `next_cube` is an integer that determines which element of `Boundary_List` contains information about the next cube. Using this data structure we can devise an algorithm to translate pixelwise data into a boundary method representation, see Fig. 13.

5   Summary

Using the algorithm given in Fig. 13 will result in an array with a structure such as (1). This storage is inefficient since there is, as was argued earlier, no need to store the coordinates of every cube. The relative coordinates are important and can be determined from the storage structure. Giving the coordinates of a single cube will allow the absolute coordinates of any other cube to be determined. There is also no need to store the complete variable since this was only used for construction purposes. The storage can also further be reduced by noting that if we pass through face 1 of cube A into cube B, then this is equivalent to stating that we pass through face 6 of cube B into cube A. Hence every connection is stored twice, causing redundancy. A simple trimming operation can halve the storage required for the edges.

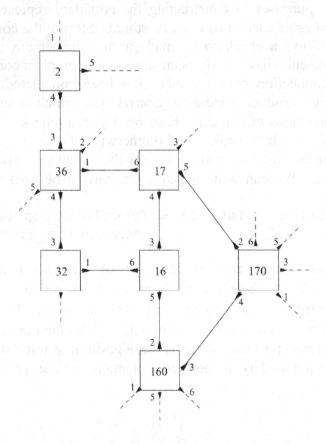

Fig. 12. *A section of a graph depicting the boundary of a three-dimensional object.*

In this chapter we have presented algorithm that take pixelwise representations of two- or three-dimensional objects as input and output a representation of the boundary. Storing information about an object with reference to its boundary requires far less storage than storing every pixel. Using the methods presented herein does in no way lead to a loss of information. We do not approximate the object any more than the original input, and as such do not suffer the shortcomings highlighted in [7]. Storing the boundary as a graph rather than the coordinates of boundary pixels allows the use of graph-matching algorithms to determine common boundary segments in different objects. The storage of the graph of the boundary is optimal.

References

1. Donafee, A., Maple, C., (2000) Rotating Squares, preprint.
2. Heiden, W., Goetze, T., Brickmann, J., (1993) Fast generation of Molecular-Surfaces from 3D Data Fields with an Enhanced Marching Cube Algorithm, Journal of Computational Chemistry, **14**, 2, 246-250.
3. Howie, C.T., Blake, E.H., (1994) The Mesh Propagation Algorithm for Isosurface Construction, Computer Graphics Forum, **13**, 3 SICI, C65.
4. Levitt, D.G., Banaszak, L.J., (1992) Pocket - A Computer-Graphics method for Identifying And Displaying Protein Cavities and Their Surrounding Amino-Acids, Journal of Molecular Graphics, **10**, 4, 229-234.
5. Lorensen, W.E., Cline, H.E., (1987) Marching Cubes: A High Resolution 3D Surface Construction Algorithm, Computer Graphics, **21**, 4, 163-169.
6. Nagae, T., Agui, T., Nagahshi, H., (1993) Interpolation of CT Slices for Laser Stereolithography, IEICE Transactions on Information Systems, **76**, 8, 905-911.
7. Rosin, P.L., (1997), Techniques for Assessing Polygonal Approximation to Curves, IEEE Transactions on Pattern Analysis and Machine Intelligence, **19**, 6, 659-666.
8. Zhou, C., Shu, R.B., Kankanhalli, M.S., (1994) Handling Small Features in Isosurface Generation Using Marching Cubes, Computers & Graphics, **18**, 6, 845-845.
9. Zhou, Y., Chen, W.H., Tang, Z.S., (1995) An Elaborate Ambiguity Detection Method For Constructing Isosurfaces Within Tetrahedral Meshes, Computers & Graphics, **19**, 3, 355-365.
10. Zhou, C., Shu, R.B., Kankanhalli, M.S., (1995) Adaptive Marching Cubes, Visual Computer, **11**, 4, 202-217.

Chapter 08

A Rational Spline with Point Tension: An Alternative to NURBS of Degree Three

A Rational Spline with Point Tension: An Alternative to NURBS of Degree Three

Muhammad Sarfraz

Department of Information and Computer Science, King Fahd University of Petroleum and Minerals, KFUPM # 1510, Dhahran, Saudi Arabia

A rational cubic C^2 spline curve has been described which has point tension weights for manipulating the shape of the curve. The spline is presented in both interpolatory and local support basis form, and the effect of the weights on these representations is analysed.

1 Introduction

Interactively designing of curves and surfaces is an important area of Geometric Modeling, Computer Aided Design (CAD), and Computer Graphics. A good amount of work has been proposed by various authors in the last few years. The work has been oriented towards state of the art techniques, applications and system and tools. For brevity, the reader is referred to [1-18].

This chapter describes a parametric C^2 rational cubic spline representation, which has point tension weights. These weights can be used to control the shape of the curve. The spline can be considered as an alternative to the cubic v-spline formulation of Nielson [12], or the rational cubic spline of Boehm [3]. These splines provide shape control parameters through the use of geometric GC^2 continuity constraints and hence are C^2 with respect to a reparameterization. A result of Degen [4] indicates that, ignoring the possibility of singular reparameterizations, any piecewise defined cubic or rational cubic GC^2 curve can be reparameterized as a C^2 piecewise defined rational cubic. This result suggests the use of parametric C^2 rational cubic splines as an alternative to the use of geometric GC^2 cubic or GC^2 rational cubic spline representations. The rational spline also provides a C^2 alternative to the C^1 weighted v-spline of Foley [6, 7], since the interval and point tension weights have a remarkably similar influence on the curve to those of the weighted v-spline. (The paper by Piegel [14] on rational cubic representations should also be noted.)

The rational spline is a further study of the method introduced in Gregory and Sarfraz [9]. The solution we adopt here is to restrict the class of rational cubic representations to those having quadratic denominators. This leads to representations having sensible parameterizations with well defined and well behaved point tension weights. The use of this restricted class of rational splines has the added advantage that conic segments could easily be accommodated within the representation.

Given the volume of literature on rational splines in Geometric Modeling, CAD, and Computer Graphics, the reader might be dubious as to the merit of introducing yet another rational spline representation. However, we believe that the introduction of point weights, in a C^2 setting, gives a smoother and powerful tool for manipulating the shape of the curve representation and hence will be useful in Geometric Modeling, CAD, and Computer Graphics applications. We were also motivated to develop the spline in the context of surface design, see Sarfraz [15, 17], where a surface is to be constructed through a general network of spline curves similar to tensor products. The method proposed in [15, 17] involves variable parameters, which lead to the local shape control. We are not, however, proposing to describe this surface construction here, and hence will assume the methodology for the spline representation.

One of the most attractive feature of the proposed spline method is that it can be represented as a stronger alternate to NURB (nonuniform rational B-spline) of degree three. This strength lies in the nature of Point Tension parameters, which are completely independent of its neighbor counter parts. This feature is not found in traditional NURBS as the shape parameters cancel out when they are taken equal globally or in a specific range of neighborhood. This happens due to their canceling out in the numerator and denominator of the NURBS description. The proposed method has another interesting feature that it can also be extended to have a generalization of NURBS for keeping the NURBS features if necessary. This feature is not discussed here and will be discussed in a subsequent work.

The organization of the manuscript is as follows. In Section 2, the basic rational cubic form is described. Then, in Section 3, the interpolatory form of the rational spline is presented. The local support basis form is described in Section 4. Finally, the extension of the method, to surfaces, is proposed in Section 5. Section 6 summarizes the whole discussion.

2   The Rational Cubic Form

Let $F_i \in R^m, i \in Z$, be values given at the distinct knots $t_i \in R, i \in Z$, with interval spacing $h_i = t_{i+1} - t_i > 0$. Also let $D_i \in R^m, i \in Z$, denote first derivative values defined at the knots. Then a parametric C^1 piecewise rational cubic Hermite function $p: R \to R^m$ is defined by

$$p|_{[t_i, t_{i+1})}(t) = \frac{(1-\theta)^3 \alpha_i F_i + \theta(1-\theta)^2(2+\alpha_i)V_i + \theta^2(1-\theta)(2+\alpha_{i+1})W_i + \theta^3 \alpha_{i+1}F_{i+1}}{(1-\theta)^2 \alpha_i + 2\theta(1-\theta) + \theta^2 \alpha_{i+1}}$$

(2.1)

where

$$\theta|_{[t_i, t_{i+1})}(t) = (t - t_i)/h_i,$$

(2.2)

and

$$V_i = F_i + \frac{\alpha_i}{2+\alpha_i} h_i D_i \; , \; W_i = F_{i+1} - \frac{\alpha_{i+1}}{2+\alpha_{i+1}} h_i D_{i+1}.$$

(2.3)

For simplicity, we assume positive weights

$$\alpha_i > 0, \, i \in Z,$$

(2.4)

and have made use of a rational Bernstein-Bezier representation, where the 'control points' $\{F_i, V_i, W_i, F_{i+1}\}$ are determined by imposing the Hermite interpolation conditions

$$p(t_i) = F_i, \text{ and } p^{(1)}(t_i) = D_i, \, i \in Z$$

(2.5)

The scalar weights in the numerator of (2.1) are those given by degree raising the denominator to cubic form, since

134

$$(1-\theta)^2\alpha_i + 2\theta(1-\theta) + \theta^2\alpha_{i+1} =$$

$$(1-\theta)^3\alpha_i + \theta(1-\theta)^2(2+\alpha_i) + \theta^2(1-\theta)(2+\alpha_{i+1}) + \theta^3\alpha_{i+1} \qquad (2.6)$$

Since the denominator is positive, it follows from Bernstein-Bezier theory that the curve segment $p|_{[t_i,t_{i+1})}(t)$ lies in the convex hull of the control points $\{F_i, V_i, W_i, F_{i+1}\}$ and is variation diminishing with respect to the 'control polygon' joining these points. It should also be observed that the weights are very similar to those in NURBS and are associated with control points.

The rational cubic Hermite function can also be written in the Ball like form, Ball [1],

$$p|_{[t_i,t_{i+1})}(t) = \frac{(1-\theta)^2\alpha_i F_i + 2\theta(1-\theta)^2 V'_i + 2\theta^2(1-\theta)W'_i + \theta^2\alpha_{i+1}F_{i+1}}{(1-\theta)^2\alpha_i + 2\theta(1-\theta) + \theta^2\alpha_{i+1}}, \qquad (2.7)$$

where

$$V'_i = F_i + \frac{\alpha_i}{2}h_i D_i, \quad W'_i = F_{i+1} - \frac{\alpha_{i+1}}{2}h_i D_{i+1}. \qquad (2.8)$$

The rational cubic form then degenerates to a rational quadratic form (that is a conic curve) when $V'_i = W'_i$.

Remarks: If $p(t)$ is the interpolant for scalar data $F_i \in R$, with derivatives $D_i \in IR$ $i \in Z$, then $(t, p(t))$ can be considered as the interpolation scheme applied in R^2 to the data (t_i, F_i), with derivatives $(1, D_i)$, $i \in Z$. This is a consequence of the property that the interpolant is able to reproduce linear functions. In particular, for scalar data $F_i := t_i$ and derivatives $D_i = 1$, $i \in Z$, the interpolant reproduces the function t. The reproduction property also indicates a well behaved parameterization. For example, with $D_i = D_{i+1} = (F_{i+1} - F_i)/h_i$ the interpolant reproduces the straight line segment with parameterization $(1-\theta)F_i + \theta F_{i+1}$ for all choices of the shape parameters.

For the practical implementation we will write

$$\alpha_i = 1/\lambda_i. \qquad (2.9)$$

This leads to a consistent behavior with respect to increasing weights and avoids numerical problems associated with evaluation at $\theta = 0$ and $\theta = 1$ in the (removable) singular cases $\alpha_i = 0$. We now have

$$V_i = F_i + \frac{1}{2\lambda_i + 1} h_i D_i, \quad W_i = F_{i+1} - \frac{1}{2\lambda_{i+1} + 1} h_i D_{i+1}. \quad (2.10)$$

The following 'tension' properties of the rational Hermite form are now immediately apparent from (2.1), (2.9), and (2.10), see Figure 1:

(i) Point Tension

$$\lim_{\lambda_i \to \infty} V_i = F_i \text{ and}$$

$$\lim_{\lambda_i \to \infty} p \big|_{[t_i, t_{i+1})} (t) = \frac{(1-\theta)^2 (2F_i + \theta(1-\theta)(\gamma_i + \beta_i)W_i + \theta^2 \lambda_{i+1} F_{i+1}}{2(1-\theta) + \theta \lambda_{i+1}}.$$

$$(2.11)$$

$$\lim_{\lambda_{i+1} \to \infty} W_i = F_{i+1} \text{ and}$$

$$\lim_{\lambda_{i+1} \to \infty} p \big|_{[t_i, t_{i+1})} (t) = \frac{(1-\theta)^2 (\lambda_i F_i + \theta(1-\theta)(2 + \lambda_i)V_i + 2\theta^2}{(1-\theta)\lambda_i + 2\theta}.$$

$$(2.12)$$

This behavior accentuates the point tension property, since the behavior of the curve is controlled by limiting processes from both right and left of the parametric point t_i. One thus has a point tension parameter λ_i controlling the curve in the neighborhood of t_i. It is this case which, in the spline discussion which follows, mirrors the behavior of the polynomial v-spline.

(ii) Interval Tension

The interval tension property is recovered by letting $\lambda_i, \lambda_{i+1} \to \infty$. Since the behavior of the curve is controlled by limiting processes from both right and left of the parametric interval $[t_i, t_{i+1}]$. One thus has an interval tension controlling the curve on the interval $[t_i, t_{i+1}]$. It is this case which, in the spline discussion which follows, mirrors the behavior of the polynomial weighted spline as well as the rational cubic spline with interval tension [9].

3 Rational Cubic Spline Interpolant

We now consider the problem of constructing a parametric C^2 rational cubic spline interpolant on the interval $[t_0, t_n]$, using the rational cubic Hermite form of Section 2. This is the situation where $F_i \in R^m$, $i = 0, ..., n$ are the given interpolation data at knots t_i, $i = 0, ..., n$, and the derivatives $D_i \in R^m$, $i = 0, ..., n$, are degrees of freedom to be determined by the imposition of C^2 constraints on the piecewise defined rational Hermite form. We will assume that D_0 and D_n are given as end conditions, although other well known end conditions can be applied, as for the particular case of cubic spline interpolation. For example, we leave the case of periodic end conditions, required for a closed parametric curve, as an exercise for the reader.

The C^2 constraints

$$p^{(2)}(t_i^+) = p^{(2)}(t_i^-), \; i = 1, ..., n-1, \tag{3.1}$$

give the tri-diagonal system of 'consistency equations'

$$h_i \alpha_{i-1} D_{i-1} + 2(h_i + h_{i-1})D_i + h_{i-1}\alpha_{i+1}D_{i+1} =$$
$$h_i(2 + \alpha_{i-1})\Delta_{i-1} + h_{i-1}(2 + \alpha_{i+1})\Delta_i, \; i = 1,...,n-1, \tag{3.2}$$

where

$$\Delta_i = (F_{i+1} - F_i)/h_i. \tag{3.3}$$

Thus, in terms of the reciprocal weights (2.9), the tri-diagonal system is

$$(h_i / \lambda_{i-1})D_{i-1} + 2(h_i + h_{i-1})D_i + (h_{i-1} / \lambda_{i+1})D_{i+1} =$$
$$h_i(2 + 1/\lambda_{i-1})\Delta_{i-1} + h_{i-1}(2 + 1/\lambda_{i+1})\Delta_i, \; i = 1,...,n-1. \tag{3.4}$$

Suppose that $2 > 1/\lambda_i$, $i = 1,...,n-2$, that is

$$2\lambda_i > 1, \; i = 1,...,n-2. \tag{3.5}$$

Then, in (3.4), we have a diagonally dominant, tri-diagonal system of linear equations in the unknowns D_i, $i = 1, ..., n-1$. Thus (3.5) provides a sufficient condition for the existence of a unique, easily computable solution.

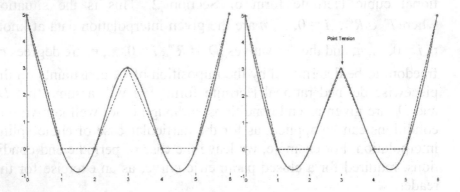

Fig. 1. *(a) Cubic Spline (sold curve) with control polygon (dotted lines), (b) Point Tension behavior shown at the middle point.*

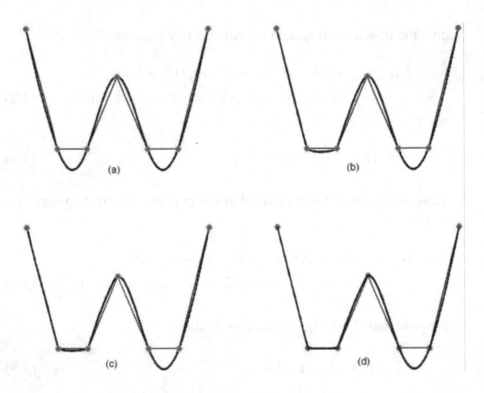

Fig. 2. *(a) Cubic Spline, (b-d) Gradually increasing Interval Tension behavior shown at the second and third points.*

We conclude this section with the examples shown in Figures 1 and 2, which illustrate the effect of the point and interval tension weights described above. The point tension behavior is confirmed by the example curves of Figure 1. The curve (a) shows a cubic B-spline curve with its defining control polygon. The rest of the curves (b-d) then show the effect of increasing one point tension weight with values 3, 10 and 50 at the middle point of the figure. The interval tension behavior is demonstrated by the example curves of Figure 2. The curve (a) shows a cubic B-spline curve with its defining control polygon. The rest of the curves (b-d) then show the effect of increasing two consecutive point tension weights with values 3, 10 and 50.

4 The Local Support Basis

We now seek a local support basis representation for the space of C^2 rational cubic splines on the knot partition $\{t_i \in R : i \in Z\}$. Thus assume that there exists a local support basis $\{N_j(t)\}_{j \in Z}$, where

$$N_j(t) = 0, \quad t \notin (t_{j-2}, t_{j+2}), \quad (\textit{local support}) \tag{4.1}$$

and

$$\sum_{j \in Z} N_j(t) = 1 \quad (\textit{partition of unity normalization}) \tag{4.2}$$

Then, given *control points* $\{C_j \in R^m : j \in Z\}$, we consider the parametric curve representation

$$p(t) = \sum_{j \in Z} C_j N_j(t), \tag{4.3}$$

that is

$$p\big|_{[t_i, t_{i+1})}(t) = \sum_{j=i-1}^{i+2} C_j N_j(t) \tag{4.4}$$

Following the approach of Boehm [3] and Lasser [11], we wish to represent the curve in the rational Bernstein Bézier form (2.1), with control points $\{F_i, V_i, W_i, F_{i+1}\}_{i \in Z}$. The existence of the transformation to rational Bernstein Bézier form will, in fact, demonstrate the existence of the local support basis. The transformation also provides a convenient tool for computing and analysing the local support basis representation.

Imposing the constraint

$$p^{(1)}(t_i^+) = p^{(1)}(t_i^-), \qquad (4.5)$$

on the rational Bernstein Bézier form gives

$$F_i = (1 - \delta_i)W_{i-1} + \delta_i V_i, \qquad (4.6)$$

where

$$\delta_i = \frac{h_{i-1}}{h_{i-1} + h_i}. \qquad (4.7)$$

Also, imposing the constraint

$$p^{(2)}(t_i^+) = p^{(2)}(t_i^-) \qquad (4.8)$$

gives

$$a_{2,i}F_i - (a_{2,i} + a_{1,i})V_i + a_{1,i}W_i = b_{2,i-1}F_i - (b_{2,i-1} + b_{1,i-1})W_{i-1} + b_{1,i-1}V_{i-1},$$
$$(4.9)$$

where

$$a_{1,i} = \frac{2}{h_i^2}(2\lambda_i + \lambda_i / \lambda_{i+1}), \ a_{2,i} = \frac{2}{h_i^2}(4\lambda_i^2 - \lambda_i / \lambda_{i+1}), \qquad (4.10)$$

$$b_{1,i} = \frac{2}{h_i^2}(2\lambda_i + \lambda_i / \lambda_{i+1}), \ b_{2,i} = \frac{2}{h_i^2}(4\lambda_{i+1}^2 - \lambda_{i+1} / \lambda_i), \qquad (4.11)$$

Eliminating F_i using (4.6) then gives

$$(1 - \tau_{i-1})W_{i-1} + \tau_{i-1}V_{i-1} = (1 - \sigma_i)V_i + \sigma_i W_i, \qquad (4.12)$$

where

$$\sigma_i = \frac{-a_{1,i}}{a_{2,i}(1 - \delta_i) + b_{2,i-1}\delta_i}, \quad \tau_i = \frac{-b_{1,i}}{a_{2,i+1}(1 - \delta_{i+1}) + b_{2,i}\delta_{i+1}} \qquad (4.13)$$

Equation (4.12) represents the equation for the intersection of the two lines through W_{i-1}, V_{i-1} and V_i, W_i respectively and we will see that this intersection is the control point C_i of the local support basis representation. Thus, given $\{C_i\}_{i \in Z}$ we have

$$\left.\begin{aligned}
(1 - \sigma_i)V_i + \sigma_i W_i &= C_i, \\
\tau_i V_i + (1 - \tau_i)W_i &= C_{i+1},
\end{aligned}\right\} \quad i \in Z, \tag{4.14}$$

and hence

$$\left.\begin{aligned}
V_i &= [(1 - \tau i)/\Delta_i]C_i - [\sigma_i/\Delta_i]C_{i+1}, \\
W_i &= -[\tau_i/\Delta_i]C_i + [(1 - \sigma_i)/\Delta_i] = C_{i+1},
\end{aligned}\right\} \quad i \in Z, \tag{4.15}$$

where

$$\Delta_i = 1 - \sigma_i - \tau_i. \tag{4.16}$$

Equations (4.15) and (4.6) define the transformation from the control points $\{C_i\}_{i \in Z}$ to those of the piecewise defined Bernstein-Bézier representation. The existence of this transformation implies the existence of the local support basis, since with scalar data

$$C_j = \delta_{i,j} = \begin{cases} 1, j = i, \\ 0, j \neq i, \end{cases} \tag{4.17}$$

the transformation generates the rational Bernstein-Bézier representation of the basis function $N_i(t)$. The condition

$$\gamma_i^2 > \frac{1}{\lambda_i \mu_i}, i \in Z, \tag{4.18}$$

is a sufficient one for negative σ_i and τ_i, in (4.15). This implies that the transformation is obtained by a corner cutting process on the *control polygon* joining the points $\{C_i\}_{i \in Z}$, see Figure 3. (The diagonal dominance condition $\lambda_i \gamma_i > 1$, $\mu_i \gamma_i > 1$ of the previous section is a sufficient condition for this.) This guarantees that the local support basis is *non-negative; that representation is variation diminish-*

ing; and that the curve $p(t)$, $t \in [t_i, t_{i+1}]$ lies in the *convex hull* of $\left\{ C_j \right\}_{j=i-1}^{i+2}$ (more precisely, the curve lies in the *convex hull* of $\{F_i, V_i, W_i, F_{i+1}\}$ which is contained in that of $\left\{ C_j \right\}_{j=i-1}^{i+2}$.

We now consider the tension properties of the local support basis representation:

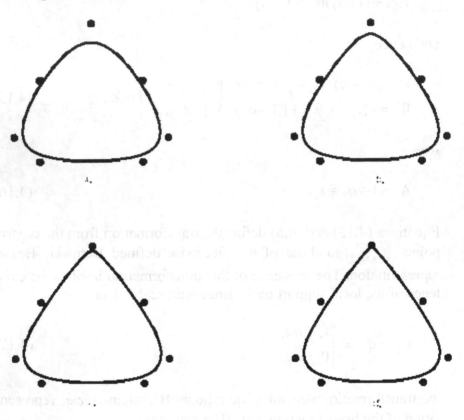

Fig. 3. *(a) Cubic Spline (sold curve) with control polygon (dotted lines), (b-d) Gradually increasing Point Tension behavior shown at the top point.*

(i)   Point tension

Consider $\lambda_i \rightarrow \infty$. Then

$$\lim W_{i-1} = \lim V_i = \lim F_i = C_i. \tag{4.19}$$

Thus the curve is pulled towards C_i, and in the limit a tangent discontinuity is introduced at C_i.

This behavior is confirmed by the example curves of Figure 3. The curve (a) shows a cubic B-spline curve with its defining control polygon. The rest of the curves (b-d) then show the effect of increasing one point tension weight with values 3, 10 and 50 at the top point of the figure.

(ii) Interval tension.

Consider λ_i, $\lambda_{i+1} \to \infty$. In this case the curve is pulled onto a straight line segment joining the points C_i, and C_{i+1}.

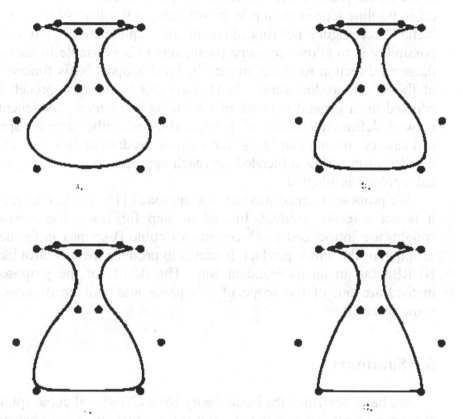

Fig. 4. *(a) Cubic Spline, (b-d) Gradually increasing Interval Tension behavior shown at the second and third points.*

This behavior is confirmed by the example curves of Figure 4. The curve (a) shows a cubic B-spline curve with its defining control polygon. The rest of the curves (b-d) then show the effect of increasing two consecutive point tension weights with values 3, 10 and 50 in the base of the figure.

5 Surfaces

The extension of the curve scheme, to tensor product surface representations, is immediately apparent. However, this representation exhibits a problem common to all tensor product descriptions in that the shape control parameters now affect a complete row or column of the tensor product array. Nielson [13] solves this problem for his cubic v-spline representation by constructing a Boolean sum, spline-blended, rectangular network of parametric v-spline curves. Another possibility is to allow the shape parameters to be variable in the orthogonal direction to, for example, the local support basis functions of the tensor product form. In [10] another blending approach is adopted for a general network of rational spline curves. A general network defines a topology of 'polygonal faces', rather than the special case of 'rectangular faces' for a tensor product or Boolean sum blended form. Thus a blended approach appropriate to this polygonal topology is adopted.

We propose a tensor product like approach [15, 17] but actually it is not a tensor product. Instead of step functions, the tension weights are introduced as C^2 continuous cubic B-splines in the description of the tensor product. It causes to produce local control like NURBS but in an independent way. The details of the proposed method are out of the scope of this paper and will be discussed somewhere else.

6 Summary

We have described the basic theory for a C^2 rational cubic spline curve representation, which has point tension weights, which behave in a well-controlled and meaningful way. The method is a stronger alternate to the NURBS of degree three. The proposed method has another interesting feature that it can also be extended to have a generalization of NURBS for keeping the NURBS features if necessary. This feature is not discussed here and will be discussed in a subsequent work.

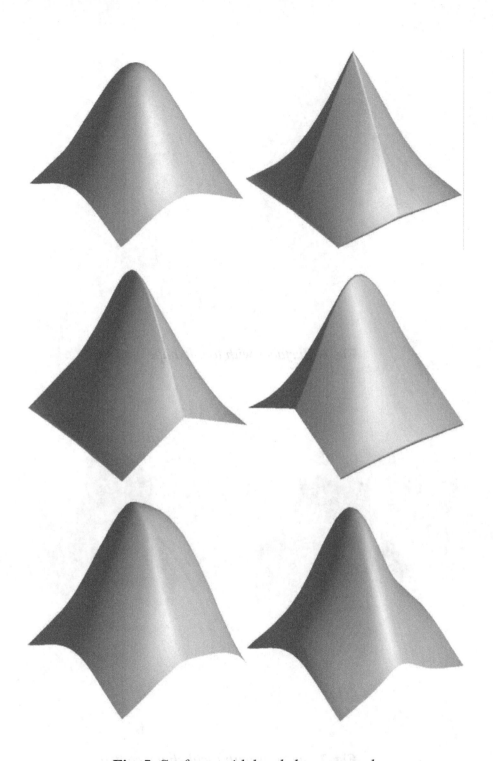

Fig. 5. *Surfaces with local shape control.*

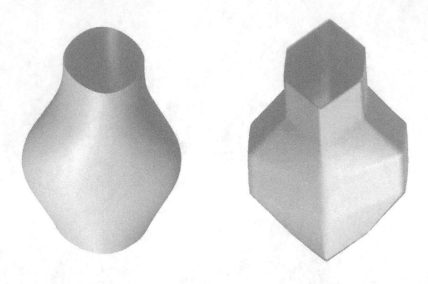

Fig. 6. *Surfaces with local shape control.*

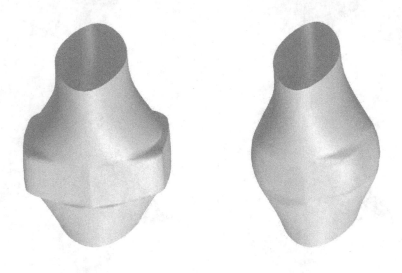

Fig. 7. *Surfaces with local shape control.*

The extension to tensor product surface representations is immediately apparent. However, this representation exhibits a problem common to all tensor product descriptions in that the shape control parameters now affect a complete row or column of the tensor product array. Nielson [13] solves this problem for his cubic v-spline representation by constructing a Boolean sum, spline-blended, rectangular network of parametric v-spline curves. Another possibility is to allow the shape parameters to be variable in the orthogonal direction to, for example, the local support basis functions of the tensor product form. In [10] another blending approach is adopted for a general network of rational spline curves. A general network defines a topology of 'polygonal faces', rather than the special case of 'rectangular faces' for a tensor product or Boolean sum blended form. Thus a blended approach appropriate to this polygonal topology is adopted.

We have not considered here the use of the shape parameters for the control of convexity or number of inflections of the curve representation, particularly for the interpolatory case. (The local support case is obviously controlled by the behavior of the contol polygon.). The case of shape preserving C^1 rational cubic interpolatory curves is a straightforward exercise, since, given derivative vectors consistent with the shape of the interpolatory data, the parameters can be chosen to give a piecewise defined rational

Bernstein-Bezier polygon defined by (2.1) of the required shape. The automatic choice of the parameters for the C^2 case is not such a simple matter, however, and this problem is now being investigated.

Acknowledgment

This work has been supported by King Fahad University of Petroleum and Minerals under the Project No. FT/2001-18.

References

1. Ball, A.A. (1974), Consurf I, CAD 6, 243-249.
2. Barsky, B.A. (1981), The beta-spline: A local representation based on shape parameters and geometric measures, Ph.D. dissertation, Dept. Computer Science, the University of Utah.
3. Boehm, W. (1985), Curvature continuous curves and surfaces, Computer Aided Geometric Design 2, 313-323.

4. Degen, W. (1988), Some remarks on Bezier curves, Computer Aided Geometric Design 5, 259-268.
5. Farin, G. (1990), Curves and Surfaces of Computer Aided Geometric Design, 2nd edition, Academic Press.
6. Foley, .a. (1987), Interpolation with interval and point tension controls using cubic weighted v-splines, ACM Trans. Math. Software, 13, 68-96.
7. Foley, T.A. (1988), A shape preserving interpolant with tension controls, Computer Aided Geometric Design 5, 105-118.
8. Goodman, T.N.T. (1988), Shape preserving interpolation by parametric rational cubic splines, in: International Services of Numerical Mathematics, Vol. 86, Birkhauser Verlag, Basel, 149-158.
9. Gregory, J.A. and Sarfraz, M. (1990), A rational cubic spline with tension, Computer Aided Geometric Design 7, 1-13.
10. Gregory J.A. and Yuen, P.K. (1992), An arbitrary mesh network sheme using rational splines, in: T. Lyche and L.L. Schumaker (eds.), Mathematical Methods in Computer Aided Geometric Design II, Academic Press, 321-329.
11. Laser, D. (1988), B-Spline-Bezier representation of Tau-Splines, NPS technical report 53-88-006, Naval Postgraduate School, Monterey, CA 93943.
12. Nielson, G. M. (1974), Some piecewise polynomial alternatives to splines under tension, in: R.E. Barnhill and R.F. Riesenfeld, eds., Computer Aided Geometric Design, Academic Press, New York, 209-235.
13. Nielson, G. M. (1986), Rectangular v-splines, IEEE CG&A February 86, 35-40.
14. Piegl, L. (1987), Interactive interpolation by rational Bezier curves, IEEE CG&A April 87, 45-58.
15. Sarfraz, M. (1995), Curves and Surfaces for CAD Using C^2 Rational Cubic Splines, *Engineering with Computers*, 11(2), 94-102.
16. Sarfraz, M. (1994), Cubic Spline Curves with Shape Control, *Computers & Graphics*, 18(5), 707-713.
17. Sarfraz, M. (1993), Designing of Curves and Surfaces Using Rational Cubics, Computers and Graphics, **17**(5), 529-538
18. Sarfraz, M. (1992), A C^2 Rational Cubic Spline Alternative to the NURBS, Comp. & Graph. 16(l), 69-78.

Section II

Applications

Chapter 09

Muscle Modeling and Rendering

Muscle Modeling and Rendering

F. Dong and G.J. Clapworthy
Department of Computer & Information Sciences, De Montfort University, Milton Keynes MK7 6HP, United Kingdom

This chapter introduces techniques used in muscle Modeling and rendering. An anatomically based approach to muscle Modeling is presented in which the muscle models are generated from anatomical data. These models provide a good visual description of muscle form and action and represent a sound base from which to progress further towards medically accurate simulation of human bodies. Deformation of these models is also performed on the basis of their anatomical structures. The result is an efficient, anatomically accurate, muscle representation that is specifically designed to accommodate the particular form of deformation exhibited by each individual muscle.

Muscle rendering is addressed in two radically different ways. In the photorealistic approach, the muscle texture is generated from anatomical data, and a texture synthesis method is used to simulate the fibre patterns on the muscle surface. In the non-photorealistic method, the muscles are rendered directly from volumetric data using a pen-and-ink styles. This provides a useful alternative to traditional visualisation techniques.

1 Introduction

Human simulation is a prominent area in computer graphics, and many researchers have addressed different aspects of the problem, including Cohen [1], Hodgins & Pollard [2], Gleicher [3], Laszlo [4], Funge et al.[5], Maurel & Thalmann [6], Scheepers et al.[7] and Wilhelms & Van Gelder [8]. In most of this work, the authors adopted greatly simplified models that were far from representing the functioning of a human body.

As technology has improved, anatomically and biomechanically accurate Modeling has become increasingly demanded. A recent example is that of Savenko et al.[9] who adapted work by Van Sint Jan et al.[10,11] on knee kinematics to create an improved joint model for figure animation that is more in keeping with newly available forms of data from biomechanics.

In this chapter, we shall focus on muscle Modeling, which is one of the most important elements of the human anatomy. Muscles have complicated individual structures and layout, complex dynamic properties and elaborate deformation actions, so any attempt at realistic muscle simulation should address 3 major, related, problems:

- muscle geometry
- muscle deformation
- muscle rendering.

This chapter introduces an anatomy-based approach to muscle Modeling in which each muscle is modelled individually according to its anatomical characteristics. The main purpose is to bring muscle models one step closer to the real anatomical structure. The work involves three aspects.

Firstly, the geometric surface models for muscles are constructed from anatomical data. The individual surface models thus obtained are then divided into several categories on the basis that all models in one category should share the same representation and deformation action.

Secondly, deformation models are used to produce realistic changes in the shapes of the muscle models produced in the first step. The movement of the muscle action line and tendons and the changes of the muscle belly shape are taken into account, as is interaction between muscles during the motion.

Thirdly, two different approaches for muscle rendering are presented – texture synthesis and non-photorealistic rendering.

2   Related Work

Much attention has been paid in recent years to realistic human Modeling. Nowadays, representations of the human figure tend to use a layered model, rather than a simple line-segment skeleton. Normally, such a model has four layers: skeleton, muscle, fatty tissue, skin, and its introduction represented an important step in a direction towards anatomically-based Modeling.

Examples of such work include Chadwick et al.[12], who presented a method for the layered construction of flexible animated characters; Kalra et al.[13], who built a single system for simulating a virtual human that allowed real-time animation; and others such as Gourret et al.[14] and Turner & Thalmann [15]. Chen & Zeltzer [16] developed a finite element model to simulate the muscle force and visualise its deformation. However, none of these authors used models constructed in an anatomically-based way, which has the capacity to achieve the best results in terms of realism.

Our work is thus closest to that of Scheepers et al.[7] and Wilhelms & Van Gelder [8], who attempted to model human muscle according to its anatomical structure. Scheepers at al. were motivated by the artistic study of anatomy – they used ellipsoids as the graphics primitives for the muscle belly and used this as a basis on which to build a more sophisticated model. Wilhelms & Van Gelder introduced a general muscle primitive model, a "deformed cylinder", to model various sorts of muscles. Human intervention was needed to manipulate the default muscle type into the form necessary for each particular muscle. The muscle deformation was greatly simplified and used just to produce some good visual effects.

However, our Modeling differs in purpose from the work of these authors as they wished only to achieve good visual results for the figure as a whole. Muscle Modeling constituted just one layer of their layered model and it was, therefore, greatly simplified, as further attention had to paid to the other layers, such as skin. In contrast, our work attempts to perform Modeling that is much closer to the real human anatomy. We focus here on muscles and attempt to perform the geometric Modeling, deformation and muscle Modeling using features of each individual muscle captured from real anatomical data.

3   Modeling and Classifying

Normally, a muscle is attached to two locations through tendons on to the bones. One attachment is mostly fixed and is called the ORIGIN, while the other is considered to be the more movable part and is termed the INSERTION. For simplicity, most of our muscle models contain both the muscle bodies and their tendons in one geometric surface. This is because, from a geometric viewpoint, the tendons can be regarded as extensions of the main muscle bodies.

To create the geometric models, we first find the geometric features of the muscle from well-segmented Visible Human data and build the models based on these features. Two advantages of this approach are that the resulting shape is very realistic and that little human intervention is required. In this section, we discuss the two key techniques relating to this issue:

- analysing and extracting the features from the muscles
- Modeling the muscle geometry via B-spline surfaces.

3.1 Analysis of the features

We analyse the geometric features of a muscle via its cross-sectional contours on the 2D slices in the data. Within the slice, we shall require the straight line $p = x\cos\theta + y\sin\theta$, which is defined relative to the local x-y co-ordinate system within the slice. Here, θ defines the line direction.

For each contour, we compute the area, direction radius, circumference, direction concavity and general concavity and use these as its features – please refer to Dong et al.[17] for more details.

These features of a contour form a multi-dimensional feature vector. By considering the feature vector of each contour, we select a set of key contours, which is used for the geometric Modeling at the next step – the "distance" between the feature vectors of two successive key contours must be beyond a pre-defined threshold.

3.2 Modeling

The geometric models are represented by cubic B-spline surfaces, the input for which are the key contour sets that were built in Section 3.1. The problem is then as follows: given several contours, find a B-spline control polygon mesh to fit the data.

Our solution is similar to that of Hoppe et al.[18] and is based on subdivision. However, we use a B-spline subdivision scheme rather than the Loop scheme, since our goal is a B-spline representation.

As we know, subdivision can be used to create uniform and non-uniform B-splines: starting from the mesh of control points, we obtain the B-spline surface by subdividing the control mesh repeatedly. This subdivision scheme is used here to fit a B-spline surface to the key contours. The algorithm starts by initialising the mesh of control points to an ellipsoid that approximates the muscle shape, then successively employs the following steps:

- *Subdivision*: for a fixed control-point mesh, compute the B-spline subdivision to produce the subdivided points

- **Matching**: for each data point on the contours, find the closest point among the subdivided points.
- **Optimisation**: since the subdivision can be seen as a linear combination of control points, we can set up a system of linear equations, and create a new control-point mesh using least-squares fitting.

We now return to the first step and resume the whole procedure; this continues until the distance between the data points and the subdivided points is below a pre-defined threshold.

3.3 Classifying

The muscle models created in Section 3.2 are classified into the following categories:

- **Basic Representation and Single Head Muscles**

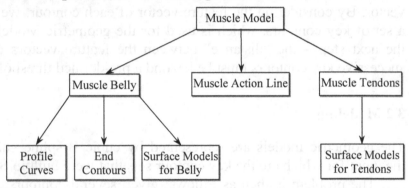

Fig. 1. *Basic representation of a muscle model*

Figure 1 shows the basic representation of a muscle model. This representation covers all of the single-head muscles, which connect to the underlying bones via a single head at each end, either directly or by means of a tendon or aponeurosis (a wide, flattened tendon). Many muscles are single headed, for example, the semimembranosus, the semitendinosus, the gracilis, etc.

- **Multiple-Head Muscles**

Multiple-head muscles attach to the bones via two or more heads at one end. A multiple-head muscle is represented as a combination of several single-head muscles. Each head is modelled separately using the basic representation described above.

The following muscle models have to be dealt with individually, because of their special shapes or particular deformation actions.

- **_Quadriceps Femoris._** Quadriceps femoris is a group containing four distinct muscles – vastus medialis, vastus lateralis, vastus intermedius and rectus femoris. For each muscle in the quadriceps femoris we use the basic representation illustrated in Figure 1. However, they all connect to a common tendon model, which requires the basic muscle representation to be modified.

- **_Sartorius._** The physical representation of the sartorius is the same as that of a single-head muscle. However, its deformation is handled differently (see Section 4), which is why it is modelled individually.

- **_Adductor Magnus._** The adductor magnus is a special case because of its three end contours, one at the origin, and the other two at the two insertions separately. It is not treated as a multiple-head model since it is not split into two single-head components. We produce the initial surface directly from the slice contours using the method for contour-based meshing described by Christiansen & Sederberg [19].

- **_Hip Muscles._** The hip muscles include the tensor fascia lata, the gluteus medius, the gluteus maximus, and the iliopsoas. All of these muscle representations are single-headed. However, the gluteus maximus and the tensor fascia lata share a large tendon, and the iliopsoas is made up of the large, flat, triangular iliacus and the fusiform psoas major.

4 Muscle Deformation

Here, we present the deformation of the muscle models generated in Section 3. The way in which each model contracts is based on its features, such as its shape and its attachments to the bone. This is conducted on the same category basis as above.

4.1 Individual Deformation

We first describe a general deformation model for single-head muscles and then give examples of a few more complex cases.

- **General Deformation Model**

The following steps are involved in deforming a general model.

i) Compute the new position of the bones resulting from the motion data of the joints.

ii) Re-position the action lines of the muscles – the muscle action line indicates the basic position, orientation and length of a muscle during the motion. The movement of a straight action line follows the joint. If an action line consists of several line segments, a cubic curve is used to fit the action line.

iii) Move the tendons and hence the end contours on the tendons.

iv) Deform the profile curves in response to the movement of the end contours.

v) Amend the shape to conserve muscle volume. Step (iv) causes a change in the muscle volume. To conserve the volume, we adjust the shape of the deformed muscle by appropriate scaling. The volume of the muscle belly formed by meshing the deformed profile curves is calculated on a cross-sectional basis.

- **Deformation of Multiple-Head Muscles**

A multiple-head muscle is modelled by combining its single-head components. Interaction among the components is considered in Section 4.2. Figure 2 is an example of multiple-head muscle deformation: biceps femoris.

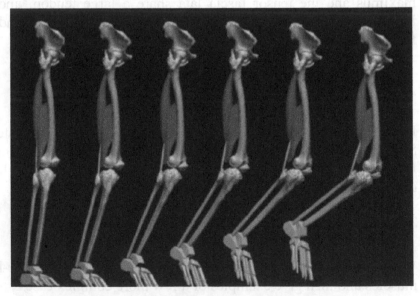

Fig. 2. *Deformation of the Biceps Femoris*

- **Deformation of the Adductor Magnus**

The adductor magnus model is special due to the three end contours of its muscle belly, one at the origin and two at the insertion. The shape of the model is deformed as a whole. We model two action lines for the magnus model; these remain straight throughout the movement. The motion of the tendon follows one of the action lines.

- **Deformation of the Quadriceps Femoris**

The four muscles in the quadriceps femoris work together and contribute mainly to knee motion. The action line of the rectus femoris curves at the front of the femur and has 3 line segments. It is fitted by a cubic curve. The vastus lateralis, vastus medialis and vastus intermedius models all have curved action lines in 3 line segments, which are fitted by cubic curves.

The compound tendon of the quadriceps moves in accordance with the motion of the action lines of the muscles and pulls the end contours of all the muscles in this group. This gives rise to the reshape of the mesh.

The Modeling of the Quadriceps Femoris is illustrated in Figure 3, together with the Sartorius.

- **Deformation of the Sartorius**

The action line of the sartorius model curves across the front of the upper leg. During contraction, this curve is strongly affected by the shapes of underlying muscles. To model this, we take some sample points along the action line and adjust the shape of the action line and the muscle according to the movement of the underlying muscles.

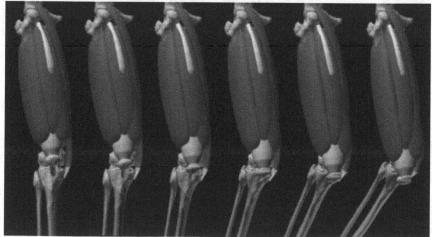

Fig. 3. *Deformation of the Quadriceps Femoris and the Sartorius*

4.2 Combining the Models of Individual Muscles

So far, we have discussed deformations of single muscle models. In fact, the deformation of a muscle model is also affected by the surrounding tissue. In this section, we discuss the interaction between muscle models, after which we put them together to produce a whole leg deformation.

Computing the interaction between muscle models is important, not only for deformation, but also to prevent penetration between models during deformation. The algorithm to compute the interaction is as follows.

After the muscle Modeling, produce a table associating with each vertex V_i on the muscle surface, the following information:

- the identity of the neighbouring muscle model
- the face on this neighbouring muscle model with which V_i is in contact.

Figure 4 illustrates two surfaces from neighbouring muscle models A and B. V_i from model A is in contact with face f_j of model B. We record model B and face f_j into the table, associated with V_i.

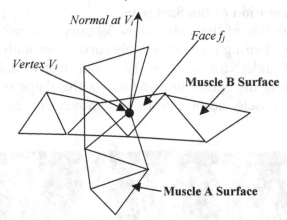

Fig. 4. *Computing the muscle interaction*

The final deformation of a muscle model relies both on its own deformation and influences from other models. We simulate the interaction between muscle models by scaling transformations.

The detailed procedure is as follows:

- compute the deformation of all the individual models
- detect penetration between neighbouring models by checking each vertex of the muscles; the information stored at the vertex is used to find which face on which muscle should be checked

- if there is an interaction between 2 models, scale the models by a pre-defined factor along the normal to the surface, which is the major direction of the common surface between the two muscles, then return to the previous step
- if there is no interaction, terminate the algorithm.

This approach efficiently describes the interaction between the muscle models and has been proved to work well in practice. Figure 5 is a demonstration of deformation of the whole leg model during knee flexion.

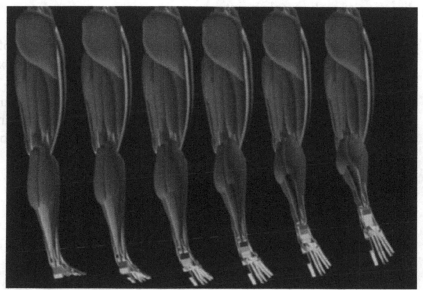

Fig. 5. *Deformation of leg muscles during knee motion*

5   Muscle Rendering

The result of muscle rendering should illustrate muscle texture, which is an important feature of the muscle and is created by the arrangements of muscle fibres. Here we shall present two different approaches, the first is by texture synthesis, and the second is by pen-and-ink illustration.

5.1 Texture Synthesis

To render the muscle models by texture mapping, we first create texture images which can identify muscle fibre orientations from the

Visible Human Data, and then synthesise the texture on the surface models. This approach is markedly different from the standard approach in which a 2D texture is simply mapped on to the surface.

The process used involves the following steps:

- **Finding the Orientation of Muscle Fibres**

The fibre orientations are sampled at points of a *u-v* iso-parametric grid on the muscle surface and expressed in terms of vector directions at these sample points. 3D edge detection is applied to the Visible Human data to detect the edge points, which indicate the directions, and it is followed by the Hough transform to identify the direction at each sample point.

- **Texture and Colour**

Using the direction vectors generated above, and the colours of the muscle fibres from the anatomical data, we synthesise the texture with colours on the surface model. The method adopted, line integration convolution (LIC) is a flow visualisation technique described by Cabral & Leedom [20]. Starting from each sample point, we integrate along the vectors and assign the result to the point as its colour. An example is given in Figure 6.

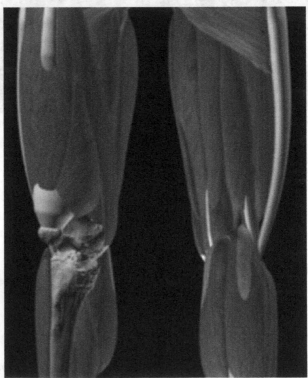

Fig. 6. *Texture of the leg muscle models*

5.2 Pen-and-Ink Illustration

Much attention has been paid to non-photorealistic rendering (NPR) in recent years. Instead of placing an emphasis on realism in 3D images, NPR seeks to produce images in the manner of traditional artistic styles, such as painting, pen-and-ink illustration, etc.

For many years, technical and medical illustration has tended to use pen-and-ink as the main medium. Typically, these pictures leave out a great deal of detail and use simple lines to demonstrate technical concepts or object shapes and features. Apart from being cheaper to print, they are frequently able to convey a higher level of information and provide a greater focus than glossy images.

The widespread adoption of pen-and-ink illustrations in the medical area, and their subsequent familiarity to medical professionals, suggests that they could be a readily accepted alternative to conventional rendering in medical visualisation.

In this section, we shall introduce a novel technique for producing muscle images in pen-and-ink style from medical volume data, which we have named *volumetric hatching*. During volumetric hatching, the interior data makes a significant contribution to the final results, but it differs from volume rendering as follows:

- there is a need to compute volume silhouettes
- the final results are vectorised lines rather than pixel colours – the contributions from data points are strokes, so we have to decide stroke orientations for each of the data points
- only data on, or a small distance beneath, the exterior surface makes a contribution; this is because hatching is to produce illustrations that represent the shape, whereas volume rendering sometimes uses data deep inside the volume to visualise another object by making the outer one transparent.

The pipeline of *volumetric hatching* is similar to that of conventional volume rendering and is illustrated in Figure 7. The input is volume data. The significant difference lies in the output, in which strokes are produced rather than pixel intensities.

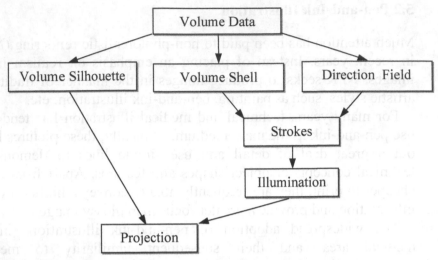

Fig. 7. *Overview of volumetric hatching*

- **Volume Silhouette**

The silhouette conveys the most important information about the subject. It is identified from volume data using an algorithm similar to Marching-Cubes, Lorensen & Cline [21], for a fixed viewpoint.

The algorithm starts by finding the first segment of the silhouette. It checks each cube in the volume, considering its eight data points for their intensity values and their normals. If the values do not span the threshold, or if all of the normals face in the same direction with respect to the viewpoint, the cube is eliminated. Otherwise, the cube is split into five tetrahedra; this can be done in either of two ways, and adjacent cubes must take different ways to maintain coherence.

From the intensity values and normals at the vertices, the tetrahedron that contains the silhouette can be identified. The iso-face inside this tetrahedron is constructed, and the first segment of the silhouette across the face is generated by linear or bilinear interpolation.

Once the first segment of the silhouette is found, it is traced into the neighbouring tetrahedra or cubes to find the remainder of the silhouette. Each time the algorithm marches into a new cube, this cube is split into five, and the same operation is applied to find the line segments inside the cube. This continues until all of the volume cubes have been visited.

- **Stroke Directions**

When illustrating muscles, the orientation of the strokes should follow the directions of the muscle fibres. The method consists of finding the initial approximate directions, 3D-edge detection and Hough transforms. For more details, please refer to Dong et al.[22].

- **Producing Strokes**

In volumetric hatching, data some distance beneath the iso-surface can contribute significantly to the results. We first build a shell beneath the surface of the subject. Next, a 3D stroke is created for each data point inside the shell; strokes are short lines that reflect the shape of the shell. This is achieved by fitting the data in the shell with surfaces and producing the strokes as lines running on the surfaces.

The fitting is based on a local scheme. At each data point of the shell, a surface patch is used to fit the data locally, and a plane is built which contains the normal of the surface patch and the stroke direction given above. The stroke is defined as the intersection between the surface patch and the plane, as shown in Figure 8.

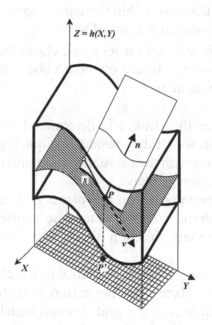

Fig. 8. *Generating strokes*

In Figure 8, the object in bold lines is the shell, with z its primary direction. The xy plane is the base on which resides a grid, which is the domain of the height field, $h(x,y)$. The shaded surface within the shell is the fitting surface. The plane containing the surface normal *n* and stroke direction *v* intersects the fitting surface to produce a stroke *s*. The process is described in greater detail below. As seen from Figure 8, the surface patches are represented in terms of height fields. They are produced by fitting local data inside the shell.

- **Illumination**

For a fixed area in a pen-and-ink illustration, including more strokes produces a darker effect, while using fewer strokes gives a lighter appearance. Thus, illumination relates to the number of strokes in a particular area.

The illumination model counts the number of strokes passing through a volume cube and removes some, if there are too many. In detail, the following steps are taken:

- for a volume cube, calculate the average lighting intensity at the eight data points of the cube – the number of strokes N is proportional to the intensity
- count the number of strokes in the volume cube – if this is larger than N (over toned), select a stroke and remove the segment of that stroke that lies within the cube; repeat this until the number of strokes in the cube is equal to N
- if a stroke is selected for removal, check the other cubes through which it passes – if any of these cubes are over toned, cut the stroke from them, too.

- **Visibility**

The visibility of the strokes is determined by the visibility of the volume cubes in which they reside. We quickly find the visibility of the volume cubes with the assistance of an intermediate plane, which is parallel to the image plane.

A grid is superimposed on this plane, and each square of the grid records the volume cubes that are the shortest distance from the plane along the view direction (that is, the "visible" cubes) and their states.

To compute the visibility of a stroke, we check its projection on the intermediate plane. The projection is recorded in terms of the collection of squares in the grid through which it passes. Then we make visibility marks on the projection, according to whether the cubes within which the stroke resides are visible in the squares. The stroke could be totally visible, invisible, or partially visible, depending upon the marks on the projection.

- **Results**

Here we present two pictures generated by volumetric hatching. The left image of Figure 9 is the front view of the muscles on a human upper leg, and the right image is the back view.

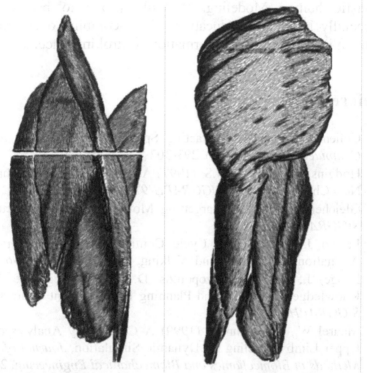

Fig. 9. *Pen-and-ink illustration from volume data*

6 Summary

This chapter has discussed anatomy-based approaches to Modeling and rendering human muscle. We have attempted to produce shapes, behaviours, and colours of muscle models that are based on their anatomical features and biomechanical properties, so as to be representative of the real morphological structures.

The geometric Modeling phase produces accurate surface models from well-segmented anatomical data. These models are classified into a number of types so that muscles with common representations and deformation actions are in the same category. The work on muscle deformation simulates muscle actions based on their anatomical structures, using this same categorisation. Two different approaches to muscle rendering were also discussed; they are based on texture synthesis, and the pen-and-ink style of non-photorealistic rendering.

We believe that the work described brings us one step closer to realistic human Modeling. It could prove to be very useful, especially in medical education, if we combine our current work with an intuitive muscle deformation control interface.

References

1. Cohen M.F. (1992) Interactive Spacetime Control for Animation, *Computer Graphics*, 26(2), 293-302
2. Hodgins J.K., Pollard N.S. (1997) Adapting Simulated Behaviors For New Characters, *Proc. SIGGRAPH'97*, 153-162
3. Gleicher M. (1998) Retargetting Motion to New Characters, *Proc. SIGGRAPH'98*, 33-42
4. Laszlo J. (1996) Limit Cycle Control and its Application to the Animation of Balancing and Walking, *Proc. SIGGRAPH'96*, 155-162
5. Funge J., Tu X., Terzopoulos D. (1999) Cognitive Modeling: Knowledge, Reasoning and Planning for Intelligent Characters, *Proc. SIGGRAPH'99*, 29-38
6. Maurel W., Thalmann D. (1999) A Case Study Analysis on Human Upper Limb Modeling for Dynamic Simulation, *Journal of Computer Methods in Biomechanics and Biomechanical Engineering*, 2(1), 65-82
7. Scheepers F., Parent R.E., Carlson W.E., May S.F. (1997) Anatomy-Based Modeling of the Human Musculature, *Proc. SIGGRAPH'97*, 163-172
8. Wilhelms J., Van Gelder A. (1997) Anatomically Based Modeling, *Proc. SIGGRAPH'97*, 173-180
9. Savenko A., Van Sint Jan S.L., Clapworthy G.J. (1999) A Biomechanics-Based Model for the Animation of Human Locomotion, *Proc GraphiCon '99*, Dialog-MGU Press, 82-87
10. Van Sint Jan S.L., Clapworthy G.J., Rooze M. (1998) Visualisation of Combined Motions in Human Joints, *IEEE Computer Graphics & Applications*, 18(6), 10-14
11. Van Sint Jan S.L., Salvia P., Clapworthy G.J., Rooze M. (1999) Joint-Motion Visualisation using Both Medical Imaging and 3D-Electrogoniometry", *Proc 17th Congress of International Society of Biomechanics*, Calgary (Canada)

12. Chadwick J.E., Haumann D.R., Parent R.E. (1989) Layered Construction for Deformable Animated Characters, *Computer Graphics*, 23(3), 243-252
13. Kalra P., Magnenat-Thalmann N., Moccozet L., Sannier G., Aubel A., Thalmann D. (1998) Real-Time Animation of Realistic Virtual Humans, *IEEE Computer Graphics and Applications*, 18(5), 42-55
14. Gourret J.P., Magnenat-Thalmann N., Thalmann D. (1989) Simulation of Object and Human Skin Deformations in a Grasping Task, *Computer Graphics*, 23(4), 21-30
15. Turner R., Thalmann D. (1993) The Elastic Surface Layer Model for Animated Character Construction. In: Magnenat-Thalmann N., Thalmann D. (eds), *Communicating with Virtual Worlds*, Springer Verlag, 399-412
16. Chen D.T., Zeltzer D. (1992) Pump It Up: Computer Animation of a Biomechanically Based Model of Muscle Using the Finite Element Method, *Computer Graphics*, 26(2), 89-98
17. Dong F., Clapworthy G.J., Krokos M.A., Yao J.L. (2002) An Anatomy-Based Approach to Human Muscle Simulation, IEEE Transactions on Visualization and Computer Graphics, 8(2), 154-170
18. Hoppe H., DeRose T., Duchamp T., Halstead M., Jin H., McDonald J., Schweitzer J., Stuetzle W. (1994) Piecewise Smooth Surface Reconstruction, *Proc. SIGGRAPH'94*, 295-302.
19. Christiansen H.N., Sederberg T.W. (1978) Construction of Complex Contours Line Definition into Polygonal Element Mosaics, *Computer Graphics*, 12(3)
20. Cabral B., Leedom L. C. (1993) Imaging Vector Fields Using Line Integral Convolution, *Proc SIGGRAPH' 93*, 263–270
21. Lorensen W.E., Cline H.E. (1987) Marching Cubes: A High Resolution 3D Surface Construction Algorithm, *Computer Graphics*, 21(4), 163-169
22. Dong F., Clapworthy G.J., Krokos M. (2001) Volume Rendering of Fine Details within Medical Data, *IEEE Visualisation 2001*

Chapter 10

Skeletal Implicit Surface Reconstruction and Particle System for Organ's Interactions Simulation

Skeletal Implicit Surface Reconstruction and Particle System for Organ's Interactions Simulation

M. Amrani, B. Crespin, and B. Shariat
LIGIM, University of Lyon I, Villeurbanne, France

We describe an automatic surface reconstruction technique from a set of planar data points organised in parallel sections. The reconstruction employs skeletal implicit surfaces. Two key points in this study are:
- Calculation of the 3D skeleton by establishing a correspondence between each pair of 2D Voronoi skeleton of two neighbouring sections.
- Use of a uniform field function necessitating the introduction of the notion of "weighted skeleton".
Another key point of this work is the proposition of an animation methodology by transforming the skeleton into a deformable mass/spring system.
We also describe the multi layer particle system model, which is used for reconstruction and simulation of deformable objects. The last part of this work consists in defining an hybrid approach by using the two previous model in the same scene, improving the simulation.
This work has been used in the context of a medical project to simulate the dynamic behaviour of organs during the conformal radiotherapy treatment.

1 Introduction

In this chapter, we describe a general methodology for reconstruction and animation of volumetric deformable objects using skeletal implicit surfaces. These surfaces are defined by a skeleton and a field function. They allow the modeling of shapes of arbitrary topology at low computation cost and can easily account for plastic or elastic behaviour of deformable objects. With this model it is also possible to introduce several kinds of geometrical constraints such as volume conservation to reproduce the behaviour of cancerous tissues and organs.

This work has been used in the scope of a conformal radiotherapy application consisting in the delivery of a lethal ionising radiation dose to cancerous regions causing a minimum of damage to surrounding healthy tissues. To achieve this goal, the use of multi-leaf collimators permits to have a precise control on the beams' shape as well as their shooting angles and intensities. Classically, the patient's initial positioning is supposed to be exactly known. Moreover, the patient as well as his organs are supposed rigid and motionless during the treatment. However, in practice, this supposition is not true.

In order to enhance the quality of the dosimetric calculation, we have to take into account the natural movement of the organs as well as their interactions. For this we have created geometrical models of the patient's organs. The animation and interaction of these models permit to have a better understanding of the quality of the therapy.

In this application, we have used two representations to model human organs (hybrid approach): implicit surfaces for the organs subject to small deformations and particle systems [2] for the others. This allows us to simulate interactions between organs in a body.

In the following sections, after the state of the art we describe our implicit surface reconstruction methodology as well as the deformation process. Then we give a short description of the particle system model and we define an hybrid approach for organs' interactions modeling. We finally present examples of simulation using this hybrid approach.

2 State of the Art of the Most Relevant Reconstruction Methods

Numerous reconstruction methods have been developed over the past two decades to fit a 3D surface to a set of points in space. The surface model usually depends on the application. In the area of deformable objects for example, the most common way for modeling elastic materials is to use finite difference or finite element techniques by approximating the body with a mesh of nodes of fixed topology [20]. However, this model is not well adapted for materials able to accept large inelastic deformations.

Other formulations for the resulting surface include parametric or implicit definitions, which allow complex algorithms such as collision detections to be applied more efficiently. We choose to focus on a particular model, namely skeletal implicit surface. A skeleton is

173

a set of geometric primitives (usually points, segments or polygons), represented either by a tree or a graph. Skeletal implicit surfaces are a natural choice for modeling and simulation applications, since each primitive can have its own motion while preserving the overall smooth, free-form shape of the generated surface [7].

Therefore, this preliminary study of existing reconstruction methods is restricted to those producing implicit surfaces, following a short presentation of implicit surfaces themselves.

2.1 Implicit surfaces

An implicit surface is defined as the set of points P in space that satisfy $f(P)=0$, where P is a point in space and f is a scalar function of $\Re^3$. For most types of implicit surfaces such as algebraic surfaces (quadrics and superquadrics), f is a polynomial.

The main property of implicit surfaces is the *point membership classification*: for any point P in space, it is possible to know whether it lies inside, outside or on the surface depending on the scalar $f(P)$. Thus, collision detection between implicit surfaces is easy since it only means testing the inclusion of one surface into another.

In the case of skeletal implicit surfaces (also known in the literature as *blobs, meta-balls, potential surfaces, convolution surfaces or soft objects* [5]), the implicit equation has the form:

$$\sum_{i=1}^{n} f_i(P) - ISO = 0 \tag{1}$$

where f_i is a potential (or field) function associated to each primitive of the skeleton, ISO is the global isovalue, and n is the number of primitives. Each f_i is defined as a gaussian decreasing smoothly to 0 according to the distance between P and that primitive. Therefore, the overall sum leads to a smooth blend between each primitive.

2.2 Implicit surface reconstruction

Most implicit surface reconstruction methods rely on an unorganised (or unstructured) set of points as an input. These methods can also be employed for reconstructing from planar sections anyway, but the connectivity of the vertices in a section won't be used.

In the first category, an algebraic surface is deformed by external forces to fit the data points: superquadrics [18], deformable super-quadrics [3] and hyperquadrics [8] were studied. But these methods are limited by the shape of the initial basic surface.

In the second category, smooth implicit surfaces are computed directly from the data points. Different formulations were proposed: simplicial surface defined from signed distance [11], R-function [17], function sampling on a 3D grid [21], variational implicit surface [14], each having its own specific properties, but no significant reduction of the size of input data is obtained.

Other methods, which were specifically designed to produce skeletal implicit surfaces, address the problem of data reduction. Indeed, the skeleton structure is a way to represent the shape of a more complex object, and finding the best surface that fits a set of points "only" requires the computation of the skeleton and the associated potential function. In [15], the skeleton is a random set of points which is refined iteratively through an energy-minimisation process by adding or deleting some skeleton points. The convergence is optimised in [4] and [9] by using the medial axis of the input data as an initial skeleton. Nevertheless, the computation is still expensive, especially for large datasets.

Finally, a few other techniques take advantage of the planar sections organisation. In practice, they can usually be decomposed in two steps: reconstruct a 2D implicit function for each section, then extend this set to a 3D implicit function that preserves the continuity between each plane. In the method proposed in [12], the 2D function is sampled in a 2D grid using the minimal distance to the contour. The set of 2D grids is then considered as a global 3D grid, thus resembling to some methods described in the previous section. Other methods provide continuous 2D functions [17] [10] that are interpolated in the second step by set-theoretic operations such as sweeping [19].

3 Methodology for Implicit Surface Reconstruction

As we are interested in reconstructing a skeletal implicit surface from planar sections, none of the methods described previously seems well adapted. Indeed, considering the input data as an unorganised set of points is not satisfying because the reconstruction process is not optimised. On the other hand, specific methods to reconstruct an implicit surface from sections do not make use of a skeleton structure that could be computed from the contours set.

Therefore, we propose a method that takes advantage of both approaches by computing a skeleton from a set of planar sections. A field function is then combined with the skeleton; there is no blend, and no iterative process is required to refine the reconstruction. Thus, the model is simplified and the deformation process is easier.

One should note that a field function affects a potential to a point according to its distance to the skeleton. Thus, a potential corresponds to a distance and an iso-potential surface is a uniform offset of the skeleton. Then, if a uniform field function is employed, the reconstruction of complex shapes needs to have complex skeletons which are obtained by erosion of the data points.

The use of such a skeleton is not realistic, because both its computation and storage are expensive. Therefore, we introduce "weighted skeletons", composed of a set of non intersecting triangles for which a scalar is affected to each vertex. These values are distances and define a non-uniform offset of the skeleton. They are computed such that the offset surface interpolates the points cloud: the distance between the data points and the weighted skeleton is always the same. Then, it suffices to combine the weighted skeleton with a uniform field function to reconstruct the object.

3.1 3D weighted skeleton computation

The computation of a 3D skeleton is usually based on a 3D Voronoi diagram. However, a direct generation of a 3D Voronoi diagram is prohibitive. Therefore, we prefer a two-steps approach: generate the 2D skeleton of each section, then deduce the 3D skeleton.

Computation of the 2D skeletons

Algorithms that generate a 2D skeleton from a contour are well-known [6]; the vertices are the nodes of the 2D Voronoi diagram. But this skeleton may contain too many branches, which increases the complexity of the model. This is due to the sensitivity of Voronoi diagram to noise and to small perturbations on the contour. To simplify the skeleton, we have to suppress unnecessary branches by the deletion of the corresponding Delaunay triangles. These are the triangles of the contour whose area is smaller than a given threshold, defined according to a precision criteria (see Fig. 1) [16].

Fig. 1. *Initial 2D contour, non simplified 2D skeleton and simplified 2D skeleton.*

Computation of the 3D skeleton

The computation of the 3D skeleton consists in establishing a correspondence between each pair of neighbouring 2D Voronoi diagrams. The result is a 3D skeleton composed of triangles whose vertices are the nodes of the 2D Voronoi diagrams.

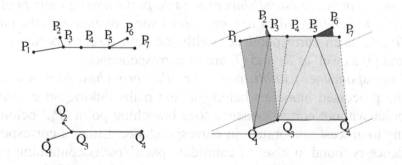

Fig. 2. *2D skeleton of two neighbouring sections and the deduced 3D skeleton*

To understand the process of the 3D skeleton computation, let us consider two Voronoi skeletons of two neighbouring sections and define some terms (see figure 2):

- A *branching point* is a 2D Voronoi vertex belonging to at least three 2D Voronoi skeleton branches (such as P_3, P_5 and Q_3).
- An *extreme point* is a 2D Voronoi belonging to one skeleton branch (such as P_1, P_2, P_6, P_7, Q_1, Q_2 and Q_4). A *corresponding point* is a 2D Voronoi vertex which is in corre-

spondence with at least one branching point or extreme point of a neighbouring skeleton (Q_2 is a corresponding point of P_2)

- A *corresponding branch* is a skeleton branch in correspondence with a branch of a neighbouring skeleton (the branch $Q_1;Q_3$ is a corresponding branch of the branch $P_1;P_3$)

The 3D skeleton calculation process is done in three steps:

1. *Correspondence establishment for the branching points and the extreme points.* This is an automatic process based on the 3D Voronoi diagram of the set of points organised in parallel sections. The correspondences of a point are the 2D Voronoi vertices of the neighbouring sections to which the point is linked by paths composed of 3D Voronoi edges. For example in Fig. 2, P_3 is linked to Q_1 and Q_3. To simplify the process, the branching points (respectively the extreme points) are in correspondence only with branching points (respectively extreme points).

2. *Correspondence establishment for two paths having their beginning and ending points in correspondence.* For example, the path (P_3,P_7) is in correspondence with the path (Q_3,Q_4) because P_3 and Q_3 as well as P_7 and Q_4 are in correspondence.

3. *Correspondence establishment for other branches.* At this stage, the processed branches belong to the paths linking an extreme point without correspondence to a branching point (P_B) belonging to at least one branch in correspondence. Either a correspondence is found in a set of candidate paths (paths containing one extreme branch without correspondence and one of the corresponding points of P_B) or no correspondence is found and the path is in correspondence with the nearest corresponding point of P_B. For example, the branch P_6,P_5 is in correspondence with Q_4.

Weighting

As mentioned earlier, the Voronoi skeleton can not be combined to a uniform field function to reconstruct complex shapes. Therefore, we need to define weights for the skeleton vertices. For this, we first compute the weights of the 2D skeletons vertices, according to their distance to the contour. The 3D skeleton is composed of triangles whose vertices are 2D Voronoi vertices of the neighbouring sections. Thus, the vertices of different triangles of the 3D skeleton are already weighted. Weights of other skeleton points are obtained by an affine combination of triangle vertices weights.

4   Animation and Deformation of Implicit Objects

For a realistic simulation of the modeled objects, we use physically based animation techniques. Our solution is based on the geometrical structure of the proposed reconstruction technique: since the skeleton is a set of triangles, we choose to transform the vertices (resp. the edges) of the triangles into masses (resp. springs). In this method, two levels are used, external (surface) and internal. When a collision occurs on a surface point, the reaction force is distributed on the skeleton's triangle that contains the projection of this point.

The general process of the simulation is defined by successive iterations. Each iteration represents the transition of the object's state during the time t to $t + dt$. The computation of the object's evolution during this "*time step*" dt is obtained in four steps:
- rigid motion according to the forces applied to the skeleton (gravity for example)
- collision detection
- deformation of the surface to approximate the exact contact surface
- deformation of the skeleton through the propagation of the constraints applied to the surface and computation of the new shape

The rigid displacement of the object consists in using the Newtonian laws of mechanics. For the collision detection of two implicit surfaces, we use a sampling method [16]: we first sample the two surfaces then, using the potential functions, we test if sampled points of one surface are included in the other one. When such points exist, a collision has occurred.

In case of collision, the surface of the object should be deformed (displacement of each sampled point P belonging to the contact zone) by applying a reaction force $\vec{F}_{R_P}$. The amplitude of this force depends on the degree of the penetration of P in the other colliding object, and its direction is obtained from the direction of the velocity of P when the collision occurs. It is also possible to take into account the friction forces $\vec{F}_{f_P}$ on each point P. In this case the force applied to P is:

$$\vec{F}_P = \vec{F}_{R_P} + \vec{F}_{f_P} \qquad (2)$$

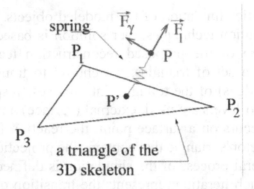

Fig. 3. *Reaction of a surface/skeleton spring*

Using the reaction force, the contact surface can be computed. In our approach, each point P is supposed to be linked to its corresponding point on the skeleton by a spring. This spring reacts to the force $\vec{F}_P$, which deforms the surface. The spring has an elastic behaviour, *i.e.* the surface should come back to its initial state when the constraint is removed. Therefore, two types of linear and angular forces should be defined (see Fig. 3). Linear forces $\vec{F}_l$ control the length and angular forces $\vec{F}_\gamma$ control the orientation of the spring with respect to the skeleton. Thus, two elasticity parameters are introduced: spring's linear elasticity k_l and spring's angular elasticity k_γ. Motion of point P is then obtained by:

$$\vec{F} = \vec{F}_l + \vec{F}_\gamma \qquad (3)$$

where:

$$\begin{aligned} \vec{F}_l &= k_l (l_0 - l) \\ \vec{F}_r &= k_r (\gamma_0 - \gamma) \end{aligned} \qquad (4)$$

with l_0 and l the spring's initial and actual length, and γ_0 and γ the spring's initial and actual orientation. These reaction forces, as well as other constraints such as gravity, are then propagated through the skeleton, inducing the motion of different skeleton's nodes.

Recall that the skeleton is composed of triangles whose edges are springs. As stated for the surface/skeleton springs, three kinds of forces are needed so that the skeleton can recover its initial state when external and internal constraints are removed [13] (see Fig. 4):

- *linear forces* $\vec{F}_L$, to conserve the length of the edges
- *angular forces* $\vec{F}_\alpha$, to conserve the angles between edges
- *co-planar forces* $\vec{F}_{\bar{n}}$, to conserve the orientation of the triangles

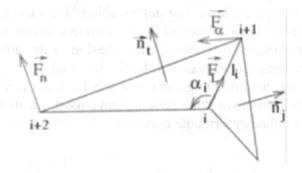

Fig. 4. *Internal forces applied to object's skeleton*

The equilibrium condition can then be written as :

$$\sum_i \vec{F}_{L_i} + \vec{F}_{\alpha_i} + \vec{F}_{\bar{n}_i} = \vec{F} \qquad (5)$$

Consequently, vertices' velocity is obtained by integrating these forces. The motion of the vertices deforms the skeleton, inducing the deformation of the object's surface. Figure 5 shows an example of the application of our methodology, with an extra parameter controlling the volume of the object.

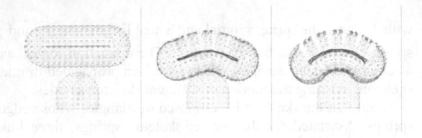

Fig. 5. *Implicit surface animation and deformation*

5 Particle Systems

Particle systems consist in a set of solid spheres whose movement follow physical laws. For deformable objects modelling, this simple model has been improved by the introduction of internal forces between particles to preserve the cohesion of the object. Hence, particle systems model seems to be the most adapted for highly deformable objects modelling since many behaviours can be simulated [1]. In this section we show how we have used this model to reconstruct and simulate deformable organs.

5.1 Reconstruction with multi layer particle system

Our reconstruction method requires to first compute a closed surface from the initial set of points. When the initial data is structured in parallel sections, we can compute an implicit surface with the method described in previous sections. The reconstruction method consists in gradually filling this volume with particles, and let them evolve to a stable state according to physical laws. Thus, we easily obtain a regular sampling which corresponds to a maximal filling. Unfortunately, if we want to increase the precision of the resulting model, we need to decrease the radius of the particles. This will dramatically increase their number. To solve this problem, we have developed a multi-layer particle model: small particles are placed where details are required and big ones elsewhere (see Fig. 6).

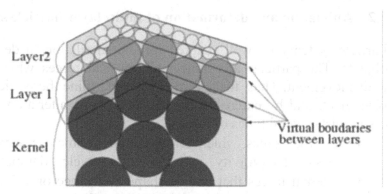

Fig. 6. *Multi-layer structure*

- the centre is composed of particles of great radius. They are the kernel of the object,
- this kernel is surrounded by one or more layers, according to the desired precision, with decreasing radius as we get closer to the boundaries.

The consistency of each layer is preserved by virtual separations between layers. The reconstruction algorithm consists in successively filling each layer with particles. However, when computing the internal forces, we have to include the interactions with the particles of the existing layers.

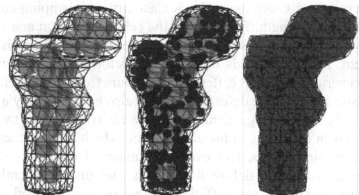

Fig. 7. *Femoral head reconstruction with multi layer particle systems*

The reconstruction process is finished when no particle can be added to any of the layers, and the system has evolved to a stable state. Figure 7 shows the reconstruction steps of a femoral head.

5.2 Animation and deformation of multi layer particle systems

Particle systems are well adapted for the simulation of deformable objects. The particles are subject to physical forces which induce their movement. Thus, these forces will determine the particles' displacement, and by extension the object's shape alteration. We can distinguish several types of forces:

- external forces, which define the environment in which particles evolve (gravity, collision with obstacles, damping...)
- internal forces that ensure shape coherence or volume preservation (attraction/repulsion, auto-collision)

It is possible to handle a wide variety of behaviours, simply by changing force formulation and its parameters. If there is no change in the objects' topology (objects do not undergo large deformations), the internal forces can be represented by a network of mass/spring elements as in section 4. This reduces greatly the complexity of the system. The interaction between particles is then simplified and modelled with the help of springs that can be written as follows:

$$\vec{f} = -\left[k_s(\|\vec{r}\| - l_0) + k_d \frac{\vec{v}.\vec{r}}{\|\vec{r}\|} \right] \frac{\vec{r}}{r} \qquad (6)$$

where k_s and k_d are the spring's elasticity and damping constants, l_0 is the initial length, $\vec{r}$ and $\vec{v}$ are the relative position and velocity of two adjacent particles. The constants define the characteristics of the organ: rigidity depends on k_s, while deformability depends on k_d. This permits to model different behaviours from rigid to elastic state. Elastic and inelastic deformations can also be handled by a combination of simple springs. Cutting may be easily simulated by removing springs in contact with the cutting tool, which is very interesting for surgery simulations. However, the fusion of objects is difficult to obtain. For highly deformable object, and in the general case, the spring system is not sufficient. We use non coupled interaction forces such as Lennard-Jones (see Fig. 8) to model internal interactions:

$$\vec{f}(r) = \frac{-mn\varepsilon}{(n-m)r_0} \left(\left(\frac{r_0}{\|\vec{r}\|} \right)^{(n+1)} - \left(\frac{r_0}{\|\vec{r}\|} \right)^{(m+1)} \right) \frac{\vec{r}}{\|\vec{r}\|} \qquad (7)$$

where m and n are called the shape parameters, r_0 is the equilibrium distance, r the distance between two particles and ε the force amplitude.

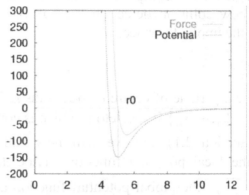

Fig. 8. *The Lennard-Jones force*

6 A Hybrid Approach

Implicit surfaces described in section 4 are able to model organs of arbitrary topology. However, they cannot simulate large deformations since this would generally involves the re-computation of the skeleton. On the other hand, particle systems described in the previous section are able to simulate large deformations, but they are more expensive in computing time and memory size. The key idea is to use both models. Therefore, particle systems will be used to model organs that are subject to large deformations, and implicit surfaces for the others.

6.1 Interaction between particle systems and implicit surfaces

In order to use and animate two models simultaneously, we have to handle possible interactions between them. To this end, we first define a collision detection method between a particle system and an implicit surface. A first problem is to quickly eliminate a particle which do not collide the surface. One approach consists in using the classical sampling based method, which needs to test the potential of all sampled points on the particle surface, and also the distances between the sampled points on the implicit surface and the particle centre. However, this method is very expensive. Indeed, for a precise collision detection we need to increase the number of sampling points, and hence the computing time. To avoid this problem, we propose a new collision detection method based on the potential function of the implicit surface.

Definitions

Let $P(c,r)$ be a particle of centre c and radius r. Let S be an implicit surface, $S_q = \{s_1, \cdots, s_m\}$ its skeleton, which is composed of m primitives (see section 2.1) that we will now call *skeletal elements*, $f_1, \cdots, f_m$ the local potential functions associated with these elements, $F = \sum f_i$ the global potential function and *ISO* the iso potential value. We also define the following:

- **Point of maximum local potential**: it is the point p_i of the particle P which has the maximum potential according to the skeletal element s_i:

$$\forall p \in P(c,r), f_i(p) \le f_i(p_i) \tag{8}$$

and it is also the particle's point closest to s_i:

$$\forall p \in P(c,r), d(p,s_i) \le d(p_i,s_i) \tag{9}$$

- **Local minimal distance**: it is the distance $d\min_i$ between the point p_i and the skeletal element s_i:

$$d\min_i = \begin{cases} d-r & if \quad d > r \\ 0 & otherwise \end{cases} \tag{10}$$

Proposal

Let $B = \sum_{i=1}^{m} f_i(d\min_i)$ then

$$\forall p \in P(c,r), F(p) \le B \tag{11}$$

Proof

$$\forall p \in P(c,r), \; \forall i \in \{1, \cdots, m\} \qquad d(p,s_i) < d\min_i$$
$$\Rightarrow \forall i \in \{1, \cdots, m\} \qquad f_i(d(p,s_i)) \le f_i(d\min_i)$$
$$\Rightarrow F(p) \le B$$

The previous proposal means that B is the maximal potential value of a point of the particle. So, if $B < ISO$, we can deduce that there is no collision between the particle and the implicit surface. In the other case, a collision can exist and we have to refine the detection. In order to do that, we first consider the implicit surface to be composed of two kinds of surface pieces:

- surface areas resulting from the potential due to only one skeletal element. Although an implicit surface is defined as the result of the blending of potentials of all its skeletal elements, in practice the contribution of a skeletal element is restricted by an action ray beyond which the potential due to this element is negligible;
- surface areas resulting from blending the potentials of the skeletal elements.

When we consider the first case, if a collision occurs in one of these areas it should be on one of the points of maximum local potential. Therefore, in order to detect the collision, we just have to test for each p_i if its local potential according to the skeletal element s_i is greater or equal to the iso potential value of the implicit surface (*i.e.* if $f_i(p_i) \geq ISO$). Then, if this condition is true for one of these points, a collision has occurred on it.

Finally, to detect a collision in the areas defined from blending we have developed an heuristic which consists in computing the weighted affine combination (p_{aff}) of the points of maximal local potential. The weights are the normalised value of the local potentials associated to these points:

$$w_i = \frac{f_i(p_i)}{\sum_{j=1}^{m} f_j(p_j)}$$

$$p_{aff} = \sum_{i=1}^{m} w_i p_i$$

(12)

Using these weights allows us to take into account that the particle is closer to the part of the implicit surface defined by a certain skeletal element but also the contributions of the other elements. Then, we project the resulting point on the particle's surface, and we compute the potential value of the projection. The comparison of this value with ISO allows us to determine whether there is a collision or not (see Fig. 9).

For collision detection of a particle system we just have to apply
this method to each particle.

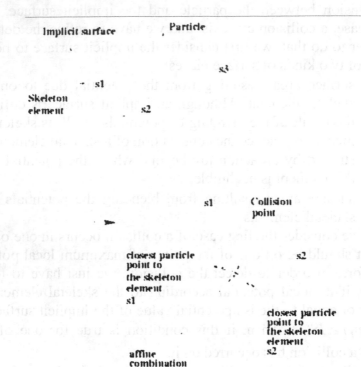

Fig. 9. *Collision detection between a particle and implicit surface*

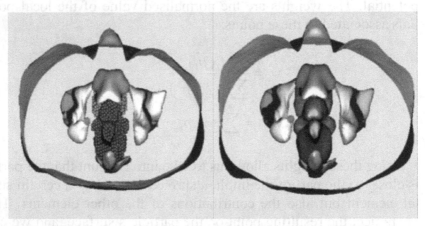

Fig. 10. *Hybrid model of the bladder, prostate and rectum system*

Figure 10 shows an example of an hybrid scene. This is a model
of the bladder, prostate and rectum system. The left picture shows

that particle systems are used to model the bladder, the prostate and the rectum, and implicit surfaces for the body shape and the bones (pubic bone and pelvises). The right picture shows the same model with a rendering method applied to the particle systems making them look more realistic.

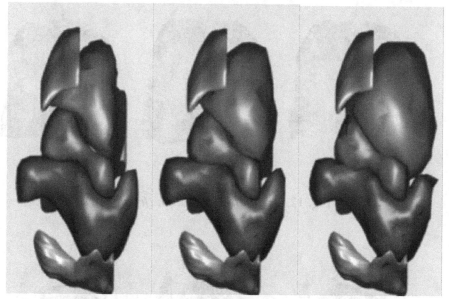

Fig. 11. *Bladder, prostate, rectum interactions simulation*

7 Simulation Examples

Figure 11 shows a simulation of prostate movement and deformation during the bladder's filling process. In this simulation, the body shape and a pelvic have been made invisible, but are still considered when computing organs' interactions. So, we can see the pubic bone acting as a constraint on the bladder, and making it moving to the back when growing. The prostate is deformed and pushed down by the bladder.

Our hybrid approach has also been used to model and animate the respiratory system depicted on Figs. 12 and 13. The mediastinum (*i.e.* trachea) and rib cage were reconstructed with skeletal implicit surfaces using the method described in this chapter, while the heart and lungs are modelled with particle systems. Three key frames of an animation are presented in these pictures showing empty to full lungs growing. All elements are in interaction with each other: see for example the deformation of the rib cage on Fig. 13.

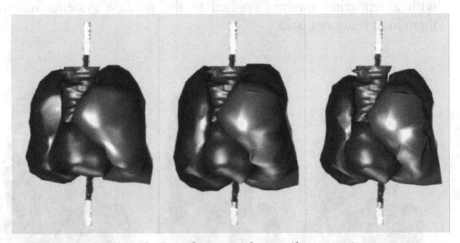

Fig. 12. *Breathing simulation (front view)*

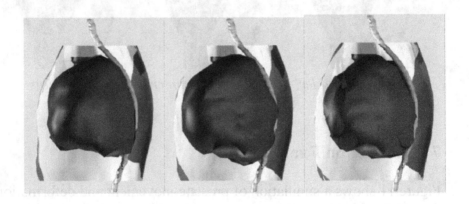

Fig. 13. *Breathing simulation (side view)*

8 Summary

In this chapter, we have presented a methodology to reconstruct objects' shape using skeletal implicit surfaces. The reconstructed objects can contain holes or have several branches.

To avoid the use of a blended field function, the objects are described with a uniform field function and a weighted skeleton.

Rather than computing the skeleton from the whole 3D cloud of points, we deduce it automatically from the 2D Voronoi skeleton of each CT scan section. We have also presented a technique to simulate the deformation of these surfaces by transforming their skeleton into a mass/spring system.

In the context of a medical application, we have used a hybrid technique. This has permitted us to model most of anatomical shapes. Our modeling algorithms is being integrated in a global treatment software. In our trials, the physical parameters that permit to define the organs' characteristics have been set intuitively. In cooperation with Christie Hospital in Manchester (UK) we are planning to find a better approximation of these parameters by long series of tests. Moreover, the computation time is actually rather prohibitive. Consequently, we are optimising our calculation algorithms.

References

1. M. Amrani, F. Jaillet, M. Melkemi, and B. Shariat. Simulation of deformable organs with a hybrid approach. *Revue Int. d'infographie et de la CFAO*, 16(2), 2001.
2. M. Amrani and B. Shariat. Deformable organs modelling with multi layer particle systems. In *Information Visualisation 2000*, pages 351–356, 2000.
3. E. Bardinet, L. D. Cohen, and N. Ayache. A parametric deformable model to fit unstructured 3d data. Technical Report 2617, INRIA, July 1995.
4. E. Bittar, N. Tsingos, and M. Gascuel. Automatic reconstruction of unstructured 3d data : Combining a medial axis and implicit surfaces. In *Eurographics'95 Proceedings*, pages 457–468, 1995.
5. J. Bloomenthal, C. Bajaj, J. Blinn, M.-P. Cani-Gascuel, A. Rockwood, B. Wyvill, and G. Wyvill. *Introduction to Implicit Surfaces*. Morgan Kaufmann, 1997.
6. J. D. Boissonnat and B. Geiger. Three dimensional reconstruction of complex shapes based on the Delaunay. Technical Report 1697, INRIA, April 1992.
7. M.-P. Cani. Implicit representations in computer animation : a compared study. In *Proceedings of Implicit Surface '99*, Sep 1999. Invited paper.
8. I. Cohen and L. Cohen. A hybrid hyperquadric model for 2-d and 3-d data fitting. Tech. Report 2188, INRIA, January 1994.
9. E. Ferley, M. C. Gascuel, and D. Attali. Skeletal reconstruction of

branching shapes. In *Implicit Surfaces'96 Proceedings*, pages 127–142, 1996.

10. E. Galin and S. Akkouche. Fast surface reconstruction from contours using implicit surfaces. In *Implicit Surfaces'98*, pages 139–144, 1998.

11. H. Hoppe, T. DeRose, T. Duchamp, J. McDonald, and W. Stuetzle. Surface reconstruction from unorganized points. In *SIGGRAPH'92 Proceedings*, pages 71–78, 1992.

12. M. W. Jones and M. Chen. A new approach to the construction of surfaces from contour data. *Computer Graphics Forum*, 13(3):75–84, 1994.

13. A. Joukhadar, F. Garat, and C. Laugier. Parameter identification for dynamic simulation. In *IEEE Int. Conf. on Robotics and Automation*, pages 1928–1933, New Mexico, US, 1997.

14. B. Morse, T. Yoo, P. Rheingans, D. Chen, and K. Subramanian. Interpolating implicit surfaces from scattered surface data using compactly supported radial basis functions. In *Proceedings of Shape Modeling International (SMI) 2001*, 2001.

15. S. Muraki. Volumetric shape description of range data using "blobby model". In *SIGGRAPH'91 Proceedings*, pages 227–235, 1991.

16. S. Pontier. *Reconstruction d'objets déformables à l'aide de fonctions implicites*. PhD thesis, University Lyon 1, April 2000.

17. V. V. Savchenko, A. A. Pasko, O. G. Okunev, and T. L. Kunii. Function representation of solids reconstructed from scattered surface points and contours. *Computer Graphics Forum*, 14(4):181–188, 1995.

18. S. Sclaroff and A. Pentland. Generalized Implicit Functions for Computer Graphics. In *SIGGRAPH'91 Proceedings*, pages 247–250, 1991.

19. A. Sourin and A. Pasko. Function representation for sweeping by a moving solid. *IEEE Transaction on Visualization and Computer Graphics*, 2(1), 1996.

20. D. Terzopolous and K. Fleicher. Modeling inelastic deformation: viscoelasticity, plasticity, fracture. In *Proceedings of SIGGRAPH'88*, pages 269–278. Computer Graphics, 1988.

21. H.-K. Zhao, S. Osher, and R. Fedkiw. Implicit surface reconstruction and deformation using the level set method. Tech. report, Group in Computational and Applied Mathematics, UCLA, 2001.

Chapter 11

Constraint Satisfaction Techniques for the Generation Phase in Declarative Modeling

Constraint Satisfaction Techniques for the Generation Phase in Declarative Modeling

O. Le Roux, V. Gaildrat, and R. Caubet
Department of Computer Science (IRIT), University of Paul Sabatier, 118 route de Narbonne, 31062 Toulouse cedex, France

Declarative modeling is an emergent research domain in computer-aided geometric design. To deal with the generation problems in declarative modeling, a recent approach consists in using constraint satisfaction techniques. After an introduction to constraint satisfaction and an overview of related works in declarative modeling, this chapter presents an object-oriented constraint solver. This generic tool is based on constraint propagation and domain reduction and supports heterogeneous parameters. It can be used in many generation systems. As an application, a declarative modeler for virtual 3D-environments planning is briefly presented.

1 Introduction

Declarative modeling is a recent and emergent paradigm in the world of computer-aided design systems. A declarative modeler allows designers to produce geometric or architectural data (for example 2D-ground plans, 3D-scenes or shapes, etc.) giving a high-level and natural description. Opposite the imperative geometric modeling, it requires neither a complete knowledge of the final result at start time nor to specify all numeric details. Furthermore, consistency of the description can be automatically and continuously maintained by the system. For example, logical relationships between entities resulting from the description can be preserved without additional user's work.

A declarative modeler is usually the combination of three modules [8]:

- The first one is the *description module*. It defines the interaction language and offers the user an interface to supply its description. An intermediate layer translates the high-level description into a well-suited internal format (for example a primitive constraint graph).

- The second module, called the *generation module*, can be considered as the kernel of the system. Its role is to produce models that match the description given by the user. The generation module must be able to generate one, several or even all the solutions if required and possible. If no solution meets the description, it can either return an error or free some constraints to give a partial solution.
- The third part of a declarative modeling system is the *insight module*. It allows presentation, navigation and refinement of valid models to help the user choose a solution.

These three main components interact in a spiral design process to progressively converge towards the expected solution [9].

Note: we only focus on the generation phase in the following of this chapter.

There are mainly four approaches to deal with the generation phase problem in declarative modeling [20]:
- Specific procedural approach: designed for a particular application, it can be very efficient but is not flexible and difficult to extend.
- Deductive approach: it is either a rules-based system or an expert system based on an inference engine [25, 26]. The difficulties to build rules and their dependencies in relation to the application are the main drawbacks.
- Stochastic approach: it mainly leans on genetic algorithms [15] or local search (tabu search, simulated annealing [16]), also known as metaheuristics. It is a powerful and very flexible approach but it suffers from incompleteness of related algorithms. However, when the search space is very large, it is the only available approach. See [11, 23].
- Search tree approach [5, 9, 19]: it is a general and flexible approach, which allows a systematic exploration of search space. It can generate all the solutions either at once or in several times. Moreover, the user can control the search process by giving tree branches to prefer or to prune. For all those reasons, it is a well-suited approach to the generation problems. Search tree approach frequently leans on constraint satisfaction techniques because of expressivity of this formalism and efficiency of related algorithms.

In this chapter we only focus on the search tree approach based on constraint satisfaction because we consider it as one of the most appropriate to build a powerful and efficient generic generation tool.

The chapter is organized as follows: section 2 introduces constraint satisfaction problems and classical solving techniques. Section 3 surveys constraint satisfaction-based declarative modeling systems. Section 4 describes the architecture and the main algorithms of a generic constraint solver based on constraint propagation and domain reduction. This part also insists on hotspots to make generation process efficient, like domain modeling and filtering algorithms. Finally, section 5 briefly presents an application: a 3D declarative modeler dedicated to space planning in virtual environments. Section 6 summarizes and concludes this chapter.

2 Constraint Satisfaction Problems (CSP) in a Few Words

This section introduces basic terminology and fundamental notions about CSP.

The origins of constraint satisfaction come from research in AI (search, combinatorial problems [24]) and Computer Graphics (Waltz line labeling [27]). This is now a topic of growing importance in many disciplines.

Definition 1: a CSP
A *CSP* is a tuple $P = <V, D, C>$ defined by:
- a finite set of variables $V = \{ v_1, \ldots v_n \}$
- a set of domains $D = \{ D_1, \ldots D_n \}$, where D_i is the domain associated with v_i; i.e. the only authorized assignment of v_i are values of D_i.
- a set of constraints $C = \{ C_1, \ldots C_e \}$; each $c \in C$ is a subset of the cartesian product $\prod_{i \in V_c} D_i$ where $V_c \subset V$ is the set of variables of c.

Constraints restrict the combination of values that a set of variables can take.

Each constraint can be defined:

- *extensionally*, for example giving sets of authorized tuples in the case of finite domains;
- *intentionally*, giving either a mathematical equation or a procedural method.

Note: the previous definition does not involve any restriction neither on constraints arity nor on reference sets of domains (each domain of D can be finite or continuous, ordered or not, etc.).

If each constraint is either unary or binary, then the CSP is said *binary*. A binary CSP can be depicted by a constraint graph (usually referred as a constraint network). In such a graph, each node represents a variable, and each arc represents a constraint between variables.

Example 1: a very simple constraint network with four variables {v1, v2, v3, v4}. Constraints are given extensionally; i.e. the authorized tuples are enumerated. For example, (v1, v2) can take the values (0; 0) or (1; 1).

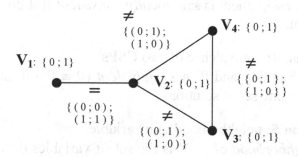

Fig. 1. *A simple constraint network (example 1).*

In this example, we can notice two important points about CSP: constraints are bi-directional (arcs are not oriented) and cycles are supported ({v_2, v_3, v_4} forms a 3-vertices cycle).

Example 2: another simple CSP where constraints are given intentionally (mathematical equations).

Variables: x, y
Domains: x, y: integer in [-5..5]
Constraints: $y <= x$, $x^2 + y^2 <= 9$

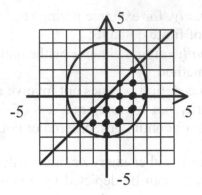

Fig. 2. *Representation of problem 2.*

Definition 2: solution to a CSP
A solution is an assignment of a value to each variable from its domain such that all the constraints are satisfied.

Definition 3: partial assignment locally consistent
A partial assignment is said *locally consistent* if it doesn't break any constraint.

Definition 4: equivalence of two CSPs
Two problems P and P' are *equivalent* (P ≡ P') if and only if they have the same set of solutions.

Definition 5: neighborhood of a variable
The *neighborhood* of $v \in V$ is the set of variables denoted $\Gamma(v)$ such that $\Gamma(v) \subset V$ and $\forall w \in \Gamma(v)$ $\exists c \in C$ $v \in V_c$ and $w \in V_c$.

Definition 6: projection of a constraint (see Fig. 3.)
Let c be a n-ary constraint on $V_c = \{V_{c_1}, \ldots, V_{c_n}\}$. The projection of c on $W \subset V_c$, that will be denoted $c{\downarrow}w$, is the constraint defined on W and equals to the set of projections on W of the values of c.

Definition 7: a disjunctive constraint
A constraint c on $V_c = \{V_{c_1}, \ldots, V_{c_n}\}$ is said *disjunctive* if and only if $\exists i \in \{1..n\}$ such as $c{\downarrow}V_{c_i}$ is non-convex.
For example, the constraint C (Fig.3.) is disjunctive because of the projection on X.

Definition 8: terminology

A CSP P is said to be:

- *under-constrained* if P has several solutions;
- *over-constrained* if P has no solution;
- *well-constrained* if P has one and only one solution.

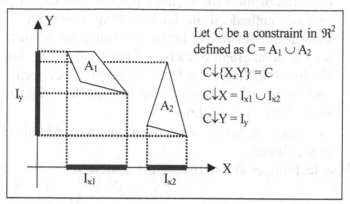

Fig. 3. *Example of 1D projections.*

Examples: problem 1 (Fig. 1.) has no solution, it is over-constrained; {x=-1, y=-1} is a solution of problem 2 (Fig. 2.) which is under-constrained.

Although it exists a few tractable subclasses, a CSP is a highly combinatorial problem (NP-complete) in the general case. When size of problems is reasonable, exhaustive search can be used. The most trivial algorithm – called *generate and test* - generates all possible combinations of values and tests whether each combination satisfies all the constraints or not. Due to its very bad running time, this algorithm is never used in practice. The well-known chronological backtracking (see Fig. 4.) is always preferred as a complete enumeration technique: a *consistency test* is performed after each instanciation of a variable.

Numerous techniques can improve computation time of the backtracking algorithm (exponential in the worst case) by several orders of magnitude:

- Specific heuristics on choice points. Heuristics can be static or dynamic; they strongly depend on the application field. The fail-first principle is often quite effective: when selecting a variable to branch on, choose the "most constrained" one (the variable with the smallest domain for example).

- Look-ahead scheme: anticipate dead end by inferring local consistencies (constraint propagation techniques). Consistency inference algorithms reason through equivalent problems (see definition 4): each step consists in narrowing down domains of the CSP to make it more explicit. A propagation method embedded in backtracking algorithm – so-called look-ahead scheme – limits thrashing by pruning some detected inconsistent branches of the search tree. The main difficulty is to find an effective trade-off between exploration and propagation. This approach is at the origin of best currently known algorithms.
- Look-back scheme: learn from past to avoid similar failures in the future.

All these techniques aim to reduce the search tree.

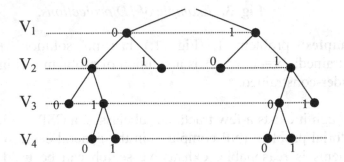

Fig. 4. *The entire backtracking search tree of problem 1 (Fig. 1.) with variables instanciated in lexicographic order. 14 nodes are generated.*

Declarative modeling is an interactive designing task. Most of the resulting generation problems are either under-constrained or over-constrained. Over-constrained problems generally arise when user makes an inconsistent description (there is no solution), whereas under-constrained problems come from an inaccurate or incomplete description. Furthermore, the structures of problems are often simple enough to be solved without many backtracks. For example: large intermixed cycles, which frequently lead to hard problems, are rare. For all those reasons, hard problems (transition phase) or exceptionally hard problems (EHP) are very singular. Consequently, a stable and efficient solving process, for a large amount of entries, can be expected when combining well-chosen domain modeling, propagation technique and heuristic. Due to its specificity, the theoretic intractability of CSP approach must not be considered as a handicap if the search space size remains reasonable.

To learn more about CSP, refer to [12-14, 21, 24].

3 Constraint Satisfaction in Declarative Modeling: A Brief Review of Related Works

[9] was one of the first to introduce CSP in declarative modeling. His goal was to speed up the generation process by using learning methods. Technically, the main algorithm is based on a back jumping process (BJ is a look-back scheme) coupled with a constraint subtrees search (CSS). The search is improved by a static variable ordering (minimal width ordering). Exhibit concepts are illustrated in a simple 3D-houses builder. This approach seems to be limited to small problems: in presented results, search space size is about 10^{13} elements. This apparently huge search space corresponds to CSP with 15 to 25 variables with domain sizes of around 5 elements; which is common in constraint satisfaction.

Some remarks:

- The search tree is entirely explored: all solutions are computed and classified;
- The description generates some n-ary geometric complex constraints hard to handle and propagate (hence the choice of a simple look-back scheme).

[19] explores a different approach. He designed a generic constraint solver ORANOS dedicated to a 3D declarative space planning application (DEM²ONS[1]). The user gives spatial relations between objects and the system automatically builds a corresponding virtual environment. The solver leans on numeric CSP; i.e. variables and constraints are numeric; the domains are real intervals discretized during the search process. Two others important features of ORANOS are hierarchical and dynamic aspects. Constraints are organized in a hierarchy (given by the user). This enables the system to always propose a solution even degraded. If the description is in-

---

[1] Declarative Multimodal Modeling System

consistent, then some constraints are relaxed according to their weight. The search process is a backtracking algorithm improved by both a propagation mechanism (forward checking or real-full look-ahead) and a dynamic variable ordering heuristic. The propagation mechanism is based on interval arithmetic and bounds propagation. The dynamic aspect allows saving time when an interactive modification occurs. It takes place in the filtering pre-processing phase and also in the search process: while a previously computed value holds for a variable, it is maintained; i.e only uninstanciated or variables that became inconsistent are processed. In the given examples, arity of the primitive constraints is one to four. The scenes contain about 10 to 30 objects (each object is defined by 6 parameters: location (x, y, z) and size (dx, dy, dz). No computation times are provided but problems seem to be easy to solve due to simple internal structure. More complex scenes can probably be handled; n-ary constraints can be added to extend the basic set as well. As the author says, this approach is very general (not limited to space planning problems) and could probably be used in numerous declarative modelers.

In recent works [4, 5] use a CLP(FD)[2] approach to design a powerful generative kernel for the XMultiForm application. This declarative platform enables the user to build complex 3D-objects or buildings by assembling 3D isothetic[3] blocks from a hierarchical high-level description. Same as [19], each block is defined by 6 parameters: location (x, y, z) and size (dx, dy, dz). The constraint solver leans on primitive constraints over finite integer domains. The canonical order relation on integers is at the origin of a finely tuned search tree algorithm (technically this sophisticated algorithm is a look-ahead scheme based on AC5 [16]). This pure finite domain approach is more restrictive than [19] (according to non-linear constraints, for example) but it turned out to be very efficient in this particular case: the resulting solver is able to process very large problems (more than 2000 variables with an average domains size of 20 elements) in a reasonable time (only a few minutes). It is probably close to the limits of the traditional CSP approach. This version of XMultiForm gives a good idea of what can be obtained when underlying problem specificities are taken into account to design the constraint solver.

---

[2] CLP(FD) = Constraint Logic Programming on Finite Domains
[3] isothetic = parallel to the reference axes (global coordinate system)

Some remarks:

- As an improvement, the search algorithm gets benefits from concurrent programming;
- Due to very large search space the entire set of solutions cannot be computed at one time: only one solution is proposed to the user. If it does not match his description, another one is proposed and so on.

4 A Generic Constraint Solver

In this section, we present a generic object-oriented constraint solver based on CSP. It can be adapted for various generation problems in declarative modeling.

4.1 Architecture

The adaptability of the solver leans on genericity of both constraints and domains. Here are the main classes and the most important associated fields and methods.

The CSP class embeds the problem description:

Table 1. *CSP class description.*

Class CSP
Main fields
V: list<Variable>
C: list<Constraint>
domains[n][i] : domain of the ith variable at the nth step
agenda: list<Constraint) : constraint to revise
Main methods to solve the CSP
boolean solve() *find the first solution (see definition 2) if it exists and others by successive calls (see algorithm in the next section)*
void collectConstraint() *fill agenda with constraints to revise*
boolean propagation() *apply revise method on each constraint of agenda; return false iff inconsistency has been detected*
void chooseValue() *choose next value to try for the current variable*

void chooseNextariable() *choose next variable to instanciate (dynamic heuristic on variable choice)*	
Main methods to modify the CSP	
void addVariable(in v:Variable, in initialDomain:Domain)	
void removeVariable(in v:Variable)	
void addConstraint(in c:Constraint)	
void removeConstraint(in c:Constraint)	

To efficiently take into account the dynamic aspects (add/remove variable/constraint), see algorithms in [19].

The Variable class describes free parameters of the generation problem:

Table 2. *Variable class description.*

Class Variable
Main fields
id: int *variable identifier*
deg: int *degree = # neighbours*
C: list<Constraint> *constraints to revise when this variable is modified*

The solver is able to manipulate a heterogeneous set of variables. That means several types of domains can be handled simultaneously in the resolution process. This permits to precisely model a wide range of complex generation problems (an example is presented on section 5). Its specificity requires the implementation of a generic support of domains. [7] already used such an idea to design effective algorithms for a 2D isothetic floor planning system (EAAS). Note it is a much more general approach than [4, 9, 19] which restrict all the domains to only one mathematical reference set.

The performance of the resolution process strongly depends on the domains internal representation. These representations must optimize the ratio *filtering quality / filtering computation time*: the more a domain data structure is precise and in accordance with its mathematical reference set the more its memory cost and filtering evaluation time are important. So, it is important to choose a relevant trade-off. In case of continuous domain, combinatorial explosion due to disjunctive constraints (see definition 7) and domain splitting should be avoided by setting a lower bound for domain size.

Table 3. *Domain class description.*

Abstract Class Domain
Main fields
`V: Variable` *variable associated with this domain*
Main methods
abstract `boolean isEmpty()` *returns true iff the domain is empty*
abstract `List<Value> discretize()` *returns a discrete representation of the domain (a list of values)*
`Domain clone()` *returns a copy of the Domain object*
abstract `void setSingleValue(in val:Value)` *reduce domain to a single value*

In the solver, constraints are generic. They can be defined according to the application.

Table 4. *Constraint class description.*

Abstract Class Constraint
Main fields
`V: list<Variable>` *variables of this constraint*
`RM : list<ReductionMethod>`
Main methods
abstract `boolean` `          testConsistency(in t: Assignment)` *returns true if the constraint is satisfied, false otherwise* *Assignment* `t` *is a list of pairs (variable, value)*
`boolean revise()` *constraint filtering method: applies reduction methods on free parameters (i.e. not yet instanciated); returns false iff inconsistency has been detected, (at least one domain is empty)*

A *reduction method* is an over-estimated projection function of a constraint on one of its variables (see definition 6). In the search process reduction methods are used to tighten domains.

205

Table 5. *ReductionMethod class description.*

Abstract Class ReductionMethod
Main fields
`C : Constraint`
`outVar : Variable`
variable on which the constraint is projected
Main methods
abstract `Domain proj()`
the projection method of `C` *on* `outVar`

Requirements: let C_E be a constraint on $E = D_1 x \ldots x D_n$. Let F a set such that $F \subseteq E$. We denote C_F the restriction of the constraint C_E to F. The method `proj` – here written proj(C, outVar) - must satisfy the two following conditions:

$(1):$ $\forall i \in \{1..n\}$ $C_E \downarrow V_i \subseteq \text{proj}(C_E, V_i)$

$(2):$ $\forall i \in \{1..n\}$ $\text{proj}(C_F, V_i) \subseteq \text{proj}(C_E, V_i)$

4.2 Main algorithm

The following algorithm is a general non-recursive backtracking scheme. It includes generic heuristics and consistency inference mechanism.

To get the first solution if it exists: (a) set n to zero, (b) call `chooseNewVariable()` to initialise the process, (c) call `solve()`. To get the next solution, (a) decrement n, (b) recall `solve()`. While `solve()` returns true, successive calls compute the whole set of solutions.

```
boolean solve() {
  while (true) {
    if (domains[n][ivar[n]].isEmpty()) {
      if (n == 0)
        return false; // inconsistency
      nivar.addFirst(ivar[n]);
      n--;   // backtrack
      continue;
    }

    // backup current domains
    for (i=0; i<N; i++)
      domains[n+1][i]=domains[n][i].clone();
```

```
// choose a value for the current variable in its
// domain (=_domains[n][ivar[n]])
// if domain is continuous, discretize() is used
// assign chosen value to current variable
// (=> domains[n+1][ivar[n]].setSingleValue(...)
chooseValue();

// collect constraint to revise
collectConstraint();

// prepare next step (n+1)
n++;

if (propagation()) {  //FC or other
   if (n == N)
     return true;   // solution found

   // heuristic on variable choice
   // assign ivar[n] (update nivar)
   chooseNewVariable();

} else {

   // local inconsistency detected: stay on step n
   n--;
 }
}
```

Glossary:

> N = # variables in the CSP
> n = current step (in 0..N)
> ivar[i] = the ith instantiated variable
> nivar = list of not yet instanciated variables
> domains[n][i] = domain of the ith variable at the nth step

Some remarks:
- This algorithm can be extended with a learning technique like backjumping or others related techniques [6];
- It is not limited to binary CSP.

propagation() is a local consistency inference mechanism (or *filtering mechanism*). It refines the domains at the nth step and returns true if and only if no inconsistency has been detected; i.e each domain is non-empty. Constraint propagation can be either limited to direct neighborhood (Forward-Checking = FC) [3] or performed on the complete CSP (Real-Full-Lookahead = RFL or MAC = Maintain Arc-Consistency). The local consistency must be relevantly chosen according to problem modeling.

The most-studied among local consistencies is arc-consistency[4] (AC). Many efficient algorithms implement this consistency: from very general scheme (like AC3 [21] or recent related AC2000/AC2001 [2]) to finely tuned AC6 and optimal AC7 [1], via specialized AC5 [16]. Consequently AC is often efficient when dealing with pure finite domains problems. On hard problems, stronger local consistencies (like restricted path-consistency (RPC), path inverse consistency (PIC), path-consistency (PC), etc.) can sometimes be better [10].

Finally a specific local consistency can be designed to get benefits from particular domain modeling. For example, [22] defines a suitable consistency on real intervals and [7] proposes an efficient one for isothetic 2D-rectangles.

5 An Application: A Declarative Planning System for Virtual Environments

This section presents a declarative modeler called DEM²ONS-NG[5].

5.1 Overview

A functionality of this application is to place 3D-objects – for example furniture – in a virtual environment according to a high-level description. Note that the goal is neither to create the most efficient space planning system, nor to propose the most complete one; but only to design a generative system to illustrate solver capabilities.

---

4 A constraint $c \in C$ is said arc-consistent iff $\forall Vi \in Vc$, $\forall d \in Di$, d is supported by c. A CSP is arc-consistent iff all its constraints are arc-consistent.

5 DEM²ONS-NG = Declarative Multimodal Modeling System – Next Generation (initial version, called DEM²ONS, was partly developed by Kwaiter in [19]).

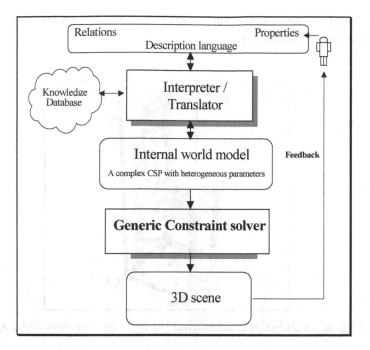

Fig. 5. Synoptic of DEM²ONS-NG

As said, the underlying generation problem is a space-planning problem. Each 3D-object is identified with its local coordinate space and defined by three different kinds of parameters:

- *Position variable*: location of the local coordinate space origin in the world coordinate space. A position variable is a 3D vector made up of three one-dimensional values [x y z];
- *Orientation variable*: an angle in degree. Three orientation parameters (the three Euler angles) are used to orientate an object;
- *Size variable*: a one-dimensional parameter. It describes a dimension along an axis. Three size parameters are needed to define the size of a 3D-object.

Consequently to describe an object in the application, seven Variable objects are required: one for position, three for orientation and three for size.

In case of 2D space planning problems, it is proven [7] that such a representation gives good results. Here, this idea is extended to 3D. One of the most important differences with [7] is the ability of the system to handle any orientation, not only isothetic ones. Note also that objects should be identified with hierarchical structures of bounding boxes - i.e. a set of local coordinate spaces - for a more accurate representation.

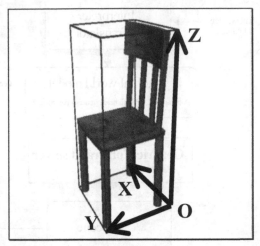

Fig. 6. *A 3D-object identified with its local coordinate space*

5.2 About domains internal representation

As parameters are heterogeneous, a different kind of domain is assigned to each other.

Orientation variable domains: in this application, consistency tests often require a fixed orientation. So associated domains are represented by a finite list of values.
Example: $D_O = \{\ 0°;\ 15°;\ 30°;\ 90°\ \}$
Classical CSP filtering technique is used.

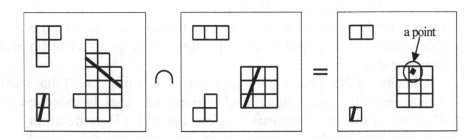

Fig. 7. *Two position domains and the resulting intersection (2D-orthographic projection)*

Size variable domains: to achieve a correct filtering even processing disjunctive size constraints, a list of independent and continuous intervals is used.

Example: D_S = { [0, 100]; [150,155]; [250;300] }

The numeric CSP technique described in [22] is used.

Position variable domains: as position is a three-dimensional parameter, the internal representation must describe a 3D-space. Various representations, either continuous or discrete can be used; for example: a list of polyhedra, a voxel array, an octree, a list of 3D-points, a set of 3D-isothetic boxes, etc. Actually a good choice is a trade-off between accuracy of modeling and efficiency of resulting reducing algorithms. As the system must hold both equality and inequality constraints, the chosen modeling is finally a voxel array coupled with a set of *geometrical elements* (3D-points, lines and planes). Voxels describe space elements generated by inequality constraints while geometrical elements are used to tighten domain in case of equality constraints. The two complementary models are maintained simultaneously. In practice each voxel owns a list of pointers towards the geometrical elements that pass through it.

5.3 About constraints

The application kernel is composed of a set of basic constraints. Some examples of significant basic constraints are:

- On position parameters (pos):
```
Position-fixed (pos, value) (unary)
```

- On orientation parameters (ort):
```
Orientation-fixed (ort, value) (unary)
Orientation-same-as (ort1, ort2) (binary)
```

- On size parameters (size):
```
Size-fixed-in (size, min, max) (unary)
Size-equal-to (size1, size2) (binary)
Size-greater-than (size1, size2, k) (binary)
```

- Heterogeneous ones (obj = {pos, ort, size}):
```
Non-overlapping (obj1, obj2) (arity is 14)
In-fixed-half-space (obj, half-space) (arity
is 7)
```

On-relative-plane (`obj1.pos`, `plane<obj2>`):
constrains position parameter of obj1 to be on a plane defined
from obj2.

Note: a half-space is defined by a 3D-plane that splits 3D-world space in two parts.

Each basic constraint has a set of appropriate reduction methods to perform domain reduction during the search phase. They can be combined to get high-level constraints:

target_object **spatial_preposition** landmark_object

Examples: obj1 **on** obj2; obj1 **in-front-of** obj2, etc.

5.4 Experimental results

Some experimental results are presented to illustrate what can be expected from such a system.

Test scene description: the scene is composed of 30 complex geometric objects (four walls, three tables, two chairs, a TV, two computers, shelves, books, etc.). The high-level description contains meta-constraints like *SCREEN$_1$ is on TABLE$_1$, dimensions of TABLE$_1$ are 2m x 0.7m x 0.8m, TABLE$_2$ is against the back wall, TABLE$_1$ is to the right of TABLE2, etc.*

Solver configuration is:

Voxel size (for position parameter domains)	20 cm
Step for position parameters	20 cm
Step for orientation parameters	15°
Step for size parameters	10 cm
Size of the room (in meters)	4.4 x 3.4 x 2.5

Translation of the description in the internal representation gives:

Number of variables	210
Number of basic constraints	245

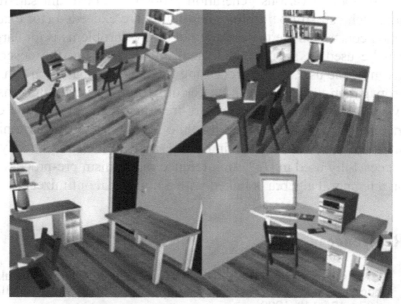

Fig. 8. *Four views of a scene generated by the application that meets the description*

Time to get the first solution – the first valid scene according to the given constraints – is approximately 3 seconds. In this example, there are too many solutions to generate them all. However, some other solutions can be obtained (required computation time is generally < 1 second).

Note: the implementation of our application is in Java (JDK 1.3) / OpenGL and runs under Windows 2000© on a PIII600 / 256Mb. With a good implementation in C++ and some heuristics adjustments, computation time should probably be highly restricted.

6 Summary

In this chapter, we presented an overview of constraint satisfaction techniques applied to the generation phase in declarative modeling. We provided some arguments and results to show that this approach is currently one of the best to deal with generation problems. We also presented a generic constraint solver based on classical CSP techniques. As shown in the last section dedicated to the DEM²ONS-NG application, the use of generic constraints and heterogeneous parameters (via generic domains) allows the adaptation of this tool to various generation problems. Constraint satisfaction approach can still be useful when limits are exceeded (for example, when constraints are too complex or size of problems is too large): it can be used as an internal formalism to represent the description given by the user. A pre-processing filtering pass based on constraint propagation techniques can be performed on the resulting constraint system. This time-polynomial pre-treatment can detect some coarse inconsistencies in order to reduce the search space before running an alternative (stochastic) solving process (This idea has already been successfully used in [11]: an inference mechanism pre-process based on a temporal algebra is linked with a numerical optimizer).

References

1. C. Bessière, E. Freuder, and J.C. Régin. Using constraint metaknowledge to reduce arc consistency computation. In Artificial Intelligence 107, p125-148, 1999.
2. C. Bessière and J.C. Régin. Refining the basic constraint propagation algorithm. In proceedings IJCAI'01, p309-315, Seattle WA, 2001.
3. C. Bessière, P. Meseguer, E. Freuder and J. Larrosa. On Forward Checking for Non-binary Constraint Satisfaction. In Principles and Practice of Constraint Programming - CP99. Alexandria, USA 1999. LNCS 1713, Springer-Verlag.
4. P-F Bonnefoi, D. Plemenos. Object Oriented Constraint Satisfaction for Hierarchical Declarative Scene Modeling. In proceedings of WSCG'99, Plzen, Czech Republic, 1999.
5. P-F Bonnefoi, D. Plemenos. Constraint satisfaction techniques for declarative scene modeling by hierarchical decomposition.. In proceedings of 3IA'2000, Limoges, France, 2000.
6. X. Chen, P. van Beek. Conflict-Directed Backjumping Revisited. In journal of AI Research 14, p53-81, March 2001.
7. P. Charman. A constraint-based approach for the generation of floor plans. In 6th IEEE International Conference Tools with Artificial Intelligence, November 1994.
8. C. Colin, E. Desmontils, J.Y. Martin, J.P. Mounier. Working Modes with a Declarative Modeler. Compugraphics'97, Villamoura, Portugal, GRASP, p117-126, 1997.
9. L. Champciaux. Declarative Modeling: Speeding up the generation. In proceedings of CISST'97, Las Vegas, Nevada, p120-129, July 1997.
10. R. Debruyne and C. Bessière. Domain filtering consistencies. In journal of Artificial Intelligence Research, volume 14, p205-230, 2001.
11. S. Donikian, G. Hegron. The Kernel of a Declarative Method for 3D Scene Sketch Modeling. In proceedings GRAPHICON'92, Moscow, Russia, September 1992.
12. R. Dechter, A. Dechter. Belief maintenance in dynamic constraint networks. In proceedings of AAAI'88, St. Paul, MN, p337-342, 1988.

13. R. Dechter, F. Rossi, Constraint Satisfaction, Survey ECS, March, 2000.
14. E. Freuder, R. Dechter, B. Selman, M. Ginsberg, and E. Tsang, Systematic Versus Stochastic Constraint Satisfaction. Panel, IJCAI-95, p2027-2032, 1995.
15. D. Goldberg. Genetic Algorithms in Search, Optimization and Machine Learning. Addison-Wesley, 1989.
16. P. van Hentenrick, Y. Deville, and C.M. Teng. A generic arc-consistency algorithm and its specializations. Artificial Intelligence 57, p291-231, 1992.
17. J-K Hao, J. Pannier. Simulated annealing and Tabu search for constraint solving. In proceedings of AIM'98. Fort Lauderdale, Florida, USA, January 1998.
18. W. Hower, W. Graf. A bibliographical survey of constraint-based approaches to CAD, graphics, layout, visualization, and related topics. Knowledge-Based Systems, Vol. 9, No. 7, p449-464, 1996.
19. G. Kwaiter, V. Gaildrat, R. Caubet. Dynamic and Hierarchical Constraints Solver with Continuous Variables. In proceedings of Conference on LP and CP, Orleans 1997.
20. G. Kwaiter, V. Gaildrat, R. Caubet. Modeling with Constraints: A Bibliographical Survey. In proceedings of IV'98, London UK, July, 1998.
21. V. Kumar. Algorithms for Constraint Satisfaction Problems: A Survey by Vipin Kumar. AI Magazine Spring 2000.
22. O. Lhomme. Consistency techniques for numeric CSPs. In Proceedings of IJCAI-93, p232-238, 1993.
23. S. Liège, G. Hégron. An Incremental Declarative Modeling Applied to Urban Layout Design. In proceedings of WSCG'97, Plzen, Czech Republic, February 97.
24. A. K. Mackworth. Consistency in Networks of Relations. Artificial intelligence, Vol. 8 (1), p99-118, 1977.
25. P. and D. Martin. Declarative Generation of a Family of Polyhedra, In proceedings of GRAPHICON' 93. St Petersbourg, September 1993.
26. D. Plemenos. Declarative Modeling by Hierarchical Decomposition. The Actual State of the MultiFormes Project, GRAPHICON'95, St Petersbourg, Russia, July 1995.
27. D.L. Waltz. Understanding line drawing of scenes with shadows. In The Psychology of Computer Vision, ed P. Winston, p19-91; 1975.

Chapter 12

Surface Reconstruction Based on Compactly Supported Radial Basis Functions

Surface Reconstruction Based on Compactly Supported Radial Basis Functions

Nikita Kojekine
Faculty of Engineering, Tokyo Institute of Technology, 2-12-1, O-okayama, Meguro-ku, Tokyo 152-8552, Japan

Vladimir Savchenko
Faculty of Computer and Information Sciences, Hosei University, 3-7-2 Kajino-cho Koganei-shi, Tokyo 184-8584, Japan

Ichiro Hagiwara
Faculty of Engineering, Tokyo Institute of Technology, 2-12-1, O-okayama, Meguro-ku, Tokyo 152-8552, Japan

In this chapter the use of compactly-supported radial basis functions for surface reconstruction is described. To solve the problem of reconstruction or volume data generation specially designed software is employed. Time performance of the algorithm is investigated. Thanks to the efficient octree algorithm used in this study, the resulting matrix is a band diagonal matrix that reduces computational costs.

1 Introduction

Traditionally, constructive solid geometry (CSG) modeling uses simple geometric objects for a base model that can be further manipulated by implementing a certain collection of operations such as set-theoretic operations, blending, or offsetting. The operations mentioned above and many others have found quite general descriptions or solutions for geometric solids represented as points (x,y,z) in space satisfying $f(x,y,z) \geq 0$ for a continuous function f. Such a representation is usually called an *implicit model* or a *function representation*. Set-theoretic solids have been successfully included in this type of representation with the application of R-functions and their modifications (see [1], and [2]). In [3], an approach to volume modeling is proposed. It combines the voxel representation and function representation. It is supposed that the two main representations (voxel/function) are given rather autonomously and a rich set of operations can be used for modeling volumes. Whatever complex operations have been applied to a geometric object that can be given by a voxel raster, by elevation data, or by a function representation, equivalent volume data, can be generated. Many practical surface reconstruction techniques such as restoration design or reverse engineering tasks based on measured data points require the solution of optimization problems in fitting surface data. We use the term *volume model* to refer to the wide class of surfaces that like other implicit surfaces descriptions can be used for CSG.

A vast volume of literature is devoted to the subject of scattered data reconstruction and interpolation. In most applications, a Delaunay triangulation is used for 3D reconstruction. Unfortunately, this method has some serious drawbacks, even with the elimination of large triangles; the reconstructed shape remains convex-looking, as noted in [4]. Another approach to surface reconstruction is skeletal. An implicit surface generated by point skeletons may be fit to a set of surface points [5], but this method is rather time consuming. One of other approaches is to use methods of scattered data interpolation, based on the minimum-energy properties [6], [7], [8]. These methods are widely discussed in mathematical literature (see [9], [10]). The benefits of modeling 3D surfaces with the help of radial basis functions (RBF) have been recognized in [11] for Phobos reconstruction. They were adapted for computer animation [12], [13] and medical applications [14], [15] and were first applied to implicit surfaces by Savchenko et al. [16]. However, the required computational work is proportional to the number of grid nodes and the number of scattered data points. Special methods to reduce the processing time were developed for thin plate splines and discussed in [17], [18].

Actually, the methods exploiting the RBFs can be divided into three groups. First group is "naive" methods that are restricted to small problems, but they work quite well in applications, dealing with shape transformation (see, for example [19]). Second group is fast methods for fitting and evaluating RBFs that allow large data sets to be modeled [20], [17], [21]. The third and last group is compactly supported radial basis functions (CSRBFs) introduced by Wendland in [22]. The benefits of modeling 3D implicitly defined surfaces with the help of CSRBFS have been recognized in [23] where authors have used a k-d tree [24] to build the resulting sparse matrix. To find the solution to the system of equations a direct LU sparse matrix solver [25] was used.

In practice, the problem of reconstruction consists of the following steps: sorting the data, constructing the system of linear algebraic equations (SLAE), solving the SLAE, and evaluating the functions. In fact, while the solution of the system is the limiting step, constructing the matrix and evaluating the functions to extract the isosurface may also be computationally expensive. In this contribution, we made an attempt to solve the problem according to the above-mentioned steps. Thus, the main goal of the ongoing project was to develop an effective library of C++ classes that can be successfully applied to computationally intensive problems of surface reconstruction using RBFs splines.

2 Method of Shape Reconstruction with RBF Splines

For a three-dimensional arbitrary area Ω, the thin–plate interpolation is the variational solution that defines a linear operator T when the following minimum condition is used:

$$\int_\Omega \sum_{|\alpha|=m} m!/\alpha! (D^\alpha f)^2 d\Omega \rightarrow \min, \tag{1}$$

where m is a parameter of the variational functional and α is a multi-index. It is equivalent to using the radial basis functions $\phi(r) = r^1$ or r^3 for m = 2 and 3, respectively, where r is the Euclidean distance between two points. Since the function $\phi(r)$ is not compactly supported, the corresponding system of linear algebraic equations (SLAE) is not sparse or bounded. Storing the lower triangle matrix requires $O(N^2)$ real numbers, and the computational complexity of a matrix factorization is $O(N^3)$. Thus, the amount of computation becomes significant, even for a moderate number of points N.

Wendland in [22] constructed a new class of positive definite and compactly supported radial functions for 1D, 3D, and 5D spaces of the form

$$\phi(r) = \begin{cases} \phi(r), 0 \leq r < 1 \\ 0, r > 1 \end{cases} \tag{2}$$

whose radius of support is equal to 1. The function $\phi(r) = (1 - r)^2$, which is an interpolated function that supports only C^0 continuity, is used in our software. This function provides positive defined and nonsingular systems of equations. However, it is also possible to apply functions that support a higher continuity. An investigation [26] of the smoothness of this family of polynomial basis functions shows that each member $\phi(r)$ possesses an even number of continuous derivatives.

The volume spline $f(P)$ having values h_i at N points P_i is the function

$$f(P) = \sum_{j=1}^{N} \lambda_j \phi\,(|P - P_j|) + \mathrm{p}(P), \tag{3}$$

where $p = v_0 + v_1 x + v_2 y + v_3 z$ is a degree one polynomial. To solve for the weights λ_j we have to satisfy the constraints h_i by substituting the right part of equation (3), which gives

$$h_i = \sum_{j=1}^{N} \lambda_j \phi\,(|P_i - P_j|) + \mathrm{p}(P_i) \tag{4}$$

Solving for the weights λ_j and v_0, v_1, v_2, v_3 it follows that in the most common case there is a doubly bordered matrix T that consist of three blocks, square submatrices A and D of size $N \times N$ and $k \times k$, respectively, and B, that is not necessarily square and has the size $N \times k$, where $k = 4$ for 3D space.

3 Algorithm for Reconstruction

3.1 Sorting scattered data

Space recursive subdivision is an elegant and popular way of sorting scattered 3D data. We propose an efficient approach based on the use of variable-depth octal trees for space subdivision, that allows us to obtain the resulting submatrix A as a band diagonal matrix that reduces the computational complexity. Because of the structure of octal trees [27], for each node of the tree, we need to store the following:

- a pointer to the parent node,
- 8 pointers to child nodes,
- a pointer to the list of points (empty if this node is not a leaf).

All memory needed to store such a tree is (memory for each node) × (number of nodes) + (memory for storing point) × N. In our software implementation, we are employing standard approach for creating the tree from an initial point data set (Algorithm I - "Octal tree creation") with an additional required parametric value K, which denotes the maximum number of points in the leaf. Afterwards we can use this tree to search for neighbors of any given point from the given N points. The neighbors are points belonging to a sphere of radius r with center at the given point. We call this sphere an r-sphere.

Algorithm II "Searching using the tree," is a "searching" function based on the fact, that for each node in the tree, we can state that this node is either:

- entirely inside the given r-sphere,
- entirely outside the given r-sphere,
- partly inside the given r-sphere, partly outside it.

To accelerate the search function octal tree is simplified after creation. If node A has only one sub-node B, we can remove node A and replace it with B. This procedure is reasonable only if K is small. For example, consider the following octal tree (numerals represent node numbers):

Initial octal tree   Octal tree after simplification

For instance, for data from the "Head" example (see, Table 1 (a)) if K is set equal to 1 this procedure removes 113 nodes of the octal tree from the initially created 2427 nodes.

The maximum complexity of the two algorithms that have been described strongly depends on the initial data and the parameter K. The depth of the tree depends on the length of the cube edge corresponding to the leaf. This length is equal to $(1/2)^M$, where M is the depth of the tree that depends on the original data. If the initial points are distributed more or less uniformly, then the tree will have sufficiently uniform filling and will be symmetric. If K is set equal to 1, at that time the tree will be close to a full octal tree with N leaves. The maximum complexity of searching the tree will be proportional to the depth of this tree, which is $\log_8 N$.

A more detailed account of these algorithms, including a C++ library description can be found in [28]. The procedure of searching for the neighbors of a point in a given r-sphere is applied several times in the application. For example, it is used for calculating the function (3) to sum up only the points that are neighbors of the specified point with coordinates (x,y,z) and thus accelerate the function calculation and surface extraction. But the first application is the construction of the band diagonal submatrix $\phi(|P - P_i|)$ that accounts for a significant portion of the computational cost. In our application, to store a band-diagonal matrix, we use the so-called profile form or a slightly modified version of the Jennings envelope scheme [29]. To store the submatrix A, an array can be used for diagonal elements; values of non-diagonal elements and correspondent indices of the first non-zero elements in the submatrix lines are placed in two additional arrays.

To make our submatrix band diagonal, we need to re-enumerate the initial points in a special way. We propose the following <u>Algorithm III</u> "Sorting data using an octal tree":

- Take a point from the initial data and put it in the list.
- Go through the list, and for each point in the list search for the neighbors from the initial data. When new points are found add them to the list.
- Remove from the initial array points that have been placed in the list.
- If there are no more points in the list (all points were appended in the second step), then take the first point remaining in the initial array and repeat the above steps.

Algorithm III can be represented as pseudo-code:

```
input_list         - input list of points
output_list        - output list of points
neighbors_id_list  - temporary list of ids

i:=0;
while (input_list.length >= 0) do
begin
    // add first element of input_list to
    // output_list, remove first element from
    // input_list
    output_list.add(input_list[0]);
```

```
      input_list.remove(0);
      while (i < output_list.length) do
      begin
          // find in input_list all neighbors
          // of output_list[i] and put their
          // indices into neighbors_ids_list.
          // this is done with octree
          neighbors_id_list=
      FindNeighbors(output_list[i],input_list);
          for j:=0 to neighbors_id_list.length
          do
          begin
              // add neighbor element of
              // input_list to output_list
              // remove this element from
              // input_list
      output_list.add(input_list[neighbors_list[j]
      ]);
      input_list.remove(neighbors_id_list[j]);
          end
      end
end
```

As a result of this algorithm a band with maximum size α_1 of the neighbors of a point is obtained. The maximum complexity of this algorithm is the complexity of searching for neighbors through the octal tree for each point, that is, $N \times$ (the maximum complexity of the algorithm II). We can reduce our computational outlays by calculating the matrix and the order of the points simultaneously.

Fig. 1. *Typical matrix with band-diagonal part created by algorithm III.*

224

After sorting we get a band-diagonal submatrix as shown in Fig. 1. Note that the half-width for selected r cannot be decreased. Considering the following unlikely event would clarify this concept. If we connect all neighboring points we will obtain a graph, and if this graph has a cycle, then the maximum size we will get is less than or equal to the cycle length. Thus, if the radius of support is quite large, then the cycle will include nearly all the points from the input data. In this case, the maximum size of the band will also be large, and we will have an expansion of the band at some point.

3.2 SLAE solution

Note that solving any sparse system has the goals of saving time and space. The advantage of Gaussian **LU** [30] decomposition has been well recognized, and many software routines have been developed. For a symmetric and positive definite matrix, a special factorization, called Cholesky decomposition, is about twice faster compared to alternative methods for solving linear equations. From the discussion in section 3.1 it follows that in the most common case there is a doubly bordered band diagonal system T that consist of three blocks, square sub-matrices A and D of size $N \times N$ and $k \times k$ respectively, and B that is not necessarily square and has the size $N \times k$. A combination of block Gauss solution and Cholesky decomposition was proposed by George and Liu [31] and in our software tools we follow their proposal.

3.3 Surface evaluation and extraction

The surface fitted to a set of surface data point forms a volume model of a geometric object. This surface can be visualized directly by using an implicit ray-tracer, and can be voxelized, or polygonized to extract a mesh of polygons. For the visualization of reconstructed volumes an implicit function modeler tool is used (see [32]).

Table 1. Processing time. Test configuration: *AMD Athlon 1000 Mhz, 128 MB RAM, Microsoft Windows 2000*.

	"Head", Fig. 2 (a), number of points $N =$	"Seashell", Fig. 2 (b), number of points $N = 915$,	"Venus", Fig. 2(c), number of points $N =$

	1487, the selected radius of support $r = 0.2$	selected radius of support $r = 0.2$	6719, selected radius of support $r = 0,13$
Tree creation	0.001 sec.	0.001 sec.	0.01 sec.
Sorting time	0.03 sec.	0.03 sec.	0.26 sec.
Matrix calculation time	0.05 sec.	0.02 sec.	0.58 sec
Memory requirement to store the band diagonal submatrix of the matrix A	1,675,800 bytes (1 MiB) (if stored traditionally it would be 8856576 bytes (8 MiB))	669,068 bytes (0 MiB) (if stored traditionally it would be 3348900 bytes (3 MiB))	21171936 bytes (20 MiB) (if stored tradianally it would be 180579844 bytes (172 MiB))
Solution time with Cholesky decomposition	0.911 sec.	0.1 sec.	39.01 sec.
Polygonal surface extraction time	0.49 sec.	0.41 sec.	2.73 sec.

Note that the approach taken in this study does not guarantee a restoration of highly topologically complex volume objects. For 3D reconstruction using cross-sectional data, in [33] it is proposed that, for m different contours in one slice, m different function descriptions of separate contours must be used, and that union of the m carrier functions calculates the description of the reconstructed 2D object. For the 3D case such an approach looks exceedingly complicated. Moreover, RBFs demonstrate excessive blending features that lead to undesirable smoothing effects. The approach taken by Turk and O`Brien in [34] of using points specified on both sides of the surface will provide successful restoration of a surface, but it involves drawbacks that leads one to suppose that it would be inefficient for applying RBFs for volume reconstruction. Since it is out of the scope of this chapter to discuss all these matters, we would like to mention the following: this approach does not produce CSG-like solids as required in CSG (see [1]); the "both sides" approach has a problem with the surface extraction (a surface extractor can jump outside the band of non-zero points). There is also a problem of constructing or specifying off-surface points along a surface normal that also leads to a doubling of the given number of surface points.

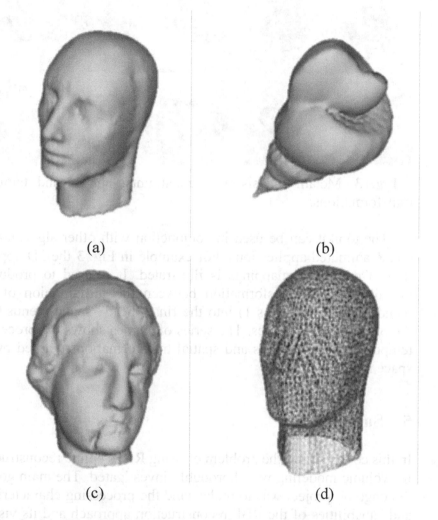

(a)   (b)

(c)   (d)

Fig. 2. *(a) "Head" reconstruction; (b) "Seashell reconstruction; (c) "Venus" reconstruction; (d) Wire-frame image of the extracted model "Head" (the number of polygons to be extracted is controlled by user-defined parameter in our software system).*

4 Results

Visual inspection (images in Fig. 2) allows us to judge the interpolation features of the algorithm that has been discussed. Benchmark results can be seen in Table 1.

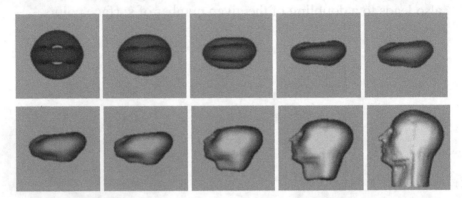

Fig. 3. Metamorphosis with constraints: spatial and temporal transformations.

The toolkit can be used in conjunction with other algorithms to create animated applications. For example in Fig. 3 the 3D application of combined mappings is illustrated. It is used to produce a visually smooth transformation between the initial union of two blended "torus" (genus 1) into the final "head" shape (genus 0) in the presence of obstacles. The series of frames shows the process of temporal metamorphosis and spatial transformations defined by the space mapping.

5 Summary

In this contribution, the problem of using RBFs spline reconstruction for volume modeling was thoroughly investigated. The main goal of the ongoing project was to understand the processing characteristics and capabilities of the RBF reconstruction approach and its visualization aspects.

Thanks to the efficient octree algorithm, the resulting matrix of size $N+k*N+k$, where $k=4$ for 3D case, has a band diagonal submatrix (not a sparse one) of size $N*N$ that reduces greatly the computational complexity.

An online reconstruction server has been established on our web page [28] that makes it possible to get a visualization of a VRML object using a web-browser. This web page also provides our reconstruction software to download and several examples.

References

1. V. Shapiro, Real functions for representation of rigid solids, *Computer Aided Geometric Design*, 11(2), 153-175, 1994.
2. A. Pasko, V. Adzhiev, A. Sourin, V. Savchenko, Function representation in geometric modeling: concepts, implementation and applications, *The Visual Computer*, 11(6), 429-446. 1995.
3. V.V. Savchenko, A.A. Pasko, A.I. Sourin, T.L. Kunii, Volume Modeling: Representations and advanced operations, *Computer Graphics Int. Conf., F. Wolter and N.M. Patrikalakis (eds.)*, Hannover, Germany, IEEE Computer Society, 4-13, June 22-26 1998.
4. S. Djurcilov, A. Pang, Visualizing Sparse Gridded Data Sets, *IEEE Computer Graphics and Applications*, 52-57, Sept/Oct 2000.
5. S. Muraki, Volumetric Shape Description of Range Data Using "Blobby Model", *Computer Graphics* vol.25 (*Proceedings of SIGGRAPH*), no. 4, 227-235, 1991.
6. J.H. Ahlberg, E.N. Nilson, J. L. Walsh, The Theory of Splines and Their Applications, *Academic Press*, New York, 1967.
7. J. Dushon, *Splines Minimizing Rotation Invariants Semi-Norms in Sobolev Spaces, Constructive Theory of Functions of Several Variables*, W. Schempp and K. Zeller (eds.), Springer-Verlag, 85-100, 1976.
8. V.A. Vasilenko, *Spline-functions: Theory, Algorithms, Programs*, Novosibirsk, Nauka Publishers, 1983.
9. R.M. Bolle, B. C. Vemuri, On Three-Dimensional Surface Reconstruction Methods, *IEEE Transactions on Pattern Analysis and Machine Intelligence*, 13(1), 1-13,1991.
10. G. Greiner, *Surface Construction Based on Variational Principles, Wavelets, Images and Surface Fitting*, P. J. Laurent et al. (eds.), AL Peters Ltd., 277- 286, 1994.
11. V. Savchenko, V. Vishnjakov, The Use of the "Serialization" Approach in the Design of Parallel Programs Exemplified by Problems in Applied Celestial Mechanics, *Performance Evaluation of Parallel Systems, Proceedings PEPS*, University of Warwick, Coventry, UK, 29-30 Nov., 126-133, 1993.
12. P. Litwinovicz, L. Williams, Animating Images with Drawing, Computer Graphics *(Proc. SIGGRAPH)*, 409-412, 1994.
13. V. Savchenko, A. Pasko, T.L. Kunii, A.V. Savchenko, Feature Based Sculpting of Functionally Defined 3D Geometric Objects, *Proceedings Multimedia Modeling Conference*, Singapore, 14-17 Nov., T.T. Chua et al. (eds.), World Scientific Pub., 341-34, 1995.

14. J.C. Carr, W.R. Fright and R.K. Beatson, Surface Interpolation with Radial Basis Functions for Medical Imaging, *IEEE Transaction on Medical Imaging*, 16(1), 96-107, 1997.
15. F.L. Bookstein, Principal Warps: Thin Plate Splines and the Decomposition of Deformations, *IEEE Transactions on Pattern Analysis and Machine Intelligence*, 11(6), 567-585, 1989.
16. V. Savchenko, A. Pasko, O. Okunev and T. Kunii, Function representation of solids reconstructed from scattered surface points and contours, *Computer Graphics Forum*, 14(4), 181-188, 1995.
17. R.K. Beatson and W.A. Light, Fast Evaluation of Radial Basis Functions: Methods for 2-D Polyharmonic Splines, Tech. Rep. 119, Mathematics Department, Univ. of Canterbury, Christchurch, New Zealand, Dec. 1994.
18. W. Light, Using Radial Functions on Compact Domains, Wavelets, *Images and Surface Fitting*, P. J. Laurent et al. (eds.), AL Peters Ltd., 351-370, 1994.
19. V. Savchenko, L. Schmitt, Reconstructing Occlusal Surfaces of Teeth Using a Genetic Algorithm with Simulated Annealing Type Selection, *6th ACM Symposium on Solid Modeling and Applications*, Sheraton Inn, Ann Arbor, Michigan, June 4-8, 2001.
20. L. Greengard and V. Rokhlin, A Fast Algorithm for Particle Simulation, *J. Comput. Phys.*, 73, 325-348, 1987.
21. J.C. Carr, T.J. Mitchell, R.K. Beatson, J.B. Cherrie, W.R. Fright, B.C. McCallumm, and T.R. Evans, Reconstruction and representation of 3D Objects with Radial Basis Functions, *Computer Graphics, Proc. SIGGRAPH*, August 2001.
22. H. Wendland, Piecewise polynomial, positive defined and compactly supported radial functions of minimal degree, *AICM*, 4, 389-396, 1995.
23. B. Morse, T.S. Yoo, P. Rheingans, D.T. Chen, and K.R. Subramanian, Interpolating implicit surfaces from scattered surface data using compactly supported radial basis functions, *Shape Modeling conference, Proc. SMI*, Genova, Italy, 89-98, May 2001.
24. J.L. Bently, Multidimensional binary search trees used for associative searching, CACM, 18(9), 509-517, 1975.
25. J.J. Dongarra, J.D. Croz, S. S. Hammarling, and I. Duff. a set of level 3 Basic Linear Algebra Subprograms, ACM Transactions on Mathematical Software, 16(1), 1-17, Mar. 1990.
26. H. Wendland, On the Smoothness of Positive Definite and Radial Functions *(Preprint submitted to Elseivier Preprint)*, 14 September 1998.
27. H. Samet, *The Design and Analysis of Spatial Data Structures*, Addison-Wesley Pub Co., 1986.
28. http://www.karlson.ru/csrbf/
29. A. Jennings, A Compact Storage Scheme for the Solution of Symmetric Linear Simultaneous Equations, *Comput. Journal* 9, 281-285, 1966.
30. W.H. Press, S.A. Teukolsky, T. Vetterling, B. P. Flannery, *Numerical Recipes in C*, Cambridge University Press, 1997.
31. A. George, J.W.H. Liu, *Computer Solution of Large Sparse Positive Definite Systems*, Prentice-Hall: Englewood Cliffs, NJ, 1981.

32. *The Visualization Toolkit Textbook and open source C++ Library, with Tcl, Python, and Java bindings.* http://public.kitware.com/VTK/, published by Kitware, 2001.
33. V. Savchenko, A. Pasko, Reconstruction from Contour Data and Sculpting 3D objects, *Journal of Computer Aided Surgery*, 1 Supl., 56-57, 1995.
34. G. Turk and J.F. O`Brien, Shape Transformation Using Variational Implicit Functions, *SIGGRAPH*, 335-342, 1999.

Chapter 13

Grid Method Classification of Islamic Geometric Patterns

Grid Method Classification of Islamic Geometric Patterns

Ahmad M. Aljamali and Ebad Banissi

Visualisation and Graphics research Unit, Department of Computing, Information Systems & Mathematics, South Bank University. London, United Kingdom

This chapter proposes a rational classification of Islamic Geometric Patterns (IGP) based on the Minimum Number of Grids (MNG) and Lowest Geometric Shape (LGS) used in the construction of the symmetric elements. The existing classification of repeating patterns by their symmetric groups is in many cases not appropriate or prudent [13]. The symmetry group theories do not relate to the way of thinking of the artisans involved, and completely has ignored the attributes of the unit pattern and has focused exclusively on arrangement formats. The chapter considers the current symmetric group theories only as arrangement patterns and not as classifications of IGP since they have a "global approach" and have failed to explore the possibilities in the construction elements of IGP. We describe and demonstrate procedures for constructing Star/Rosette unit patterns based on our proposed classification in a grid formation dictated by the final design of the unit pattern.

1 Introduction

Islamic artisans began to adorn the surfaces of palaces, mosques and minarets with IGPs more than thousand years ago [8]. This practice has developed into a highly specialised and rich system of finely executed geometric patterns, which geographically spread along with Islam itself into the wide lands of Africa, Europe and Asia [18]. The geometric designs consistently filled the surface planes with star-shaped like regions that resulted in very highly visual symmetric patterns, which would henceforth be referred to as "Islamic Geometric Patterns". To this very day, architectural landmarks in erstwhile and present Islamic fiefdoms exemplify the artistic mastery and specialisation achieved by the artisans whose efforts have adorned the surfaces of the said structures. These geometric patterns have often presented a long-standing historic awe to group theorists who have endeavoured to present a prudent classification of these structures.

The spiritual orientation that has guided the design of these structures has not been recorded or documented very much. Other than the finished works themselves, not much of information has survived about the thought process behind these highly intricate geometric patterns [16]. Many attempts that have been made to classify the Star/Rosette patterns have resulted in a wide variety of construction groups and classifications. Grunbaum and Shephard tried to decompose these geometric patterns by their symmetry groups after obtaining the base region, which they had used to arrive at the properties of the original pattern [11]. Others have argued in favour of a simple approach, which tagged the designs to the tools available to the artisans of the said time [2]. European group theorists like Dewdney have proposed a classification based on reflecting lines off to periodically placed circles [7] while African Art Investigators like Castera have presented classifications based on the construction of networks of eight-fold stars and "safts" [4]. Lee has presented simple constructions for the common features of IGPs but has failed to present a benchmark classification theorem [14]. Also one important aspect of IGPs that has failed to attract any kind of classification is the naïve extension of lines into interstitial regions. We have had a good look into the existing complexity to the inferred geometry in order to describe the relation of the extension region with the unit pattern.

S.J. Abbas and A. Salman in their landmark thesis on A Symmetries of Islamic Geometrical Patterns have written as follows where they were of a firm opinion that no worthwhile classification of IGPs has been taken up to this day with specific focus on their construction [1]. "Although recent researches like W.K. Chorbachi [5] have unearthed a few documents in a few libraries and museums, no comprehensive treatise on the subject has come down from the past. Relatively recently, starting with the publication of the pioneering work of Bourgoin [3] in 1879 several authors have published large collections of IGPs and offered their own analysis on the methods of constructions. The methods offered, however, are often unnecessarily elaborate and offer no explanation as to how the patterns have evolved. The overall impression that is created is that from the earliest of times the inventors of IGPs where dedicated geometry inspired by the classical Greek geometry. No thought or credit has been given to the practical experience of tiling with real shapes".

This chapter presents an argument that popular and existing symmetry groups like the "7-frieze groups" and the "17-wallpaper groups" are purely base models. A more finer and refined classification based on the study of the construction of the unit pattern is required with specific focus on the gridding system of the unit pattern. The accomplishment of the unit pattern by the MNG, LGS and the infinite number of possibilities this classification presents by way of permutations and combinations of the grids are huge.

The rest of the chapter is organised as follows. Section 2 presents a critical analysis of symmetry with specific orientation towards Islamic geometric patterns. Section 3 and 4 discusses the 7-frieze groups and the 17-wallpaper groups to analyse the features of IGPs i.e. Stars/Rrosettes. Section 5 describes how complete designs may be built using the proposed new classification of the IGPs by taking the grids as the basic element of construction of unit patterns, which form the symmetry. Section 6 summarises this chapter in a precise manner. Section 7 would describe he nature of future work that might be associated with this chapter.

2 Symmetry

Symmetry is often regarded as the greatest achievement of the Islamic civilisation. Since the time of evolution many civilisations have explored and admired symmetry and consistent patterns, but it was after the onset of the Islamic civilisation around 9th century that the finesse in symmetry truly developed. The concept of symmetry is even more closer to Islam since it demonstrates in a very visible manner the concept of the all pervading and all unifying God. Generally symmetry means a balance, a repetition of parts or simple uniformity of form. Symmetry simply means pattern. However, mathematically symmetry can be simply defined in terms of invariance of properties of sets under transformation [2]. This chapter presents a visual input into the powerful notions of pattern and symmetry in IGPs by analysing the construction of the individual element of the symmetric patterns. The Following sections describe existing and conventional symmetric group theories to endorse our view that they are merely arrangement patterns rather than classification theories of geometric patterns let alone IGPs. Group theory shows that in a single dimension symmetric period pattern can be analysed into seven different types and provides the information needed to identify a particular symmetry type [1]. Also, in a two dimension symmetric period pattern, seventeen different types of patterns can be generated and identified. The Single dimensional symmetric pattern is referred to as "7-frieze groups" and the double dimensional symmetric pattern is referred as the "17-wallpaper groups". However, both together can also be referred to as period groups. It might be very important to note at this stage that two eminent group theorists Polya [17] and Speiser [18] had analysed the designs and decorations of Islamic art before arriving at the Validity of the above-specified group theories. Famous authors on IGPs have chosen to approach either through Greek inspired Ruler/Compass based constructions or through grids [9]. We would prefer to call these grids as concealed grids. For the simple reason, that the most familiar shape in IGPs the eight-pointed star itself is constructed by placing a square grid on another square grid rotated by 45 degrees. The underlying grid can never be identified until and unless we decipher the grid on the beneath one. Similarly, most of the elements of symmetric patterns are nothing but Culmination's of grids of various types arranged in a unique manner to achieve the desired pattern. These grids do not require geometric specialisation. It is observed that the most fascinating patterns have been achieved by first placing the basic grid and then various geometric shapes like circles and polygons in various sizes would be placed on it in a very consistent and predictable manner to fulfil a symmetric pattern. It might be noted that the circumference of the shapes are joined and divided at marked points. In our classification the identification of the construction of the motif is the basic data of identifying the unit pattern and the larger pattern itself.

3 The 7-Frieze Groups

The most essential characteristic feature of frieze patterns is Translational symmetry. Isometric groups that keep a given straight line invariant including translations along the line are called frieze groups (Isometric may be defined as a "linear transformation of the plane (or space) that preserve distances between points"). The essential properties of symmetric groups like translation, rotation, and reflection are all isometries. Frieze patterns found their way into mathematics through classical arithmetic. When nodes are assigned integer values on lattice they tend to obey or symbolise certain rules. To illustrate the 7-frieze groups argument (Fig. 1), we present the attributes of each frieze group by Andrew Glassner [10] who has done some very inspiring research into many related topics like frieze groups, moiré patterns, mirror reflection and a periodic tiling. He has demonstrated the value of creating physical models that enabled us to stretch our visualization skills and also our perception of the subject matter. It should be very important to note that in a very definitive sense that the frieze group theories misdirect classification of IGPs. It should be very important to note that in a very definitive sense that the frieze group theories misdirect classification of IGPs. Mathematicians find it very convenient and useful to interpret regularity of a pattern in terms of its group of symmetry. In this way the results of algebra and other mathematical disciplines can be applied to the study of such patterns. However, it could be argued that this is not the concept of regularity that artisans had in mind as they were creating their art. In fact, until a century or so ago, even to mathematicians regularity of mathematical objects had a completely different meaning. The difference between the two approaches is to a large degree the contrast of the global and local points of view. Mathematicians used to define regularity of objects such as Platonic polyhedral by requirements of congruent faces, equal angles, and other local properties, now it is customary to define regularity by the transitivity of the symmetry group on the set of flags. In the same way, it seems likely that the artisans meant to create geometric patterns in which part is related to its immediate neighbours in some specific way and not by attempting to obtain global symmetries of the infinitely extended design.

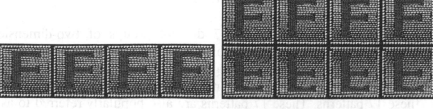

F1. Translation

F2. Horizontal reflection

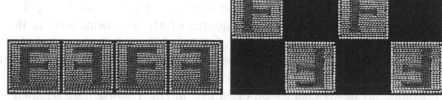

F3. Vertical reflection

F4. Corner rotation

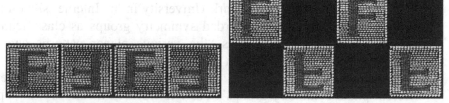

F5. Translation and center rotation

F6. Horizontal reflection and translation

F7. Vertical reflection and center rotation

Fig. 1. *The 7-frieze groups by Andrew Glassner [10]*

4 The 17-Wallpaper Groups

It has been established that 17 distinct groups of two-dimensional patterns that are periodic in two independent directions exist (Fig. 2). The laws of symmetry have long been established and also have been widely recognised by eminent mathematicians and geometricians for these 17 patterns. These 17 patterns are also popularly referred to as the "17- wallpaper groups". It might be noted that each symmetry itself involves with a peculiar combination of transformations of the plane. However, the essential nomenclature that has been assigned to the 17-wallpaper groups may be given by Xah Lee [14].

As early as 1944, Edith Muller had written a thesis on Group Theory and Symmetry Notation of the patterns of Moorish ornaments in the al-Hambra Palace. E. Muller's study remained the only one to apply Group Theory and Symmetry Notation to the study of Islamic geometric pattern. Earlier, Andreas Speiser had called particular attention to Islamic art in his chapter on ornament in *Die Theorie der Gruppen von endlicher Ordnung* as early as 1927. Her thesis work takes into account the scientific theoretical finding and reveals the value of these scientific theories to the understanding and classification of Islamic geometric patterns [5].

David E. Joyce [13] from Clark University in his Internet site on the 17 plane symmetry groups regarded symmetry groups as classification of planar patterns. He wrote as "the various planar patterns can be classified by the transformation groups that leave them invariant. A mathematical analysis of these groups shows that there are exactly different plane symmetry groups". Now if we look at the 17-wallpaper groups with the above-mentioned benchmark in section 3 it would be crystal clear that these are merely arrangement patterns and absolutely nothing else.

Now, it is very clear from the above illustrations that the frieze and the wallpaper group's theories present arrangement benchmarks, which enable us to determine the type of format pattern arrangement rather than classify the unit pattern. Also we can conclude that a viable and arguable approach has not been taken up with a holistic view to classify IGPs with specific focus on their construction.

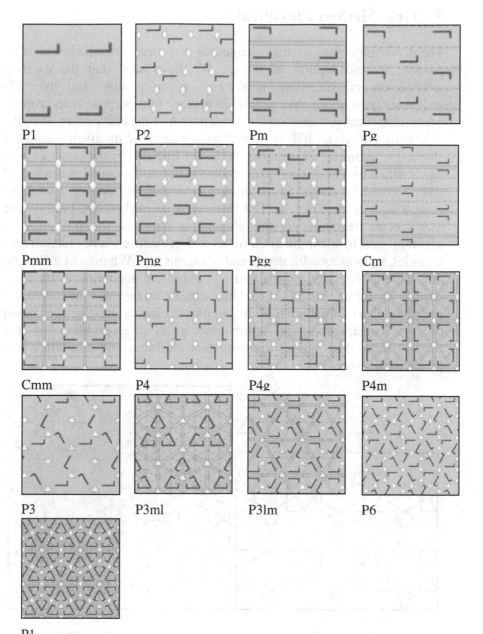

Fig. 2. *The 17-wallpaper groups by Xah Lee [14]*

5 Grid Method Classification

The basic objective of this chapter as has been stated above, is to propose a classification of IGPs. We have seen that the existing arrangement classifications like the "7-frieze groups" and the "17-wallpaper groups" are very generic in nature. Our purpose is to propose a new classification based on the construction of the unit pattern of IGP.

Usually any given IGP is named on the basis of its given geometric shape. For example, illustrations by Issam El-Said [9] the pattern could be classified as a hexagonal pattern simply because it contains hexagonal star or could be classified as an octagonal pattern because it contains octagonal star, etc. (Fig. 3). But this could be misleading, because the Star/Rosette the most popular element of most of the IGP may be accomplished by a combination of several geometric shapes like circles, triangles, squares, quadrilaterals and hexagons etc.. Where a Star/Rosette unit patterns could be normalised and classified according to its basic design. Therefore, instead of looking at these images as a hexagonal or octagonal like unit patterns, we would like to classify these images based on the construction and normalisation of the gridding of the Stars/Rosette. And look at the important attributes and properties of a given Star/Rosette.

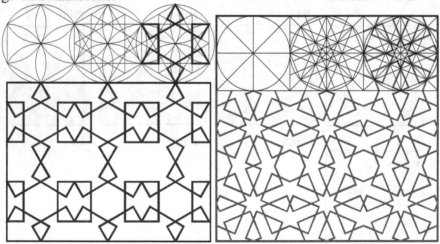

Fig. 3. *Hexagonal and octagonal patterns by Issam El-Said in "Islamic Art and Architecture, The System of Geometric Design"*

The Star/Rrosette is varying and is often found to be of different types and the differences range because of many factors like historical time gap, geographical and civilisation influences, availability of construction tools at the time of construction of the said piece of architecture etc.

We have observed that a classification that normalises the construction of the Star/Rosette or in some cases multiple Star/Rosettes with extensions that form the unit pattern are the result of a unique combination of several geometric grids.

In our method, any given Star/Rosette can be deciphered or deconstructed by normalising it. This normalisation process would be achieved by identifying individual grids that make up the Star/Rosette. Once grid elements are separated the basic geometric shape that can be used to achieve the n-gon Star/Rosette would be identified. The process of dissection of a star according to us could be taken up in the following stages:

5.1 The Planar Surface Stage

This is the basic unit circle or the planar surface on which the grids would be placed to achieve an n-gon unit star. Its radius strictly restricts the placing of the grids within the parameters of its size. Here we have a very strong difference of opinion with W. K. Chorbachi [5] who in his landmark effort "Tower of Babel: Beyond Symmetry in Islamic Design" has said as following:

".... in Geometric Concepts in Islamic Art by I. El-Said. In his introduction to it, Titus Burckhardt states that all the geometric patterns are derived by the same method of deriving all the vital proportions of a building (or a pattern) from the harmonious division of a circle ... which is no more than a symbolic way of expressing Unity (*Tawhid*) which is the metaphysical doctrine of divine unity as the source and culmination of all diversity. ... In some cases however, the authors neglected to draw in the circle, ironically revealing how unfundamental its existence is to the alleged "unique way" or "only way" of deriving all patterns" (Fig. 4) [5].

We were able to discover the same image (Fig. 5) with the base circle marked in a clear manner in El-Said's Islamic Art and

243

Architecture, The System of Geometric Design [9] which shows that El-Said did not ignore to draw the unit pattern in the circle. In this context, we can say that (Fig. 4) has been viewed from a different and convenient dimension so as to facilitate a suitable conclusion to justify the theorem that was being presented by W. K. Chorbachi. The circle in (Fig. 5) which is a replica of (Fig. 4), indeed forms the base planar surface for the design of the unit pattern, as is what is being argued by this chapter.

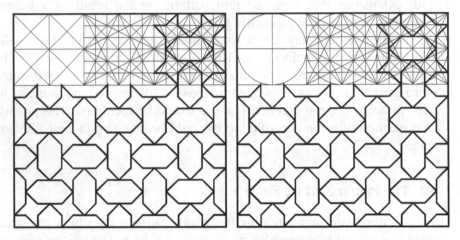

Fig. 4. *This image is Fig. 4.2 From W. K. Chorbachi, Tower of Babel, Beyond Symmetry in Islamic Design [Cho89], showing that the circle does not appear in deriving this pattern. From I. El-Said, Geometric Concepts in Islamic [8].*

Fig. 5. *The circle does appear in the unit pattern. From I. El-Said, Islamic Art and Architecture, The System of Geometric Design [9].*

5.2 The Divisional Stage

Here we divide the circle (360 degrees) by x number of points to arrive at the intended design of the Star/Rosette.

5.3 The Gridding Stage

The gridding stage would initiate the gridding process. This stage is the most important stage in the chronological stages that have been stated by this chapter. It has been observed that design formats of Stars/Rosettes found in IGPs are varied and very different to each other. Since Islam itself is spread across so many continents and each country has contributed its own artistic heritage to Islamic Art. In this background, to properly decipher a star pattern one must make a very properly guided endeavour in order to know the type of Star/Rosette. We know the complexity of this task because the very nature of Islamic art is very intricate and any intricate art is difficult to normalise. The core objective of this stage is to describe and classify the Star/Rosette with reference to the Minimum Number of Grids (MNG) and the Lowest Geometric Shape (LSG) used in achieving the design of the IGPs.

We have chosen an unusual and complicated pattern to demonstrate our method of classification (Fig. 6). The pattern is an extract from; The Mathematical Gazette *"Some Difficult Saracenic Designs, A Pattern Containing Fifteen Rayed Stars"* by E. Hanbury Hankin [12].

In any given unit pattern requires to be classified according to our method, we start by looking for the different types of Stars/Rosettes in the given unit pattern. We have taken the liberty of colouring the unit pattern given by Hankin to show the different types of Star/rosette in the given unit pattern. The given unit pattern is unusual because it consists of two types of stars which are similar in type but different in design; one is twelve rayed star and small in size and the other is fifteen rayed star and larger. What makes the pattern extra ordinary is that the two different sizes of stars beautifully connected to each other with sets of meshes of lines between them. The size of star doesn't effect our classification rather the method of design of each Star/Rosette and its gridding attributes. Therefore we conclude that the first attribute of this given unit pattern is that it consists of multiple Stars/Rosettes. If we start to classify the unit pattern by Hankin, according to the standards or norms of the conventional frieze and wallpaper group theories we would have to take the enclosed area as the primary unit pattern. By doing so we are bypassing the finite elements or finite properties of the image. The following sections elaborate in very fine detail the process of normalising the construction of IGPs based on the elements of grids (MNG) and the properties of grids themselves (LGS).

5.2.1   The minimum number of grids (MNG)

This section would initiate the first part of our naming convention, which is aimed at identifying the minimum number of set of grids mounted above each other to achieve an n-gon whose vertices bisect its edges. This section takes the end design as its core objective. An infinite loop process goes on identifying intersections and sets the correct relationship with vertices to achieve bisections of the rosette.

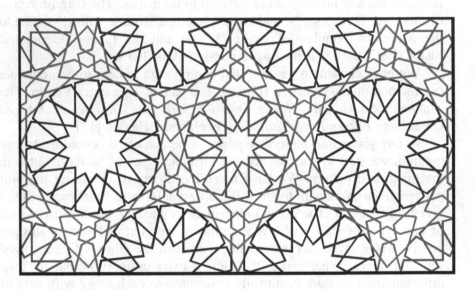

Fig. 6. *Pattern containing twelve (red) and fifteen (blue) rayed Star/Rosette by E. Hanbury, Mathematical Gazette "Some Difficult Saracenic Designs" [12].*

5.3.2   The lowest geometric shape (LGS)

This stage initiates the second part of our naming convention which is aimed at identifying the lowest possible geometric shape that is used to construct the Star/Rosette within the given unit pattern.

The following illustrations (Fig. 7) and (Fig. 8) describe the normalisation series of the pattern (Fig. 6) to achieve the classification of the twelve and the fifteen-rayed Star/Rosette respectively. The twelve-rayed Star/Rosette uses 3 Minimum Number of Grids (MNG) and a Quadrilateral as the Lowest Geometric Shape (LGS) where as the fifteen-rayed Star/Rosette uses 3 Minimum Number of Grids (MNG) and a Pentagon as the Lowest Geometric Shape (LGS) placed on a planar surface. Therefore we classify this pattern (Fig. 5) as: **Grid 3 Quadrilateral/Pentagonal Class.**

We would like to state at this point that images speak better than words regarding normalisation process. Following these sections are series of images that would illustrate the Normalization process in a very logical and visual manner that would put forth our point in a presentable manner (Fig. 9).

5.3 The Artistic Stage

This is the fourth stage, here once the core design (gridding) of the Star/Rosette is achieved we may design the final Star/Rosette by giving the necessary artistic attributes to the grid by way of presenting weights to the internal lines of the grid. This stage may also include colouring and filling the sections of the Star/Rosette.

5.4 The Extension Stage

The fifth stage is a Notional or Phantom stage because this stage might exist or might not exist. In this stage the natural extensions would evolve to accomplish the seamless mesh in the external zone within the notional boundary (usually a square or a rectangle) and beyond (Fig. 7) and (Fig. 8).

El-Said [9] has regarded the square as a definite external boundary (boundary consisting of the circle and associated mesh extensions beyond the circle) in most of the designs. W.K. Chorbachi [5] in this context has said that square cannot always be traced out in all designs. In an excerpt from his book: Tower of Babel: Beyond Symmetry in Islamic Design, he has written as following:

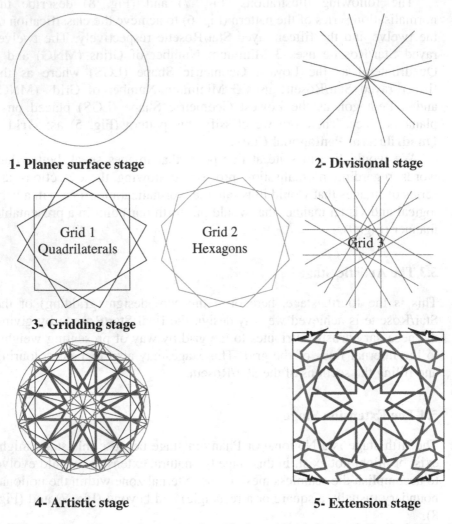

1- Planer surface stage

2- Divisional stage

Grid 1
Quadrilaterals

Grid 2
Hexagons

Grid 3

3- Gridding stage

4- Artistic stage

5- Extension stage

Fig. 7. *The 12-rayed star is classified as (Grid 3 Quadrilateral Class) because it uses minimum of 3 sets of grids and the lowest geometry is quadrilateral.*

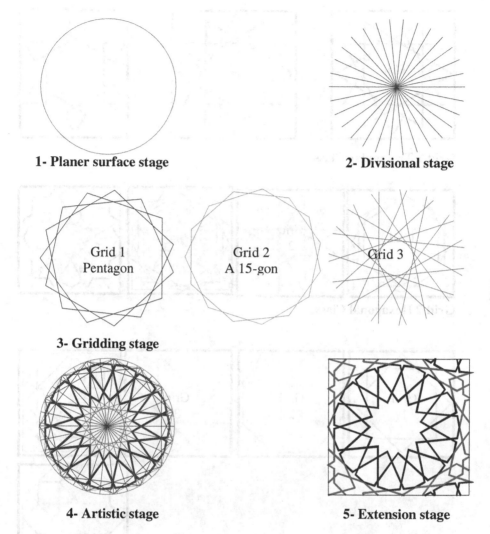

1- Planer surface stage

2- Divisional stage

Grid 1
Pentagon

Grid 2
A 15-gon

Grid 3

3- Gridding stage

4- Artistic stage

5- Extension stage

Fig. 8. *The 15-rayed star is classified as (Grid 3 Pentagonal Class) because it uses minimum of 3 sets of grids and the lowest geometry is a pentagon.*

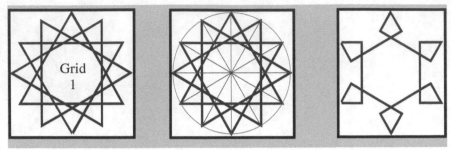

Grid 1 Triangular Class

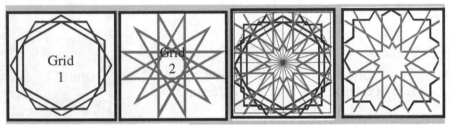

Grid 2 Hexagonal Class

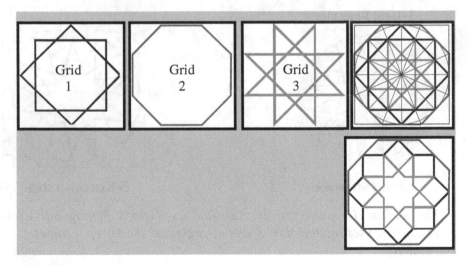

Grid 3 Quadrilateral Class

Fig. 9. *Showing the classification of some Islamic Geometric Star/Rosette.*

"... and finally, there are a few cases of designs where it was absolutely impossible to hide the fact that the analytical method did not hold. These are illustrated (Fig. 4.3) as containing a non-standard zone based as a variation. The elongated rectangular area obviously belongs to a 2-fold symmetry generalisation of the "one way" that is overwhelmingly represented in the square of a 4-fold symmetry group".

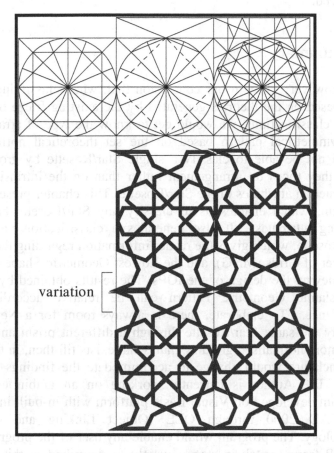

Fig. 10. *This image is Fig. 4.3 from W. K. Chorbachi [5], showing that the scheme of the circle does not fit the long rectangular unit, a variation zone is marked. From I. El-Said [8].*

251

In the illustration (Fig. 10), W.K. Chorbachi has regarded the rectangle as the external boundary for the design and proved that a square cannot always be possibly marked as the external boundary. However, he too regarded the external boundary as indispensable in a design. But, as has been proved above (Fig. 7) and (Fig. 8), the external boundary is Phantom to the Star/Rosette design and its presence cannot be always confirmed until and unless the existence of external mesh can be traced.

6 Summary

It is now very clear in our conclusion that, when classifying any given IGP based on the theories generated by the existing frieze or wallpaper group classifications, we would get to know the type of arrangement of the symmetrical pattern based on the set theoretical norms. But we would not be able to classify a single Star/Rosette by group theories since they focus on arrangement rather than on the intrinsic design or construction attributes of the Star/Rosette. This chapter presents a viable theorem, which enables us to classify any Star/Rosette based on its gridding attributes. It also generates a classification name for the Star/Rosette, which gives the reader information regarding the Minimum Number of Grids (MNG), and the Lowest Geometric Shape (LSG) used in achieving the design of the IGPs. The results obtained by classifying the Islamic Geometric Pattern can be relative according to our experience. To elaborate, there is always room for a New Critic to classify the same Star/Rosette through a different prism and come out with more normalised gridding than what exists till then, in this case the nomenclature would change to accommodate the findings of the new study. The Author is currently working on an ambitious computer program developed on Visual Basic platform with in-built integration to AutoCAD 2000 through OLE (Object Linking and embedding) technology. The program would enable any user of the program to build an IGP from scratch in stages exactly as described in this article and arrive at its classification with ease. The author hopes that the program would be an excellent tool in analysing IGPs to designers, architects, geometrists and academic fraternity working in related areas.

References

1. Abbas, S. J. and Salman, A. (1995) Symmetries of Islamic Geometrical Patterns, World Scientific.
2. Abbas, S. J. and Salman, A. (1992) Geometric and Group Theoretic Methods for Computer Graphic Studies of Islamic Symmetric Patterns. School of Mathematics, University College of North Wales, Bangor, Gwynedd, LL57 1UT, UK. Computer Graphics Forum, Volume 11, number 43-53.
3. Bourgoin, J. (1978) Arabic Geometrical Pattern and Design, Firmin-Didot, Paris, Dover, New York (1973).
4. Castera, J. M. et al. (1999) Arabesques: Decorative Art in Morocco. ACR Edition.
5. Chorbachi, W.K. (1989) In the Tower of Babel: Beyond Symmetry in Islamic Designs. Math Applic. Vol. 17, No. 751-789.
6. Critchlow, K. (1976) Islamic Pattern: An analytical and Cosmological Approach, Schocken books, New York, NY.
7. Dewdney A. K. (1993) The Tinkertoy Computer and Other Machinations, pages 222-230. W.H. freeman.
8. El-Said I., and Parman (1976) A. Geometrical Concepts in Islamic Art: World of Islam Festival. Publ. Co. London.
9. El-Said, I. (1993) Islamic Art and Architecture: The System of Geometric Design. Grant Publishing Limited, U. K.
10. Glassner, A. (1999) Andrew Glassner's Notebook: Recreational Computer Graphics. Morgan Kaufmann Publishers, San Francisco, CA.
11. Grunbaum B., and Shephard G.C. (1992) Interlace Patterns in Islamic and Moorish art. Leonardo, 25:331-339.
12. Hankin, E. H. (1934) Some Difficult Saracenic Designs Pattern Containing Fifteen Rayed Stars. The Mathematical Gazette, Vol. 18, 165-168, and Vol. 20, 318-319 (1936).
13. Joyce, D. E. (1997) The 17 plane symmetry groups Department of Mathematics and computer science, Clark University, Worcester, MA 01610.
14. Lee, X. (1998) The 17 Wallpaper Groups.
http://www.xahlee.org/Wallpaper_dir/c5_17WallpaperGroups.html

15. Mulller, E. (1944) Gruppentheoretische und Struktu-Analaytische Untersuchugen der Maurischen Ornamente aus der Alhambra in Granada. (Ph.D. Thesis, University of Zurich) Baublatt, Ruschlikon.
16. Paccard, A. (1980) Traditional Islamic Craft in Moroccan Architecture, Editors Ateliers 74,74410 Saint-Jorioz, France, Vol. 1 and 2.
17. Polya, G. (1924) Uber die analogie der Kristallsymmetrie in der Ebene, Zeitschrift fur kristallographie, 60, 278-282.
18. Speiser, A. (1927) die Theorie Der Gruppen von endlicher Ordnung, Second Edn. Springer, Berlin, Third Edn, Springer, Berlin (1937), Fourth Edn, Birkauser, Basel.

Chapter 14

Web Based VRML Modeling

Web Based VRML Modeling

Szilárd Kiss
Department of Computer Science, University of Twente, P.O. Box 217, 7500 AE
Enschede, The Netherlands

*We present in this chapter a method to connect VRML and Java compo-
nents in a web page using EAI in a way which makes possible to interactively generate and edit VRML meshes. The meshes used are based on
regular grids, to provide an interaction and modeling approach that uses
the internal semantics of such a mesh by linking the available modeling
operations either to single vertices, the vertices of a ring or column, or all
the vertices from the VRML mesh. The method permits strict mesh complexity control and scales the operations according to the mesh properties
and to the selected target vertices as well. We describe the structure of our
system and provide a few examples of the meshes that were created.*

1 Introduction

Commercial and non-commercial modeling software packages are
usually oriented for off-line rendering. Furthermore, such systems
tend to be fairly complex, thus a steep learning curve is necessary to
deal with these products, which is acceptable for a professional user,
but not for casual users who would adventure into virtual reality just
for the sake of building homepages, avatars or objects for online
communities. If somebody wants to create models that are suitable
for online rendering, he/she has to be satisfied with the export functions of complex commercial packages or with modelers that use
pre-built and non-alterable component meshes. Therefore, a 3D
mesh editor based on low-level manipulation (namely the direct manipulation of the mesh vertices) should bring a solution which is tai-
lored more to on-line deployment and real-time visualization.

Our mesh modeling system uses a mixture of editing approaches.
First of all, the editing operations can work on simple vertices of a
mesh, which means a low-level visual editing approach, but the
same editing operations are valid also on collections of vertices, giving to the editor a higher-level status. Second, additional operations

enforce this higher-level approach to editing, providing more possibilities to operate on sets of vertices. Another objective is to allow a modeling approach that gives control over the complexity of the edited mesh, as the user sets the number of vertices as needed.

2 Background

Since the requirements for the system were to be able to provide a visual, interactive, web enabled environment, we decided to use the VRML (Virtual Reality Modeling Language [15]) standard format for the system and the various players as the "graphics engine". But that was not enough, as VRML is a modeling language and not a programming language, and that's why the EAI (External Authoring Interface [11]), a set of Java classes defined for binding the VRML environment to the Java programming language were used additionally. This allowed to extend the VRML scripting possibilities with a Java applet to aid in the mesh handling processes of the modeler. The Java applet receives events from the environment, and executes the sought operations on the mesh.

When it comes to the process of generating and visualizing data dynamically in interactive systems, there are a few examples of systems that use this approach. [13] is storing behavior data in a database and generating VRML content dynamically, [3] stores data as digital documents and generates interactive content automatically.

Instead of using conventional editing operations as described in [10] and using automatically generated complex surfaces ([4], [5] and [16]) we use an approach based on vertices and sets of vertices grouped in rings and columns (regular grids), similar to the models used in [2] and [14]. We also would like to provide an easy to use interface, for which an example would be the "Teddy" system [7].

Systems with an integrated WYSWYG (What You See is What You Get) type approach can increase the usability of a system by great length. Since our modeling and target platform is essentially the same (except for the editing controls), this yields in an automatic preview method.

3 Editor Appearance

Figure 1. is a screenshot of the editor in action. The system provides
a simple, easy to use interface for all the operations possible, without
any elements that must be looked up from menus or lists. Such systems (see also [7]) could be targeted to a much broader audience, as
their use can be learned quite easily. The visual elements are
grouped and marked according to their functionality in the editor
system, and their functions are presented below.

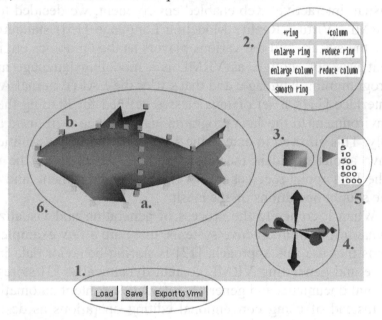

Fig. 1. *Annotated screenshot of the system.*

1. Buttons of the Java applet for loading, saving the mesh data and
exporting to VRML format. They use the familiar load/save window
of the underlying platform, provided by the Java AWT classes, and
they perform file readings and writings of the internal format and
file writings to VRML file format.

2. Editing buttons in the VRML environment. From left to right and
from top to bottom in order: add ring and add column (which increase the mesh complexity), enlarge ring and reduce ring (for the
ring-wise scaling of the mesh), enlarge column and reduce column
(for the column-wise scaling, which has a global scope, affecting the
whole mesh) and last the smooth ring button which is used for
smoothing the mesh surface ring-wise.

3. Rotation widget (box). On the figure it has been already rotated, and the elements noted with 4. and 6. follow its rotation exactly. Used for positioning the mesh in a manner that allows to perform the desired selection of vertices and to preview the mesh from different positions.

4. Editing widget. Composed of 6 arrows, with different color for each pair of arrows that represent the positive and negative direction of the main axes. A click on an arrow will displace the selection in the direction the arrow is pointing by the displacement unit selected using the unit selection widget (next widget).

5. Editing unit selection widget. The default value is set to 100 millimeters, this can be changed by dragging the triangle on the left. A wide range of values (from one millimeter to one meter) gives the possibility to handle meshes at different scales, and to perform scale transitions easily.

6. Edited mesh. a) is the current selection, a ring (the mesh was already rotated), while b) is a column which is not yet selected, but the pointing device is over the column sensor, and the column is highlighted as a possible next selection.

4   Data Format

The editor uses an internal format for the storing of mesh data. This is done for two reasons. First, there is a data parsing reason, meaning that a simple data format would allow a more comprehensible parsing method. Second, an internal format ensures no complications regarding the use of a data file with possibly non-regular structure. The latter would not be anyway a good idea (to create the mesh elsewhere and import it in the application) as the goal is to use as simple meshes as possible. The definition of the data file is as follows:

Before the first hash mark, everything is comment
#segment segment-name
rings nr-of-rings \r
points nr-of-points-in-a-ring \p
[Between these brackets, the x y z values of vertices separated by spaces, and with commas between the different vertices. Rings are separated with new line characters.]
This is the composition of a data file. However, users are not required to edit this data manually (it would be a low-level editing in-

tervention), as the application provides a visual and also low-level approach to editing. Although manual edit is certainly possible, it is more cumbersome than the possibilities offered by the editor, for instance to select a single vertex, and move it in any of 6 degrees of freedom (DOF) even with a displacement unit of one millimeter. Of course, the same operation can be effectuated also more globally, for instance on a set of selected vertices.

5   Structure of the System

We have already mentioned that the editor uses the combination of VRML, Java and EAI technologies to provide it's functionality. Since these are distinct technologies, and the EAI component handles all the interaction between the VRML and Java components, it is possible to discuss these components separately. Taking in account that the communication between the components is a two-way communication, it is best to show it in the flow diagram from Fig. 2 and explain it's mechanisms step by step.

The VRML component is a VRML file which at the beginning contains no mesh, just an anchor point for the locking of generated mesh geometry (this will be loaded and generated at user request). The file does instead contain beside the anchor point for geometry, a set of "widgets" that are tools inside the virtual environment that aid in the editing process. Depending on their functionality, these widgets act directly upon the virtual environment, or indirectly by sending their events to be handled by the Java applet, which in turn passes modified data back to the VRML component. This second component (the Java applet) is used for the data handling necessary for opening, generating, modifying and saving meshes. The applet is used also for translating into the environment the effects of user actions effectuated through some of the widget tools.

These two components that work together using EAI are packed in an HTML file for presentation in a web browser. As they are side by side in a common environment (that of a web browser), it is possible to use code that is able to make the communication possible between the Java and VRML components, and this code is the EAI.

The components of the editor are event-based, they are sending and receiving events depending on the user interaction with the environment. This means the code of the editor is not executing sequentially and it is impossible to describe it in a consequent manner. We will start off with the EAI component which explains the bindings between the other two components, the VRML component and the Java component, and then we will take a look at the VRML and Java components, in that particular order, to provide at first the description of the editing environment, and then to explain in detail how the user interaction is achieved through the Java component.

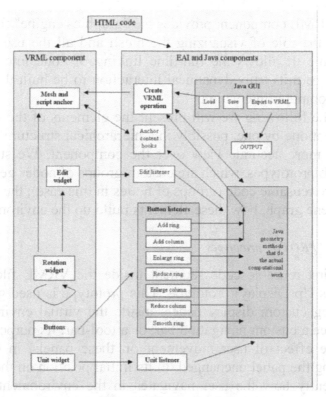

Fig. 2. *Flow graph.*

5.1 EAI component

This is the simplest component of the structure, it connects the virtual environment and the applet that are inside the same web page. It gives the possibility for the Java methods to access the VRML content, and does this by providing Java wrapper objects around the internal VRML browser, nodes and fields. As it can be observed from Fig. 2. there are a multitude of connections between the VRML and Java components, and yet these are not all. These are in fact representing only conceptual connections, and the number of real connections are much higher, since the VRML "objects" (single nodes, grouped nodes or new prototypes) usually have multiple event sending/receiving possibilities, which has to be used to provide interaction possibilities to the system.

5.2 VRML component

The VRML component provides the "graphics engine" for the editor. It has the role of visualizing the mesh and all the user controls in real-time. It allows the run-time linking of code into the system, which permits truly dynamical interaction to be built that aid in the user-computer interaction.

In the following we will present the elements of the virtual environment one by one, possibly in a hierarchical structure, this will allow a more accurate view over the component. We start with the VRML prototypes, which are custom groups of nodes generally used to store reusable code (groups of nodes in this case), then we present the scene graph, how these elements build up the environment.

VRML "PROTO" prototypes

The first reusable code segment inside the VRML file is a HUD (Heads Up Display) prototype. This prototype is used generally for creating custom display panels inside the virtual environment, for instance a custom navigation bar, or a tool-bar. Its purpose is to mirror the effect of user movement on these panels, in other words keeping the panel unchanged, in its initial position on the screen, indifferently how the user navigates in the environment. It is used generally for panels, but that is just a common usage scheme, that is certainly not a limitation. This particular HUD prototype is more general, it accepts a sub-tree of nodes as parameter, practically accepts any type of valid children node or node tree. Therefore it can be used for different purposes, as we will see when discussing the other elements of the VRML component. The only other parameter of the HUD is a starting position, which is used for positioning the visual elements of the virtual environment on the screen.

The second prototype is a geometrical assemble, representing a three-dimensional arrow. Its functionality is related for specifying the DOF (or axis) operations, which will be discussed as the prototype is used in the scene graph (next section). The parameters used for the prototype are color, translation and rotation, these specifying the appearance as well as the starting position and orientation of the object to be created. The parameters can be chosen to represent the different axes of the three dimensional space, while the arrow points to the positive or negative directions of those axes.

The last prototype of the application is a button prototype. The geometry of the prototype is very simple, consisting of a small rectangle, on top of which a picture is mapped, picture that explains in text (this is the method currently used) the functionality of that particular button. Its main purpose is however the capability to track user actions with a sensor that signals when the user's pointing device (mouse) enters the influence area of the button and when the button is actually pressed by the user, and the capability of switching the button on/off. The parameters of this prototype are related to appearance (color and texture) or initial positioning.

Scene graph

After the prototype declaration and the usual general settings code of a VRML file, the editing environment is created using the prototypes defined earlier and some additional nodes. First, an editing "widget" (or tool) is created. This is embedded in a HUD instance, thus user navigation does not affect it, since always stays is in the same position. The widget itself consists of 6 instantiations of the arrow widget, sensors attached to all of them, and a script that handles the events of the sensors. The 6 arrows represent the 6 DOFs of a three dimensional space, the +X, -X, +Y, -Y, +Z, and -Z directions. As a visual aid, X directions are colored red, Y directions green and Z directions blue (as a sort of direct mapping between the enumeration of the three dimensional coordinates and the enumeration of the three basic RGB color components). Their purpose is to collect the user's displacement commands regarding the selected mesh vertices. As the user clicks on such an arrow, the sensor attached to the arrow sends a "clicked" message to the script. Each arrow has this functionality, and the script, depending on which arrow was clicked and depending on the current displacement unit, will issue a displacement event containing the value of displacement. This value is caught by an event listener defined in the Java component, where the data will be processed. The current displacement unit is stored in a variable inside the script, and it's value can be modified by an another element of the VRML environment. The value of the unit variable is also transmitted to the Java component, as this component is responsible for all the calculations and geometry generation tasks.

The next elements of the virtual environment are seven instantiated buttons. Their functionality is to transmit their activation status to functions from the Java component that will effectuate the actions related to the buttons. The button functionalities are:

- **adding a ring**. This button triggers a Java function that adds a new ring right after the selected ring or after the ring that the selected vertex belongs to. The order of the rings is defined by the data enumeration order. This button is not enabled if a column is selected.

- **adding a column**. Same functionality as above, except that it is effectuated on columns. The new column is inserted between the selected column and the column following it, and the following order is calculated from the data using the right hand rule, with the thumb representing the ring data direction and the bent remaining fingers representing the vertex enumeration direction.

- **positive ring scale**. Selected ring or the ring of the selected vertex is enlarged triggering a Java function that adds the value of the displacement unit vector to the selected vertex values. The button is not active if a column is selected.

- **negative ring scale**. Same as above, only the displacement in this case is subtracted not added. Ring scaling introduces changes in the object thickness locally.

- **positive column scale**. Unlike the ring scale operations that are acting on a subset of mesh vertices (rings), a column scale operation acts on every vertex of the mesh. A positive scale in this case means that the object will become a longer (taller) object, while the thickness does not change.

- **negative column scale**. Object will become shorter. Scale column operations are distributed between the mesh rings based on percentages calculated from the ring positions relative to the mesh center, and the mesh will retain its original shape.

- **smooth ring**. The counterpart Java function calculates the relative distances of vertices from the center point of the mesh, and uses a fraction of it to smooth the contour line of the ring vertices. With enough activations of the smoothing button, the vertices will approximate the same distance from the center point, as if they were on the same circle.

Another element embedded in a HUD is a container element, which at start is empty, and is filled by the code generated by the Java component. The container is used only as a linking point, and contains no other functionality. The functionalities will be added at the same time with the generated geometry as the Java component generates the code simultaneously and links it to the VRML container element as one piece of code.

The next item is the rotating widget or tool. The tool is represented as a three dimensional box, which has a rotation sensor linked to it. There is a drawback to this construction, which lies in the mapping of the two-dimensional input device (the mouse) to three-dimensional manipulation, done by the VRML viewer implementation (this is a general difficulty, all implementations have this problem, mapping from 2D to 3D is a shortcoming due to input device limitations). However, as the user becomes acquainted with this particular manipulation possibility, it can provide enough control for editing purposes. The function of this widget is to enable the viewing of the mesh from all angles and also the favorable positioning of the mesh in the course of the editing process. As the user drags the rotating widget, the box itself rotates, and the same rotation is conveyed to the edited mesh container (and consequently to the edited mesh itself) and to the editing arrows. This way, not only the mesh is positioned, but the axes of the editing arrows and the actual axes of the vertices remain aligned, meaning the arrows point in the direction the vertices will be displaced by the arrow actions. The example figures from section 7 are screenshots of various objects, and it can be observed right away that all are in different orientations. From Fig. 1 it also can be observed that the mesh and arrows are facing/pointing in exact correlation, as both elements are driven by the same rotational widget.

The last element of the environment is an editing unit selection slider. The user has the possibility to select the edit unit between the 1 mm, 5 mm, 10 mm, 50 mm, 100 mm, 500 mm and 1000 mm values, allowing a wide scale of the meshes to be edited. A triangle is pointing to the current value, represented as a panel with the enumerated values written on it. By dragging the triangle, these edit values can be selected. This element consists of a HUD which incorporates the triangle slider and the values panel and consists of a script that helps mapping the slider values to editing values corresponding to the values written on the panel.

Consequences from the chosen structure

One consequence is that navigation is possible, but it does not affect the appearance of the environment at all, as all geometry in the environment is embedded in HUDs and it will move together with the user, which causes the illusion of standing still. This however does not exclude the possibility to view the edited geometry from different views, which is bundled to the rotation widget for synchronization and interface continuity purposes.

Undo operations are not implemented. However, the editing actions are effectuated by clicking on an arrow sensor, and such an action can be undone by clicking on the arrow pointing in the opposite direction. Furthermore, some of the operations represented by buttons have their symmetric counterparts, and that enables the restoring action to be effectuated (for instance, enlarging accidentally a ring can be undone by reducing the same ring in the next step).

5.3   Java component

The Java component is the part of the editor that essentially does all the work related to the interaction that occurs inside the editor. It receives the events of the VRML component, effectuates the functions/commands associated with the particular events and it returns the result in some form to the environment.

Its function is somewhat analogous to the client-server architecture. The editing data is stored by objects/variables inside the Java component, which could be considered as the server. The VRML component (the client) that tracks the user actions, sends data over these actions to the server Java component, which first calculates and accommodates the new data in its own data containers, then sends the new information back to the client VRML component for visualization.

Connections to the elements of the VRML component

To have access to the VRML component, the Java component uses code from the EAI packages. This way, it can get hold of the objects inside the virtual environment. It is not necessary to explain how exactly this has to be accomplished, a list of connections and their purpose should give enough insight to the event flow of Fig. 2.

Note: Not all the connections will be set up when the environment is created. To be able to communicate fully with the VRML component, the dynamic code created by the system must also be linked with the Java component. This is done using the EAI in the same manner as used for the static code of the VRML component, after each VRML code generating operation. Below is the list of connections made with EAI:

- connection to the browser. Without this connection, the environment cannot be accessed at all by the Java component.
- connection to the seven editing buttons, to their generated events, and to the active/inactive properties of them. For each of the buttons, there is a handling function that executes all the necessary calculations and effectuates the right back-linking steps, updating all the changes in data.
- a mesh node and a coordinate setting event. The mesh node will always hold the VRML representation of the edited mesh, and as the coordinates of the mesh change due to user actions, the coordinate setting event is used to update the VRML geometry that is visualized in the virtual environment.
- selector setting event. An event that resembles the coordinate setting event, has the same function, only it acts not on vertices, but on the positioning fields of boxes that are acting as selectors. As the coordinates change, the sensor boxes must also change to provide an accurate representation for the location of mesh vertices.
- a node that contains all the generated VRML code. The VRML code is generated when a data file is loaded, or the mesh changes complexity by adding rings or columns. In this latter case, vertices and selectors are added to the data and the VRML code is regenerated.
- other less important events such as those for removing the generated geometry (before the new geometry is transmitted to the VRML component), centering the mesh, scaling the mesh (to always provide a complete and maximal view of the mesh), or translating the mesh in the right position (also for viewing purposes).
- events to transfer vertex selection statuses (a single vertex, a ring of vertices or a column of vertices can be selected at a

time) for the Java component to know which vertices are the target vertices at all times.
- events to handle the editing unit selection process, which sets the magnitude of the editing operations.
- events to handle the rotations generated by the rotation widget. Although the rotation widget has only effect on the VRML environment itself, it is used in the Java component as a trigger for object positioning, as the rotation event is the most probable operation a user would perform after mesh modifications. It is needed as mesh modifications could cause partial occlusions of the edited mesh surface by exiting the viewable area.

The applet starting method (which also executes every time the applet gets focus, e.g. when the user returns to the applet containing window) contains all the code that makes the connections possible between the VRML and Java components. Connections are made for all the nodes and events enumerated above. To each event there is a serial number assigned, which makes it possible for a callback method to identify the source event and depending on what this event's function is, it handles the event itself, or in case the event is more complicated, it triggers the method responsible for handling that particular event.

Simpler events that are handled within the callback method are related to the selection of vertices. The selection events are of three types, a single vertex selection, a ring of vertices selection and a column of vertices selection event. For instance, if a single vertex is selected, the serial number of the vertex is stored in a variable (previous vertex selections are discarded) and all the editing buttons of the VRML component are set active. In the case of ring or column vertices selection, the selection variables are also set, but in this case only a part of the VRML buttons are set active, those effectuating ring operations in the case of ring vertices selection, and those operating on columns in the case of column vertices selection. All other events are handled by separate methods. These methods will be described in the next sections, together with the events to which they respond to.

File operations

The applet is initialized with three buttons. These are Java buttons, and they are the single Java GUI elements of the editor besides the load/save file dialog window. The buttons trigger functionalities related to file operations. These are the following functionalities: loading a data file, saving a data file and saving the data in VRML format (instead of the internally used format). The name of these buttons are "Load", "Save" respectively "Export to VRML", suggesting the functions they accomplish.

The first file handling method is the file reading (loading) method. If the data file is opened successfully, its content is read in character by character into a string variable and a parsing method is called to analyze the data. The parsing method extracts from the read in file the relevant data, and sets up the data parameters used by the Java component, like number of rings, number of columns and creates a dynamical parameter depending on the previous values to store the vertex coordinates (the number of vertex coordinates is seldom the same, therefore the variable holding this data is always created dynamically). At the end, the method calls the VRML code generation method described next, sets up initial values for a number of parameters for data consistency, and calls VRML-Java connection initializing methods for the newly generated VRML content.

The remaining two file operation methods deal with the storage of the edited mesh: one method for saving the data in the described internal format, and one method to export the mesh into VRML format (building a separate VRML file from the mesh), ready to be used.

The save method opens a window to specify the desired name and directory of the file to be saved, then it constructs the data file format and populates with the required data, extracting the data from the variables stored in the same Java component.

The last method, exporting onto VRML format resembles the previous method, only there is extra data to be generated beside the geometry (header, settings nodes), and the vertex data must be accompanied by facets indexing data, extracted also from data variables stored inside the Java component.

Generating VRML code

The VRML code creation method is the most interesting method of the Java component. It is responsible for creating the mesh geometry, the selection geometry and sensors, the script code that handles the selection process internally in the VRML component, and the events used for communicating with the Java component. It does all this by creating a single string packed with code, and sending it to the VRML component to be integrated in it. We call this VRML code dynamic because it is generated on the fly (multiple times during the editing process, when the mesh complexity changes) and it is modified constantly, with every operation committed. The dynamic code generation process executes the following tasks:

a) Creating the indexes for the mesh geometry. These indexes specify the polygons of the mesh. Since the indexes are generated automatically and not stored within the data file, it explains the reason why a regular grid of vertices is used, allowing a logical connection between the vertices, namely every item (cell) of a grid made from the vertices can be covered algorithmically by two triangle facets.

b) Creating an empty vertex coordinate node, that acts as a container for the vertex data. It is filled after the data file is parsed (also when the VRML code is re-generated), and modified later as user editing actions are performed.

c) Creating the mesh geometry string itself, an IndexedFaceSet VRML node, instantiating the data from the vertex coordinate container and using the previously created triangle indexes.

d) Next is the creation of sensors, handles and a script with functions (events) to manage the properties and events of vertex operations. After the script header is declared and stored in a string along with outgoing events that transmit selection data back to the Java component, sensors, handlers and script functions will be created for **each** of the vertices from the mesh. To be able to generate all the necessary code for the editing environment in a single pass over the array of vertices, separate string variables were used to store the generated nodes, fields, functions and routes which are concatenated after all the necessary code is generated. This task is executed in the following steps:

- First, the sizes of the sensor boxes are calculated. The exact calculation method is described in the section 'Geometry operations'.
- Next, a box is created with the calculated size, and a touch sensor is assigned to it. The boxes are created 90% transparent, and become half-transparent when the mouse is rolled over. The event receivers from the touch sensor are also declared, and their code is produced next.

- The functions that receive the sensor-over and sensor-pressed events from the sensor boxes manage the appearance of the same sensor boxes. If the pointing device is over a sensor box (e.g. close to a vertex) then the box will become half-transparent and if the pointing device exits the sensor area, the box will become transparent. Other (previous) selections are not affected, they remain unchanged. If the box sensor is activated with the input device (mouse click) then the sensor box becomes opaque, other selections will be removed and the vertex assigned to the activated sensor will become the only selection. The script is designed to be able to transmit back to the Java component information regarding the selection actions after it is placed into the VRML environment.

Note: to be able to perform ring and column operations, selection of a set of vertices that construct rings and columns also must be possible. Therefore, ring and column sensors are set up next to the vertex sensors with two iterations that go through the vertex data ring-wise respectively column-wise, described in the remaining steps of the dynamic VRML code generation process:

- A ring iteration creates the code that handles the vertex grouping into rings. The code generated contains a line-set geometry, which connects the vertices of the ring together, reusing the same coordinates the mesh uses. A touch sensor is connected to the line-set, with the same behavior as described in the case of box sensors, with the difference that the line-set sensor triggers also the behaviors of the box sensors assigned to the vertices contained within the line-set, meaning that if a line is selected, the boxes along that line are also selected.
- The column iteration has the same functionality and structure as the ring iteration, only the vertices are grouped based on columns instead of rings.
- Generating two general functions, used by all the sensor functions (there are vertex, line or column sensors). These functions are executed every time a sensor is activated and their concern is to remove the previous selections (making them transparent) and to make the new selections visible.
- The last script function generated is the initialization function, which has the task that in the case the code is created anew, it "recalls" the previous selection, and makes the selection also visible with the new code.

User actions

Next, there are the methods that handle the editing actions of the user. They are executed as the user activates the buttons described in the VRML component. Part of them act on the vertex coordinates stored in the Java component and they send the new data back to the VRML component, while the others give commands even to recreate the whole dynamic part of the VRML component, as in case of change in complexity.

Geometry operations

To be able to provide an effective manipulation environment, the Java component has to implement geometry operations that aid in the positioning and editing of the mesh, providing for instance the rotation center of the mesh, sensor box sizes, etc. We will enumerate the geometry operations below.

The most frequently used method is a method which catches the displacement event transmitted from the VRML component. In function of the current selection, applies the displacement to the data stored in the Java component (the displacement can be applied to a single vertex or to a set of vertices) and sends that data back to the VRML component. It also sends the same displacements to the sensor box(es) assigned to the selected vertex or vertices.

The size of a sensor box assigned to a certain vertex depends on the distance of the vertex from its neighbors. This way, sensors do not overlap, and remain scaled according to the complexity and the size of the edited mesh geometry. Sensor boxes are regenerated when mesh complexity changes, when the whole editing VRML code is regenerated.

Another method of the Java component is a mesh positioning method. The importance of this method lies in providing an interface where the visibility of the edited mesh is maximal, being entirely visible and without occluding by rotation the other elements of the interface. It scales and translates the geometry to a convenient position.

The ring adding method inserts a new ring in order after the selected ring (or after the ring containing the selected vertex), this in case that the selection is not the last ring, when no action is performed. The coordinates for the new vertices are calculated as the mean values of the vertices from the neighboring rings, and the dimension of the vertex data variable is changed to be able to hold the new values along the old values. Then, the dynamic code generation method is called, with all other methods necessary after such a step: vertex data setting, connections setting with the new code and mesh positioning.

The column addition method is essentially the same, only the inserted vertex data must be handled in a different order. The difference between the two methods lies in the structure of the vertices, and since after the last column comes again the first one, there is no restriction in adding a column after the last column like in the case of adding rings.

Smoothing the rings is another method assigned to a VRML button element. If activated, it calculates the center of the current ring (a ring selection or the ring containing the selected single vertex can be conceived as the current ring) and depending on the distance of vertices from the center and from the mean center distance, the vertex values are increased or decreased to achieve smoother surfaces. The new data is transmitted to the VRML component.

The ring scaling method affects the vertices of the current ring. For every vertex, distances from the center position are calculated in the horizontal (XZ) plane, and the scale value (the current editing unit is used for scaling as well) is distributed to the vertex X and Y values proportional to the previous values, such as the resulting vector from the center point overlaps with the original vector.

Column scaling is different in functionality from ring scaling. While ring scaling acts locally, only affecting one ring of vertices, column scaling affects all vertices of the mesh. It can be used for enlarging the distance between the ring vertices, thus longer/taller meshes can be obtained. A column center coordinate is calculated (from the current column) and depending on the distances of the current column vertices from the center, each ring is displaced proportionally by a certain calculated fraction of the scale value.

Updating the VRML scene

There are two types of update operations of the VRML scene, as we could see from the previous sections. The first one (which is the most time-consuming operation) consists of generating the whole mesh and control structure VRML code, and linking it to the VRML container node. This method is executed only when a data file is loaded or the structure of the geometry changes (rings or columns are added to the geometry). The second type of operation is an update operation, that can have as its target the mesh vertices itself, the mesh position or the sensor positions. These latter operations are real-time operations.

Java Component Properties

Some numbers: the simplest mesh (2 by 3 vertices) requires 2*3+2+3=11 sensors (box sensors, ring sensors and column sensors), each having two sensor functions, that is already 22 functions in the generated script code, not counting the few general functions, and the generated sensor geometry. As the size of the geometry increases, the number of functions generated increases proportionally. This might seem as too much for a simple mesh, but this complexity influences only the performance of the VRML code creation process, not the performance of the visualization and editing processes (it takes up more memory, but the extra code executes sequentially, having a minimal impact on visualization).

Regular mesh structure is important for the automatic mesh generation possibility, as well as for the ability to map higher level editing possibilities to the mesh (ring, column operations).

The vertices (single or in group) are displaced by sending new coordinate values to the coordinate container from the VRML component, without needing another VRML code generation process.

6 Symmetry

The editor was continuously tested and modified during the developing process to eliminate undesirable effects or enhance its usability. First, the construction of selection boxes was adapted to abort the production of static boxes in favor of controls scaled to the complexity of meshes, thus eliminating control boxes overlap, and permitting greater visibility of the mesh. Next, to enhance mesh size control, the editing unit selection widget was developed, allowing precisions between one millimeter and one meter. This allows the editing of objects with different magnitudes, approximately from an apple to an iceberg (but also keep in mind that all of this applies to relatively simple, structured objects).

Another enhancement made is regarding the speed of the editor. As the generated VRML code tends to be lengthy (this appears when a complicated mesh is loaded or when mesh complexity is increased and the VRML code is again generated), sectioning the code string variables into multiple pieces increases the system's performance. The sectioning of the initial string into three smaller string variables produced a doubling in the editor's speed with regard to VRML code generating.

After all these enhancements, the system still was missing something in functionality. Trying to construct different objects, finally it turned out the missing ingredient. Either it was an apple, a vase or a chess figure mesh that was created, it had symmetry, which had to be introduced by hand to the mesh, making the user's task unnecessarily complicated. To eliminate this overhead with symmetric objects modeling, the system was modified to provide this symmetric editing functionality. The mesh, according to this approach, is edited only on one side of the symmetry plane. This method gives even at the first glance the following advantages:

- Less control points. By using less control points, the size of the generated VRML code drops also, thus gain in speed can be achieved.
- Using one side of the object for editing, the other side can be used for preview purposes, as selections are not visible on that side.
- As a preview side is available, the edited side can use a semi-transparent appearance for non-selected controls, giving more comprehensibility to the mesh structure.
- Finally, if one side follows automatically the changes made to the other side of the mesh, the result will be a perfect symmetry, without the user having to worry about this.

For this functionality, changes had to be implemented to the methods from the Java component that are responsible for mesh editing and generating. These method changes are described next. Overall, the most important changes made are related to the indexes of vertices. Since the number of control points diminishes, the cycles that handle the changes run less iterations, but increase in complexity. They need to handle inside them the editing actions of the selected vertex or vertices, as well as the editing that must happen at the symmetric vertex or vertices.

An important method that is modified is the VRML code generating method. First, the indexes for the mesh faces are handled separately for the two symmetric sides, resulting in symmetric facets, even in vertex ordering. Next, the control box sizes are calculated, not using this time the symmetrical data (which in this case is redundant, the symmetric vertices have the same properties), thus a little gain in execution speed arises. Control boxes also are reduced in number. They have to connect to the vertices from only one side of the mesh, and eventually to the vertices that are on the symmetry plane. The same is true for control rings and columns, control rings become half rings, and control columns are not created for symmetric columns. For not created controls, there is no need for functions that handle their interaction, and the interaction functions have a more restricted set of actions to perform.

Another change in this method (that has no connection with symmetry) is the displacement of control points embedded in the mesh surface to a little bit above the surface, giving more visibility to the mesh and the controls at the same time.

The mesh displacement handling method suffered also changes. In the case when a single vertex is selected and displaced, if that vertex has a symmetric counterpart, the inverse displacement (relative to the YZ plane) is applied to the counterpart also, otherwise just the vertex and its control is displaced. In case of (half) ring selection, the right displacements are applied to the symmetric half ring. In case of column selections, if the column has a symmetric column, symmetric changes are applied to both columns, and if there is no symmetric column (column is on the symmetry plane) then only the column is displaced.

Due to symmetry, the column adding method in this case actually means the addition of two columns, the column that was asked, and its symmetric counterpart. The method of inserting new vertices is the same as before, except that double as much vertices are added.

Contrary to the column adding method, the ring adding method needs no changes, as the rings used in this case are symmetrical, and the differences of this approach and the previous non-symmetrical one are resolved by the VRML code generating method.

The ring smoothing method was also modified to reflect conceptual changes. Center point calculations require fewer equations (due to symmetry) and fewer displacements must be calculated, the symmetric ones come automatically.

Other methods remain unchanged, or bear only little notational changes. Only the ring scale method is modified to reflect the new symmetric structure approach, applying calculated changes symmetrically.

The editor in its current state gives a high level of control over a loaded mesh, making possible precise editing, new mesh construction, instant visibility and it is web deployable. However, its functionality and manipulation ease can be improved with the enhancements described in section 8.

7   Examples

For illustration of the possibilities offered by the system, Fig. 3 presents a few examples of the models created using the system. Textures were not applied to this objects, to reproduce exactly the output of the system. However, a texture applying utility inside the system would enhance the appearance of models and the system's usability.

Fig. 3. *Example models.*

8 Enhancements

Here is a list of enhancements that can be introduced to the application. These enhancements are mentioned for completeness, and are ordered in no manner of preference or importance:

- Rotation widget is not precise enough, a different representation or modified functionality may function better.
- Crease-angle support for interactively setting the smoothness of the edited mesh.
- The possibility to delete rings and columns from the mesh. This would mean extra buttons in the VRML component and their corresponding methods in the Java component, methods that would have to adjust the stored data (remove a set of vertices from the data set) and call the VRML code generating method for the new data set.

9 Further work

The editor in itself provides reduced functionality regarding combined shapes. It is limited to single mesh editing. To be able to edit combined shapes, it needs a wrapper application and a coequal module that takes care of the complexity of combined shapes.

Primarily, the combined shapes that are aimed by further development are the ones conforming to the H-anim VRML humanoid proposal [6]. The proposal describes a virtual humanoid in terms of joints representing bone connections and segments representing body surfaces by building a hierarchy out of these segments. Similar approach to human modeling and animation based on segments and bone structure is presented in [1].

In our approach, the mesh editor is capable of handling single segments and there is clearly a need for a wrapper application which allows the selection of single segments in order for the mesh editor to be able to provide it's functionality. This way, the mesh editor would be able to modify one by one the appearance of combined, complex shapes that can be used in the virtual environments already developed and being developed by our group [9, 12].

But the mesh surface of H-anim models are only a part of the whole model. In such a circumstance, the hierarchy of the models would be static, not trivial to change. Thus, a coequal module which would handle the model's hierarchy would be beneficial to the editor, if not necessary taking in account that the H-anim humanoid proposal gives the possibility to use different sets/levels of the same hierarchy. Results of work in this direction can be found in [8].

10 Summary

The purpose of the work presented in this chapter is to provide an easy to use tool for the visual editing of three dimensional objects. This approach allows dynamic handling of data, as the visual representations are created from a small amount of actual data, and permits as well a speed-up of the editing process using the combination of low level editing with the possibility to select groups of target vertices for higher-level, faster modeling. As a result of using a combination of VRML, Java and EAI in the application, the editing process can be viewed instantly, in real-time, and it is deployable on the world wide web.

References

1. N.I. Badler, C.B. Phillips, and B.L. Webber. (1993) Simulating Humans: Computer Graphics, Animation, and Control. Oxford University Press.
2. C. Babski and D. Thalmann. (1999) A Seamless Shape For H-anim Compliant Bodies. Proc. Fourth International Conference on the Virtual Reality Modeling Language and Web 3D Technologies. 21-28. ACM Press.
3. P. Cubaud and A. Topol. (2001) A VRML-based User Interface for anOnline Digitalized Antiquarian Collection. Proc. Web3D 2001 Symposium. 51-59. ACM Press.
4. Y. Goto and A. Pasko. (2000) Interactive Modeling of Convolution Surfaces with an Extendable User Interface. EuroGraphics Short Presentations. 37-42.
5. H. Grahn, T. Volk and H.J. Wolters. (2000) NURBS in VRML. Proc. Web3D-VRML Fifth Symposium on Virtual Reality Modeling Language, 35-43. ACM Press.
6. Humanoid Animation Working Group, Web3D Consortium. (1999) H-anim: Specification for a Standard VRML Humanoid, version 1.1. http://www.h-anim.org/Specifications/H-anim1.1/
7. T. Igarashi, S. Matsuoka and H. Tanaka. (1999) Teddy: A Sketching Interface for 3D Freeform Design. Proc. SIGGRAPH. 409-416. ACM Press.
8. Sz. Kiss. (2002) 3D Character Modeling in Virtual Reality. Information Visualisation Conference. To appear.

9. M. Kragtwijk, A. Nijholt and J. Zwiers. (2001) An Animated Virtual Drummer. International Conference on Augmented, Virtual Environments and Three-Dimensional Imaging. Mykonos, Greece.

10. G. Maestri. (1999) Digital Character Animation 2, Volume 1: Essential Techniques. New Riders Publishing.

11. C. Marrin, SGI. (1997) External Authoring Interface Reference. http://www.graphcomp.com/info/specs/eai.html. Proposal for a VRML 2.0 Informative Annex.

12. A. Nijholt and H. Hondorp. (2000) Towards Communicating Agents and Avatars in Virtual Worlds. EuroGraphics Short Presentations. 91-95.

13. J.F. Richardson. (1998) VRML Based Behaviour Database Editor. Virtual Worlds First International Conf., Volume 1434 Lecture Notes in Artificial Intelligence, 49-62. Springer.

14. P. Volino and N. Magnenat-Thalmann. (2000) Virtual Clothing. Springer.

15. Web3D Consortium. (1997) VRML97: The Virtual Reality Modeling Language. http://www.web3d.org/technicalinfo/specifications/vrml97/index.htm

16. A. Wakita, M. Yajima, T. Harada, H. Toriya and H. Chiyokura. (2000) XVL: A Compact And Qualified 3D Representation With Lattice Mesh and Surface for the Internet. Proc. Web3D-VRML Fifth Symposium on Virtual Reality Modeling Language. 45-51. ACM Press.

Chapter 15

A Rational Spline with Inflection Points and Singularities Tests on Convex Data

A Rational Spline with Inflection Points and Singularities Tests on Convex Data

Zulfiqar Habib
Department of Mathematics & Computer Science, Graduate School of Science and Engineering, Kagoshima University, Kagshima 890-0065, Japan

Muhammad Sarfraz
Department of Information & Computer Science, King Fahd University of Petroleum and Minerals, KFUPM # 1510, Dhahran 31261, Saudi Arabia

A smooth curve interpolation scheme for convex data has been developed. This scheme uses piecewise rational cubic spline functions with both unit and chord-length parametrization test. The necessary and sufficient conditions on inflection points and singularities for a convex data have been presented. Shape preserving parameters are automatically generated. The degree of smoothness attained is C^2 which is more powerful than a previous C^1 method

1 Introduction

Spline interpolation is a useful and powerful tool for curve and surface design. Smooth curve representation, to visualize the data without undesirable inflection points and singularities (loops and cusps), is of great significance. In computer aided geometric design (CAGD) applications, it is often desirable to find the conditions such that a curve may or may not have cusps and inflection points. The detection of cusps and inflection points are addressed in [1, 4-7] and there is a considerable literature on numerical methods for generating shape preserving interpolation; see for example, [2, 3, 8-12]. We show, in this chapter, that curvature continuity and high accuracy can be achieved by a simple algorithm which is based on the geometric characterization of C^2 continuity. The objectives of this chapter are as follows:

- to obtain the distribution of inflection points and singularities on the planar rational Bezier cubic curve.

282

- to produce C^2 interpolant for high degree of smoothness which provides an automatic procedure to obtain the derivative parameters instead of the approximation choices. No additional points (knots) are required.
- to visualize the difference between unit and chord-length parametrization.

The rational spline curve representation is unique in its solution. The imposition of C^2 constraints give rise to a linear system which is solvable using an efficient tri-diagonal linear system solver. C-type and S-type shapes are preserved by single interpolant with no inflection point and one inflection point respectively. Also there is no singularity (loop and cusp) in both types of shape.

This chapter begins with a definition of the rational function in next section where the description of rational cubic spline, which does not preserve the shape of a convex data, is made. This section derives the constraints, on derivative parameters, which lead to a linear system of equations and seek for the existence and uniqueness of their robust solution. The distribution of inflection points and singularities are discussed in section 3. In section 4, we describe the generation of a C^2 spline which can preserve the shape of C-type and S-type data. This section also demonstrates the output of the scheme for various data sets in literature. Summary and concluding remarks are given in last section.

2 Cubic Spline with Shape Control

Let $F_i \in R^2$, $i = 1,...,n$, be a given set of data points at the distinct knots $t_i \in R$, where $t_1 < t_2 < ... < t_n$. Then a parametric C^1-piecewise rational cubic Hermit function $P: R \to R^2$ is defined by

$$P(t) \equiv P_i(t)$$
$$= \frac{(1-\theta)^3 F_i + \theta(1-\theta)^2 v_i V_i + \theta^2 (1-\theta) w_i W_i + \theta^3 F_{i+1}(= p(t))}{(1-\theta)^3 + \theta(1-\theta)^2 v_i + \theta^2 (1-\theta) w_i + \theta^3 (= q(t))} \quad (1)$$

where

$$V_i = F_i + h_i D_i / v_i, \quad W_i = F_{i+1} - h_i D_{i+1} / w_i$$
$$\theta = (t - t_i) / h_i, \quad h_i = t_{i+1} - t_i \quad \quad (2)$$

We use this Hermite function to generate an interpolatory parametric curve which preserves the shape of locally convex data. D_i denotes the tangent vector to the curve at the knot t_i. It can be noted that $P_i(t)$ interpolates the points F_i and the tangent vectors D_i at the knots t_i. The choice of parameters $v_i \geq 0$ and $w_i \geq 0$ ensures a strictly positive denominator in the rational cubic (1). Thus from Berstein Bezier theory, the curve lies in the convex hull of the control points $\{F_i, V_i, W_i, F_{i+1}\}$ and is variation diminishing i.e., the arc (1) cuts any given straight line no more often than does the polygonal arc $F_i V_i W_i F_{i+1}$, see [2]. The parameters v_i, w_i and D_i are to be chosen such that the C-type and S-type shapes are preserved by single interpolant with no inflection point and one inflection point respectively. Also there is a guaranty of no singularity in both types of shape.

2.1 Determination of derivatives

In most applications, the derivative parameters D_i are not given and hence it must be determined either from the given data or by some other means. In this article, they are computed exactly from the system of equations which arises after applying the splining constraints.

The smoothness of the interpolant (1) is C^1. This article has been attempted to achieve higher degree of smoothness to C^2. For this, it is required to impose the second derivative constraints at the knot positions of the interpolant (1). After some simplifications, the second derivatives at t_i and t_{i+1}, respectively, are as follows

$$P^{(2)}(t_i) = \frac{2}{h_i}\left[h_i w_i S_i - (v_i - 1)D_i - D_{i+1}\right]$$

$$P^{(2)}(t_{i+1}) = \frac{2}{h_i}\left[-h_i v_i S_i + D_i + (w_i - 1)D_{i+1}\right]$$

(3)

where $S_i = (F_{i+1} - F_i)/h_i$. The C^2 constraints

$$P^{(2)}(t_{i-}) = P^{(2)}(t_{i+}), \quad i = 2, 3, \dots, n-1$$

(4)

then lead to the consistency equations

$$h_i D_{i-1} + [h_{i-1}(v_i - 1) + h_i(w_{i-1} - 1)]D_i + h_{i-1}D_{i+1}$$
$$= h_{i-1}h_i v_{i-1}S_{i-1} + h_{i-1}h_i w_i S_i, \quad i = 2, 3, \dots, n-1$$

(5)

For given appropriate end conditions D_1 and D_n, system of equations (5) is a tri-diagonal linear system. This is also diagonally dominant for v_i, $w_i \geq 2$, $i = 2,3,...,n{-}1$ and hence has a unique solution for D_i. As far as the computation method is concerned, it is much more economical to adopt the LU-decomposition method to solve the tri-diagonal system (5). The solution is also bounded with respect to the v_i and w_i since, for v_i, $w_i \geq \alpha > 2$, it can be shown that

$$\max|D_i| \leq [\alpha/(\alpha - 2)]\max|S_i| \tag{6}$$

Therefore, we can conclude the above discussion in the following

Theorem 1 *For v_i, $w_i \geq \alpha > 2$, $\forall i$, the spline solution of interpolant (1) exists and is unique.*

The readers are referred to [9] for the detailed analysis and demonstration of the parametric representation of the rational spline method

2.2 Demonstration

Let us consider unit parametrization i.e., $h_i = 1$. For the demonstration of C^2 rational cubic curve scheme, we will choose shape parameters $v_i = 3 = w_i$ and the derivatives will be computed from the system of equations (5) to generate the initial default curve which is actually the same as a cubic spline curve. Further modification can be made by changing these parameters. Figures 2, 5 and 9 are the default curves with their curvature plots. It can be seen from curvature plots that the ordinary spline curve does not guarantee to preserve the shape. It is required to assign appropriate values to the shape parameters so that it generates a data of preserved shape.

One way, for the above spline method, to achieve the shape preserving interpolant is to play with shape parameters v_i's and w_i's, on trial and error basis in those regions of the curve where the shape violations are found. This strategy may result in a required display as can be seen in Figures 3 and 6 for the shape parameters selection in Tables 2 and 5, respectively. But this is not a comfortable and accurate way to manipulate the desired shape preserving curve. It is really time consuming and is not recommended even for practical applications.

Another way, which is more effective, useful and the objective of this article, is the automated generation of shape preserving curve. This requires an automated computation of suitable shape parameters and derivative values. To proceed with this strategy, some mathematical treatment is required to remove unwanted inflections and singularities, which will be explained in the following section.

3   Inflection Points and Singularities

In this chapter, we assume that the tangent vectors D_i and D_{i+1}, $i = 1,2,\ldots,n\text{-}1$, are not parallel. Note that S_i ($= (F_{i+1} - F_i)/h_i$) can be represented as $S_i = \lambda D_i + \mu D_{i+1}$ where (λ, μ) are easily determined from the given data, i.e., $(\lambda, \mu) = (S_i \times D_{i+1}, D_i \times S_i)/(D_i \times D_{i+1})$. Here "$\times$" stands for the two-dimensional cross product, $(x_0, y_0)^{\mathrm{T}} \times (x_1, y_1)^{\mathrm{T}} = x_0 y_1 - x_1 y_0$. We introduce an auxiliary plane with the coordinates λ and μ. For more detail readers are referred to [7].

Inflection points: Using $S_i = \lambda D_i + \mu D_{i+1}$ and $s = 1\text{-}t$, we obtain

$$q(t)^2 P'(t) = a(t) D_i + b(t) D_{i+1} \tag{7}$$

where

$$\left. \begin{aligned} a(t) &= s^4 - 2s\theta^3 - s^2\theta^2 w_i + \lambda s\theta(2v_i\theta^2 + 2w_i s^2 + 3s\theta + v_i w_i s\theta) \\ b(t) &= \theta^4 - 2s^3\theta - s^2\theta^2 v_i + \mu s\theta(2v_i\theta^2 + 2w_i s^2 + 3s\theta + v_i w_i s\theta) \end{aligned} \right\} \tag{8}$$

Taking derivative of (7) w.r.t. t, we get

$$q(t)^3 P''(t) = \\ \{a'(t)q(t) - 2a(t)q'(t)\}D_i + \{b'(t)q(t) - 2b(t)q'(t)\}D_{i+1} \tag{9}$$

For number of inflection points on (1), we set $P'(t) \times P''(t) = O$, $0 < t < 1$ which implies that $a(t)b'(t) - a'(t)b(t) = 0$. Putting $t = 1/(1+\sigma)$, $0 < \sigma < \infty$, we get

$$\left[(\mu w_i - 1)\sigma^3 + 3\mu\sigma^2 + 3\lambda\sigma + (\lambda v_i - 1)\right]\frac{q(\sigma)}{h_i} = 0 \qquad (10)$$

Since $q(\sigma)/h_i > 0$, therefore

$$(\mu w_i - 1)\sigma^3 + 3\mu\sigma^2 + 3\lambda\sigma + (\lambda v_i - 1) = 0 \qquad (11)$$

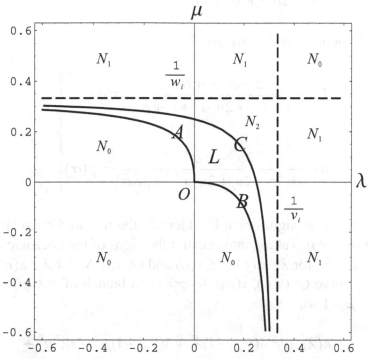

Fig. 1. *Distribution of inflections and singularities.*

The number of inflection points being equal to the number of simple positive roots of the cubic equation (11).

We have for N_i, $0 \leq i \leq 2$ representing the regions for which the curve has i inflection points and no singularity as shown in Fig. 1.

(a) $\lambda \geq 1/v_i$, $\mu \geq 1/w_i$: $(\lambda, \mu) \in N_0$

(b) $(\lambda - 1/v_i)(\mu - 1/w_i) < 0$ or $\lambda = 1/v_i$, $\mu < 1/w_i$ or $\lambda < 1/v_i$, $\mu = 1/w_i$: $(\lambda, \mu) \in N_1$

(c) $\lambda < 1/v_i$, $\mu < 1/w_i$: Descartes' Rule of Signs implies that the number of the positive roots of (c) is either zero or two, counting any double root twice. Remark that (λ, μ) is on the boundary between these cases if a double root occurs. At a positive double root σ, the cubic equation (11) and its derivative must vanish, which gives two equations that are linear in λ and μ

$$\lambda(3\sigma + v_i) + \mu(3\sigma^2 + w_i\sigma^3) = 1 + \sigma^3$$
$$\lambda + \mu(2\sigma + w_i\sigma^2) = \sigma^2$$

(12)

Solving for λ and μ, we get:

$$\left.\begin{array}{l} \lambda = \dfrac{2 - \sigma^3 + w_i\sigma}{3\sigma + 2v_i + 2w_i\sigma^2 + v_iw_i\sigma} = x(\sigma) \\[1.5em] \mu = \dfrac{2\sigma^3 + v_i\sigma^2 - 1}{\sigma(3\sigma + 2v_i + 2w_i\sigma^2 + v_iw_i\sigma)} = y(\sigma) \end{array}\right\}$$

(13)

Thus it is straightforward to identify the required boundary: $(\lambda, \mu) = (x(\sigma), y(\sigma))$. Taking into account the signs of the coefficients of (11), $(\lambda, \mu) \in N_0$ for $\lambda = x(\sigma)$, $\mu \le y(\sigma)$ and $(\lambda, \mu) \in N_2$ for $\lambda = x(\sigma)$, $\mu > y(\sigma)$. The curve C: $(x(\sigma), y(\sigma))$, $0 < \sigma < \infty$ is a branch of $k(\lambda, \mu) = 0$ limited by $\mu < 1/w_i$, $\lambda < 1/v_i$

$$k(\lambda, \mu) = 4(w_i\mu - 1)\lambda^3 + 4(v_i\lambda - 1)\mu^3 - 3\lambda^2\mu^2 +$$
$$[(v_i\lambda - 1)(w_i\mu - 1)]^2 - 6(v_i\lambda - 1)(w_i\mu - 1)\lambda\mu$$

(14)

$k(\lambda, \mu) = 0$ has two straight lines $\lambda = 1/v_i$ and $\mu = 1/w_i$ as its asymptotic lines. So region N_0 contains the boundaries A and B.

Singularities: A loop occurs if $P(\alpha) = P(\beta)$ for $0 < \alpha < \beta < 1$. Since D_i and D_{i+1} are independent, letting the coefficients of the two vectors in $\{P(\alpha) - P(\beta)\}$ be zero gives

$$\lambda = \{-(1-\alpha)^2(1-\beta)^2 + \alpha\beta(\alpha + \beta - 2\alpha\beta)$$
$$+ w_i\alpha\beta(1-\alpha)(1-\beta)\}/X$$

288

$$\mu = \{(1-\alpha)(1-\beta)(\alpha + \beta - 2\alpha\beta) - \alpha^2\beta^2 \\ + v_i\alpha\beta(1-\alpha)(1-\beta)\}/X \tag{15}$$

where

$$X = \beta^2(1-\alpha)^2 + \alpha\beta(1-\alpha)(1-\beta) + \alpha^2(1-\beta)^2 + v_i\alpha\beta(\alpha + \beta \\ - 2\alpha\beta) + v_iw_i(1-\alpha)(1-\beta)\alpha\beta + w_i(1-\alpha)(1-\beta)(\alpha + \beta - 2\alpha\beta) \tag{16}$$

Hence, we consider the image of (λ, μ) by (15) under $0 < \alpha < \beta < 1$ to get the necessary and sufficient conditions for the existence of the loop on (1). First the image of the boundary of the region determined by inequalities $0 < \alpha < \beta < 1$, is given by

(i) $\quad \alpha = 0, 0 < \beta < 1 \Rightarrow \mu^2 = \lambda(w_i\mu - 1)$

(ii) $\quad 0 < \alpha < 1, \beta = 1 \Rightarrow \lambda^2 = \mu(v_i\lambda - 1)$

(iii) $\quad 0 < \alpha = \beta < 1$ or $\alpha = \beta = 1/(1+\sigma), 0 < \sigma < \infty \Rightarrow (\lambda, \mu) = (x(\sigma), y(\sigma))$

Jacobian matrix of (λ, μ) w.r.t. (α, β) is nonsingular for $(\alpha, \beta) = (1/(1+c), 1/(1+d)), 0 < d < c$, as follows

$$\{w_icd(c+d) + (1+v_iw_i)cd + v_i(c+d) + (c^2 + d^2)\}^3 (\lambda_\alpha \mu_\beta \\ - \lambda_\beta \mu_\alpha) = (d-c)\{(1+c)(1+d)\}^2 (c^3 + v_ic^2 \\ + w_ic + 1)(d^3 + v_id^2 + w_id + 1)(< 0) \tag{17}$$

A cusp occurs if $P'(\alpha) = 0$ for $0 < \alpha < 1$. Letting the coefficients of D_i and D_{i+1} in $P'(\alpha)$ be zero give

$$\left. \begin{array}{l} (1-\alpha)^4 - 2(1-\alpha)\alpha^3 - (1-\alpha)^2\alpha^2 w_i + \lambda(1-\alpha)\alpha Y = 0 \\ - 2(1-\alpha)^3\alpha + \alpha^4 - (1-\alpha)^2\alpha^2 v_i + \mu(1-\alpha)\alpha Y = 0 \end{array} \right\} \tag{18}$$

where

$$Y = 2v_i\alpha^2 + 2w_i(1-\alpha)^2 + 3(1-\alpha)\alpha + v_iw_i(1-\alpha)\alpha \tag{19}$$

Then, $\alpha = 1/(1+\alpha),\ 0 < \alpha < \infty \Rightarrow (\lambda, \mu) = (x(\sigma), y(\sigma))$

Theorem 2 [7] *Assume that $S_i = \lambda D_i + \mu D_{i+1}$ with $D_i \times D_{i+1} \neq 0$. Then, Fig. 1 gives the distribution of inflections and singularity on the curve of the form (1) with respect to (λ, μ), where (i) $N_i,\ 0 \leq i \leq 2$, represent the regions for which the curve has i-inflection points and no singularity, (ii) C (or L limited by A, B, C) means the region for the curve to have a cusp (or a loop) and no inflection point. The region N_0 contains the boundaries A and B; and N_1 contains the two straight lines: $\lambda = 1/v_i,\ \mu < 1/w_i$ and $\lambda < 1/v_i,\ \mu = 1/w_i$.*

4   Preserving Local Convexity

The rational spline method, described in section 2, has deficiencies as far as shape preserving issue is concerned. It does not preserve the shape of the convex data. Curvature plots in Figures 2 and 5 clear that the curve is not preserving the shape of the data. It is required to assign appropriate values to the shape parameters so that it generates a data of preserved shape. So we will automate computation of suitable shape parameters and derivative values. In the light of discussion in previous section we can set $P'(t) \times P''(t) > O,\ 0 < t < 1$, for convex curve. This implies, $v_i > 1/\lambda_i$ and $w_i > 1/\mu_i$. A number of choices of the values of v_i and w_i can be adopted for graphical demonstration but for visually very pleasant results, We propose to choose

$$v_i = \max(3, 1 + 1/\lambda),\quad w_i = \max(3, 1 + 1/\mu) \tag{20}$$

as shape parameters which implies that $\lambda > 1/v_i,\ \mu > 1/w_i$, i.e., (λ, μ) $\lambda > 1/v_i,\ \mu > 1/w_i \Rightarrow (\lambda, \mu) \in N_0$

Proof (i) $3 > (\lambda+1)/\lambda,\ (\lambda > 1/2)$; then $v_i = 3$, from which follows: $\lambda > 1/v_i$. (ii) $3 < (\lambda+1)/\lambda,\ (\lambda < 1/2)$; then $v_i = (\lambda+1)/\lambda$, from which follows: $\lambda - 1/v_i = \lambda - \lambda/(\lambda+1) = \lambda^2/(\lambda+1) > 0 \Rightarrow \lambda > 1/v_i$. Similarly, $\mu > 1/w_i$.

For the case $\lambda\mu < 0$ (S-type data), the Bezier points V_i and W_i lie on the opposite sides of the line joining F_{i+1} and F_i. By the variation diminishing property, the curve segment will have a single point of inflection. In the case $\lambda\mu = 0$, the shape preserving parameters v_i, w_i $\rightarrow \infty$ and the curve segment degenerates into a line segment. When tangent vectors are parallel, if D_i and D_{i+1} possesses the same direction, then the curve segment should have only one inflection point. If D_i and D_{i+1} possesses opposite direction, then the curve segment $P_i(t)$ is convex. Since $D_i \times D_{i+1} = 0$, so from (20), we see that the parameters $v_i = 3 = w_i$.

4.1 Computation and demonstration

This section will discuss the computational aspects of the scheme and then design an algorithm to practically demonstrate the results. Some robust and automated method, for the determination of suitable derivatives, is one of the fundamental requirements of the scheme under discussion. To proceed with this strategy, Let us substitute the sufficient conditions for convexity from equation (20) into (5), we get $D_{i-1} + a_i D_i + D_{i+1} = b_i$, $i = 2,3,\ldots,n-1$, where $a_i = 2 + 1/\lambda_i + 1/\mu_{i-1}$ and $b_i = (1+1/\lambda_{i-1})S_{i-1} + (1+1/\mu_i)S_i$. Given D_1 and D_n, the set of equations (18) gives a system of $n-2$ linear equations for the unknowns $D_1, D_2,\ldots,D_{n-1}$. This is diagonally dominant tri-diagonal system and hence has a unique solution. Thus, we have the following

Theorem 3 *Given the end conditions D_1, D_n, constraints in (20) are sufficient to provide a unique spline which preserves the shape of a convex data.*

Although any end conditions D_1, D_n will work for the scheme, however three-point difference approximations $D_1 = (3S_1 - S_2)/2$ and $D_n = (3S_{n-1} - S_{n-2})/2$, which depend on the given data, have been found to produce visually pleasing results. Figures 4 and 7 demonstrate the effect of convexity preserving condition. Their curvature plots show high degree of smoothness at knots. The corresponding automatic outputs of λ, μ and shape parameters, for the shape preserving curves in Figures 4 and 7, are given in Tables 3 and 6, respectively. Figures 8 and 10 are demonstration of chord-length parametrization to compare it with examples of unit parametrization in Figures 7 and 9 respectively. Next, is an example of closed convex shown in Figure 11 with its curvature plot.

Table 1. *Data set.*

i	1	2	3	4	5	6	7
x_i	-4	-4	-2	0	2	4	4
y_i	5	0	-3	-3	-3	0	5

$$(v_i = 3 = w_i)$$

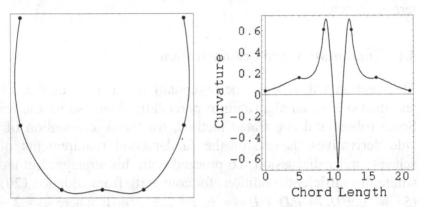

Fig. 2. *The default rational cubic spline (left) for data in Table 1 with its curvature plot (right).*

Table 2. *Suitable shape parameters for the data set in Table 1.*

I	1	2	3	4	5	6
v_i	3	3	15	15	3	3
w_i	3	3	15	15	3	3

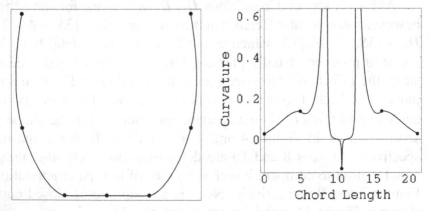

Fig. 3. *The rational cubic shape preserving spline (left) for the data in Table 2 with its curvature plot (right).*

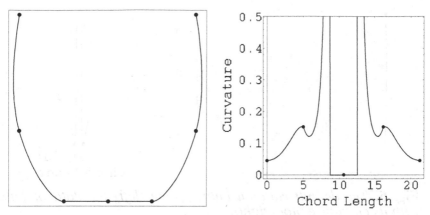

Fig. 4. *The rational cubic shape preserving spline (left) for the data in Table 2 with its curvature plot (right).*

Table 3. *Calculated shape parameters for the data set in Table 1.*

I	1	2	3	4	5	6
λ_i	0.5251	0.5251	0	1.0612	0.6076	0.4404
μ_i	0.4404	0.4404	1.0612	0	0.5407	0.5251
v_i	3	3	∞	∞	3	3.2708
w_i	3.2708	3.2708	∞	∞	3	3

Table 4. *Data set.*

I	1	2	3	4
x_i	0	0	1	10
y_i	10	1	0	0

$$(v_i = 3 = w_i)$$

293

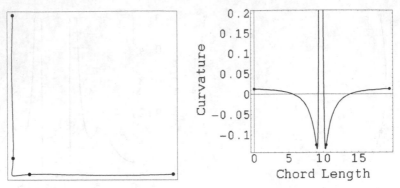

Fig. 5. *The default rational cubic spline (left) for data in Table 4 with its curvature plot (right).*

Table 5. *Suitable shape parameters for the data set in Table 4.*

I	1	2	3
v_i	3	4	3
w_i	3	4	3

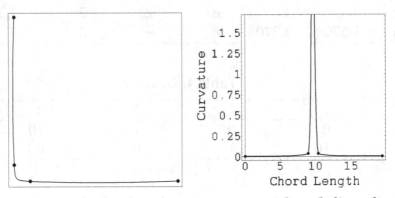

Fig. 6. *The rational cubic shape preserving spline (left) to the data in Table 5 with its curvature plot (right).*

Table 6. *Calculated shape parameters for the data set in Table 4.*

i	1	2	3
λ_i	4	0.2439	-10
μ_i	-10	0.2439	4
v_i	3	5.1	3
w_i	3	5.1	3

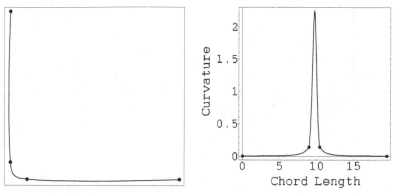

Fig. 7. *The rational cubic shape preserving spline curve (left) to the data in Table 6 with its curvature plot (right).*

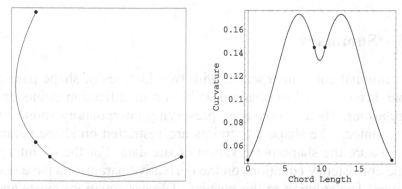

Fig. 8. *The rational cubic shape preserving spline curve (left) using chord-length parametrization with its curvature plot (right).*

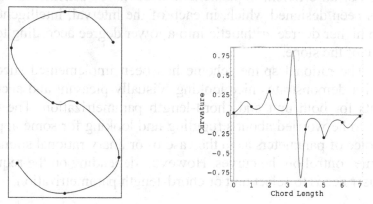

Fig. 9. *The default rational cubic spline (left) using unit parametrization with its curvature plot (right).*

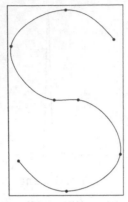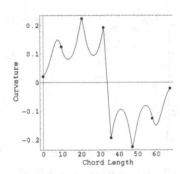

Fig. 10. *The default rational cubic spline (left) using chord-length parametrization with its curvature plot (right).*

5 Summary

A rational cubic interpolant, with two families of shape parameters, has been utilized to obtain distribution of inflection points and singularities. Then C^2 convexity preserving interpolatory spline curve is presented. The shape constraints are restricted on shape parameters to assure the shape preservation of the data. For the C^2 interpolant, the consistency equations on the derivative parameters have been derived. The solution to this system of linear equations exists and provides a unique solution. A robust solution, using the LU decomposition method, has been recommended.

For an efficient implementation of the C^2 scheme, an algorithm has been designed which, in each of the interval, intelligently saves the higher degree arithmetic into a lower degree according to the nature of the slope.

The rational spline scheme has been implemented successfully and it demonstrates nice looking, visually pleasant and accurate results for both unit and chord-length parametrization. The user has not to be worried about struggling and looking for some appropriate choice of parameters as in the case of ordinary rational spline having some control on the curves. However, depending on the requirement, a user needs to select unit or chord-length parametrization.

Acknowledgements

The authors would like to thank Prof. Manabu Sakai for helpful and valuable comments led to a considerably improved representation of the results. The authors also acknowledge the financial support of Monbusho (Ministry of Education, Japan) as well as King Fahd University of Petroleum and Minerals under Project No. FT/2001-18.

References

1. deBoor, C., Hollig, K. and Sabin, M. (1987) High accuracy geometric Hermite interpolation. Computer Aided Geometric Design, **4**, 269-278.
2. Farin, G. (1993) Curves and Surfaces for Computer Aided Geometric Design – Practical Guide. Academic Press, San Diego.
3. Habib Z. and Sakai M. (2002) G^2 two-point Hermite rational cubic interpo-lation. International Journal of Computer Mathematics **79**(11), 1225-1231.
4. Li, Y. M. and Cripps, R. (1997) Identification of inflection points and cusps on rational curves. Computer Aided Geometric Design, **14**, 491-497.
5. Sakai M. and Usmani R. (1996) On fair parametric cubic splines. BIT **36**, 359-377.
6. Sakai, M. (1997) Inflections and singularity on parametric rational cubic curves. Numer. Math., **76**, 403-417.
7. Sakai, M. (1999) Inflection points and singularities on planar rational cubic curve segments. Computer Aided Geometric Design, **16**, 149-156.
8. Sarfraz, M. (1992) Convexity preserving piecewise rational interpolation for planar curves. Bull. Korean Math. Soc., **29**(2), 193-200.
9. Sarfraz, M. (1992) Interpolatory rational cubic spline with biased, point and interval tension. Computers & Graphics, **16**(4), 427-430.
10. Sarfraz, M., Hussain, M. and Habib, Z. (1997) Local convexity preserving rational cubic spline curves. Proceedings of IEEE International Conference on Information Visualization-IV'97-UK: IEEE Computer Society Press, USA, 211-218.
11. Sarfraz M. (2002) Automatic outline capture of arabic fonts. Journal of Information Sciences, Elsevier Science Inc. **140**(3), 269-281.
12. Sarfraz M. (2002) Fitting curves to planar digital data. Proceedings of IEEE International Conference on Information Visualization IV'02-UK: IEEE Computer Society Press, USA, 633-638.

Section III

Systems & Tools

Chapter 16

Automatic Curve Fairing System using Visual Languages

- **Introduction**
- **Quantitative Analysis Method and Visual Language**
- **Key-line Extraction and Data Fitting by GA Spline**
- **Simulation**
- **Summary**
- **References**

Automatic Curve Fairing System using Visual Languages

Toshinobu Harada
Department of Design and Information Sciences, Wakayama University, 930 Sakaedani Wakayama, Japan

Fujiichi Yoshimoto
Department of Computer and Communication Sciences, Wakayama University, 930 Sakaedani Wakayama, Japan

Generally, in the field of industrial design, measurement data of a clay model are faired, and its three dimensional model is made in CAD. However, it is a problem that this fairing work take a long time and require a great deal of labor. We propose an automatic fairing system of a curve for the application to reverse engineering technology. The system is constructed by the following five steps. 1) Key-line data are extracted from a clay model by a 3D-digitizer. 2) The extracted key-line data are approximated by a spline with a genetic algorithm. 3) The spline curve (key-line) is divided into curves of monotone curvature. 4) "Characteristic" of the curve is analyzed for each curve of monotone curvature. 5) Each curve of monotone curvature is substituted by a visual language based on the result of the analysis mentioned above, and each key-line is reconstructed by the visual language. By this system, we can easily obtain an aesthetic curve intended by a designer in a short time.

1. Introduction

An important theme at present not only in the manufacturing industry (i.e. the production of product model designs) but also in the entertainment industry (i.e. the CG production of movies, TV programs, and CMs), is how to efficiently create three-dimensional (3D) models with a CAD system. CAD systems are occasionally used to directly create more realistic 3D models, if the shapes of the models are simple. However, the most general approach is to make an actual model like a clay model. In this approach, a layout machine (contact type measurement device) or a 3D-digitizer (non-contact type measurement device) makes measurements, and data of

302

high accuracy are obtained by fairing. "Fairing" is work done to correct the curvature vectors of ogee parts in the same normal direction, since measured curves do have ogee parts unintended by the designer from the measurement errors, etc. However, it is a big problem of fairing that it takes a long time and requires a great deal of labor.

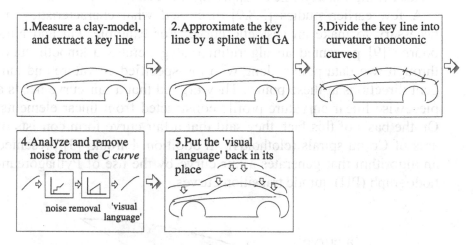

Fig.1. *Flow chart of the automatic fairing of a key-line.*

In this chapter, we propose an automatic curve fairing system using visual languages. This system is composed of the following five parts (see Figure 1).

1) Extraction of key-line data from a clay model by a 3D-digitizer. Here, key-line means a crucial line or an accent line that rules characteristic of a curved surface.

2) Approximation of the extracted key-line data by a spline curve obtained with a genetic algorithm.

3) Division of the spline curve (key-line) into curvature monotonic curves.

4) Analysis of the "Characteristic" of the curve for each curvature monotonic curve. Here, the "Characteristic" of the curve is called *C curve*.

5) Substitution of each curvature monotonic curve by a visual language [7] based on the result of the analysis mentioned above, and reconstruction of each key-line by the visual language. Here, the visual language is an elemental curve having an aesthetic rhythm (The details are given later.).

We generically call the operations from 1) to 5) "automatic fairing of the key-line". Moreover, the range of this research encompasses the development of the automatic fairing technology. As a result, we are now able to obtain the curve (key-line) that a designer intends in an extremely short period of time compared with past fairing work. In addition, we believe that we can also create an aesthetic curved surface from this key-line by applying the technology.

A few earlier studies [3,8,9] discussed what characteristics are necessary to make an aesthetic curve. For example, Pal and Nutbourne [9] presented an algorithm that generates a smooth curve through two data points along with the specified curvature and tangent directions at these points. They argued that a fair curve needs a piecewise linear curvature profile constructed from linear elements. On the basis of this fact, they said that a fair curve form consists of arcs of Cornu spirals (clothoids). In addition, Farouki [3] presented an algorithm that generates a fair curve by the use of Pythagorean-hodograph (PH) quintic transition curves.

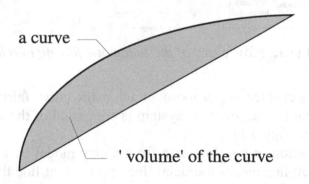

Fig. 2. *Volume of a curve.*

In this chapter, we develop a method for analyzing what characteristics are necessary to make an aesthetic curve. This method uses a relation between the length frequency of the curve and the radius of curvature in a log-log coordinate system. By using this method, we analyze many sample curves on products (drafts) made and drawn by designers. As a result, we clarify that there are a lot of curves unable to be represented by only arcs of Cornu spirals and PH spirals, and also that the designer controls the curvature changes with self-affine properties. On the basis of this finding, we develop an algorithm that generates curves having self-affine properties. Then, by using this algorithm, we get not only 'decelerating' curves like spirals but also 'accelerating' curves with the same parameters. To date, there have been no algorithm for getting 'accelerating' curves and 'decelerating' curves uniformly. We also get a new criterion for the fairness of a curve with self-affine properties.

2. Quantitative Analysis Method and Visual Language

2.1 Quantitative Analysis Method

In this section, we give an outline of a quantitative analysis method for the characteristics of a curve [5]. We assume that the curve treated in our study, satisfies the following four conditions:
1) plane, 2) open, 3) non-intersecting, and 4) curvature monotonic.

We analyzed the design words expert car designers use to find what characteristics of a curve they pay attention to. We found some of the words to express the characteristics of a curve. Some words were concerned with curvature changes and the volume of a curve from the viewpoint of mathematics. Here, the term 'volume' is defined as the area bound by the curve and the straight line binding the starting point and the ending point of the curve (see Figure 2). Therefore, we define that the characteristics of a curve indicate its curvature changes and volume in this study. In fact, when the curvature changes and volume are defined, the curve is fixed reversely.

We developed a method to analyze curvature changes and the volumes of curves mathematically, simultaneously, and intuitively. In this method, we first interpolate a sample curve using a Bézier curve on a computer. Then, we express the relation between the radius of curvature at every constitutional point on the interpolated curve and the "length" of the curve showing the total length of segmental curves corresponding to this radius of curvature in a log-log coordinate system. This relation is called the "logarithmic distribution diagram of curvature". Now, let us explain it in detail in the following.

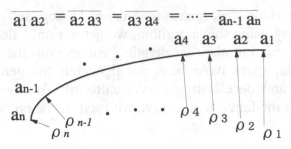

Fig. 3. *A method of extracting constitutional points.*

First, let us denote the total length of the curve by S_{all}, the length of a segmental curve by s_j, the radius of curvature at a constitutional point a_i by ρ_i, and the interval of the radius of curvature by $\overline{\rho_j}$ (the units are mm in all cases.). The radius of curvature ρ_i at constitutional point a_i on the curve is obtained by extracting constitutional points ($a_1, a_2, ..., a_n$) at equal intervals as shown in Figure 3 (e.g., at S_{all} =100mm, the constitutional points were extracted at 0.1mm intervals [in actual dimension], and therefore, $n = 1000$), and by calculating the respective radius of curvature ($\rho_1, \rho_2, ..., \rho_n$) at the respective constitutional points.

Second, let us denote the interval of the radius of curvature $\overline{\rho_j}$ by the interval corresponding to the quotient obtained by dividing the common logarithm [-3 , 2] of the value ρ_i / S_{all} [0.001, 100] by 100 equally (determined by surveying the range of curves adopted into actual cars). In other words, $\overline{\rho_m}$ = [-3+0.05(m-1), -3+0.05m] (m is an integer between 1 and 100, i.e., $1 \leq m \leq 100$).

Third, we sum the numbers of occurrences of the common logarithm values of $\rho_1 / S_{all}, \rho_2 / S_{all}, \cdots, \rho_n / S_{all}$ in each interval of $\overline{\rho_j}$. From this value ($= N_j$), we calculate the length of segmental curve s_j (= distance between neighboring constitutional points $\times N_j$) in which $\overline{\rho_j}$ appears. This means that $S_{all} = s_1 + s_2 + ... + s_{100}$. In addition, we define the "length frequency" $\overline{s_j}$[= $\log_{10}(s_j / S_{all})$] representing the ratio of the length of the segmental curve to the total length of the curve S_{all}.

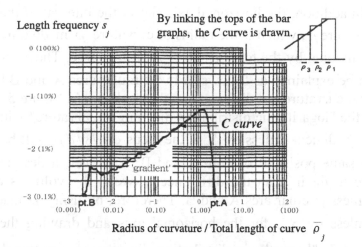

Fig. 4. *Logarithmic distribution diagram of curvature.*

The "logarithmic distribution diagram of curvature" defined above can be obtained by taking $\overline{\rho}_j$ for the horizontal axis and $\overline{s}_j$ for the vertical axis, as shown in Figure 4. In this figure, the numeric values of ρ_i / S_{all} (magnification) and s_j / S_{all} (%) are additionally shown in parentheses for convenience of understanding. To draw such a

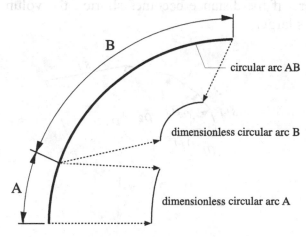

Fig. 5. *Two dimensionless circular arcs A and B.*

"logarithmic distribution diagram of curvature" means to mathematically obtain the locus of $d\overline{s} / d\overline{\rho}$ in terms of the interval of the radius of curvature $\overline{\rho}$ and the length frequency $\overline{s}$.

In addition, the horizontal axis shows the interval of the radius of curvature $\overline{\rho}$, which is the radius of curvature ρ made dimensionless by dividing by the total length S_{all} of the curve. The reason for this can be explained as follows. If two circular arcs A and B having the same curvature but different lengths as shown in Figure 5 are shown on the "logarithmic distribution diagram of curvature" without making ρ dimensionless by S_{all}, the locus of $d\overline{s}/d\overline{\rho}$ will be shown at the same position for both A and B. However, a designer will be able to distinguish the difference between the volumes of dimensionless circular arcs A and B. Therefore, by using ρ made dimensionless by S_{all} for the horizontal axis and drawing the locus of $d\overline{s}/d\overline{\rho}$, "logarithmic distribution diagram of curvature" positions can be made to differ according to the total lengths of the curves, even if they have the same curvature. The difference between the volumes of both curves is shown visually.

In this "logarithmic distribution diagram of curvature", the way the curvature changes is shown by the locus of the *C curve*, as shown in Figure 4, and the volume of the curve is shown by point A and the distance between points A and B as shown in the same figure. Here, if the distance becomes shorter, the volume of the curve becomes larger.

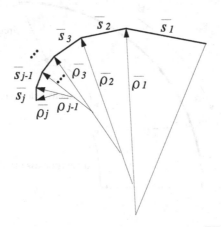

Fig. 6. *Relation of $\overline{\rho}_j$ and $\overline{s}_j$.*

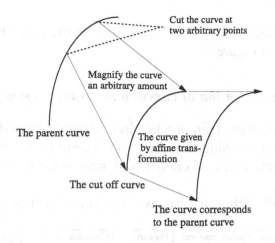

Cut the curve at
two arbitrary points

Magnify the curve
an arbitrary amount

The parent curve

The curve given
by affine trans-
formation

The cut off curve

The curve corresponds
to the parent curve

Fig. 7. *Self-affine property.*

Depending on the volume of the curve, if the curve is interpolated by Bézier curves, the exact value of the volume can be obtained directly by integrating the interpolated curve. Therefore, this value is used to obtain an accurate value of the volume. In the quantitative analysis described in this chapter, however, the definition of the volume mentioned above is used as an index of the relative changes of the volume, showing the volume changes among the sections when the plural cross sections of a curved surface are to be analyzed simultaneously.

Furthermore, the "gradient" of the C *curve* in Figure 4 is defined by

$$\text{"gradient"}=d\bar{s}/d\bar{\rho}=dY/dX=\lim_{\Delta x \to 0} \Delta Y/\Delta X$$

$$=\lim (Y_{j-1}-Y_j)/(X_{j-1}-X_j)=\lim (\bar{s}_{j-1}-\bar{s}_j)/(\bar{\rho}_{j-1}-\bar{\rho}_j).$$

In this chapter, the term "gradient" means the gradient after transforming the coordinate system to an X-Y rectangular coordinate system with the horizontal axis representing $X=\bar{\rho}_j$ and the vertical axis representing $Y=\bar{s}_j$. When the "gradient" is a, the relation of the interval of the radius of curvature $\bar{\rho}_j$ and the length frequency $\bar{s}_j$ is defined by

$$\lim (\bar{s}_1-\bar{s}_2)/(\bar{\rho}_1-\bar{\rho}_2)=\cdots =\lim (\bar{s}_{j-1}-\bar{s}_j)/(\bar{\rho}_{j-1}-\bar{\rho}_j)= a$$

(as shown in Figure 6).

Here, if the "gradient" is constant, then the curve has a self-affine property (see Figure 7).

2.2 Systematization of curves and visual languages

We analyzed more than one hundred sample curves drawn by expert car designers by using this quantitative analysis method, and studied what characteristics of curves designers control to make aesthetic curves.

As a result of this analysis, we could classify these curves into five

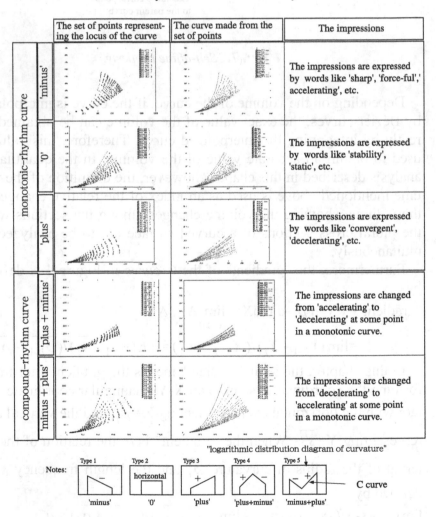

Fig. 8. *The systematization of curves and their impressions : the five basic types of curves.*

typical types. We adopted the automobile in this study, because it is known that the appearance of a car is important for potential buyers, and hence, car designers are very sensitive to curves.

Therefore, we could systematize curves for car design on the basis of our results above. Concerning the impressions of these curves, we confirmed that they differed according to what the "gradient" sign was (i.e., 'plus' or 'minus' or '0') mainly by analyzing many samples. In fact, the impressions of two curves were the same irrespective of the "gradient", if the "gradient" signs of both curves were 'plus'.

Consequently, we could systematize and classify curves into the five types shown in Figure 8. Figure 8 also shows the impressions of curves. Here, we shall call a curve whose *C curve* has one straight line a 'monotonic-rhythm curve' and a curve whose *C curve* has two straight lines a 'compound-rhythm curve'. We also developed a method to get any curve (any set of points representing the locus of a curve) among the five types having any volume, mathematically [6]. The curves shown in Figure 8 were made by this method. We were unable to abstract a *C curve* with more than three straight lines of some constant "gradient". On the basis of this fact, we considered that these curves were not aesthetic. Actually, the curvature changes were complex.

Accordingly, we consider the aesthetic curves (limited to curvature monotonic curves) that designers aim to draw as belonging to one of the five types. Furthermore, we propose a new method for curve design, where the five types of curves shown in Figure 8 are used as 'visual languages'. From this, an aesthetic curve is made by using these 'visual languages'.

3. Key-line Extraction and Data Fitting by GA Spline

3.1 Shape Measurement by a 3D-digitizer and Extraction of a Key-line

For non-contact type shape measurements, there are active methods and passive methods. Inspeck3D-M (made by Inspeck Company, Canada) (see Figure 9), which was used in this research, is a measurement device that uses an active method. This method projects a fringe pattern light on the target object to be measured while shifting the fringe pattern light where a halogen lamp is the light source, and

Fig. 9. *Inspeck 3D-M (left).*

then takes in a picture of four sheets. Next, the fringe of this picture of four sheets is analyzed and a distance value is computed. The "phase shift method" is adopted in the extraction of this distance image.

The following five characteristics characterize the Inspeck3D-M.
1) A short measurement time (0.3sec).
2) A small influence on the target object to be measured.
3) A high resolution (the pitch=0.5mm).
4) Difficulty in measuring lustrous objects and black objects.
5) Inability to measure objects exceeding the camera frame.

The necessary data in this research include the section data that

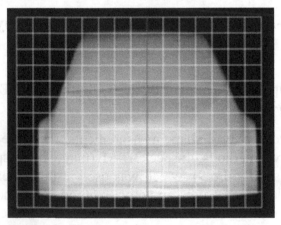

Fig. 10. *Lattice arrangement measurement.*

show the key-line of the target measurement object. However, the measurement data by Inspeck3D-M is output as numeric data in the lattice points composing the curved surface. Therefore, we have to arrange a row, with a set of lattice points, on the section we want to extract (hereafter, this is called the lattice arrangement measurement). Concretely speaking, the position of the target measurement object and that of Inspeck3D-M are determined such that the section we want to extract consists of the row with a set of lattice points, as shown in Figure 10. Accordingly, the curved surface data of the target measurement object is measured.

Next, the method of extracting the section curve we want to analyze from the measured data, that is, the key-line, is described. First, we extract the locus of the curve as a set of points forming the key-line by performing the lattice arrangement measurement. Figure 11 shows the extracted locus of a curve by a set of points from the side direction.

3.2 Data Fitting Using a Spline Curve with a Genetic Algorithm

3.2.1 What is Data Fitting Using a Spline Curve?

Splines are piecewise polynomials constructed by polynomial pieces joined together [1, 10-22]. By using splines, we can obtain smooth

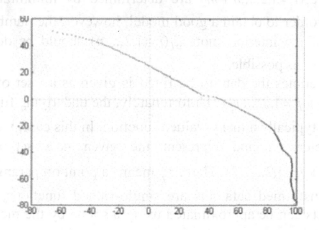

Fig. 11. *A set of points forming a key-line (side view).*

313

and flexible geometric representations. Let us assume that the data to be fitted is given in the interval $[a,b]$ of the x-axis and is written by

$$F_j = f(x_j) + \varepsilon_j (j = 1,2,..., N). \tag{1}$$

In this equation, $f(x)$ is the underlying function of the data (unknown), and ε_j is the measurement error. Let $\xi_i (i = 1 - m, 2 - m,..., n + m)$ be the knots of a spline for data fitting, where n is the number of knots $\xi_i (i = 1,2,...,n)$ located in the interval (a,b) and m is the order (degree + 1) of a B-spline $N_{m,i}(x)$. At the ends of the interval $[a,b]$, we set

$$\left. \begin{array}{l} a = \xi_{1-m} = \cdots = \xi_0, \\ b = \xi_{n+1} = \cdots = \xi_{n+m}. \end{array} \right\} \tag{2}$$

In this setting, the model function for $f(x)$ is given by a spline

$$S(x) = \sum_{i=1}^{n+m} c_i N_{m,i}(x), \tag{3}$$

where c_i is a B-spline coefficient [1].

Equation (3) is fitted to the data given by equation (1) by the method of least squares. Then, the sum of the squares of residuals Q is written by

$$Q = \sum_{j=1}^{N} w_j \left\{ S(x_j) - F_j \right\}^2, \tag{4}$$

where w_j is the weight of the data and $N > n + m$. B-spline coefficients $c_i (i = 1,2,...,n + m)$ are determined by minimizing equation (4). In order to obtain a good model, however, the number and locations of the interior knots $\xi_i (i = 1,2,...,n)$ should be determined as precisely as possible.

Sometimes the data to be fitted is given as a set of plane data, $(x_j, y_j)(j = 1,2,..., N)$. Unfortunately, the underlying function of the data is typically a many-valued function. In this case, we introduce a parameter t, and represent the given data set by (t_j, x_j), $(t_j, y_j)(j = 1,2,..., N)$. Here, t_j means a point on parameter t. Since the transformed data sets are single-valued functions, each of the data sets can be approximated using a spline by the method of least squares.

3.2.2 Segmentation of a Curve Using Data Fitting with a Genetic Algorithm

As we mentioned above, in order to obtain a good approximation using a spline, we have to place the knots involved as precisely as possible. This naturally demands dealing with the knots as variables, and we also have to solve a multivariate and multimodal nonlinear optimization problem [2]. Therefore, it is difficult to obtain the global optimum.

Recently, Yoshimoto et al. proposed a method to solve this problem using a genetic algorithm (GA) [26,27]. In [27], by using real numbers as genes, they treat not only data with a smooth underlying function but also data with an underlying function having discontinuous points and/or cusps. In this chapter, we call the method GA spline fitting. The algorithm is as follows.

Step 1: Input the interval of fitting $[a, b]$ and the data to be fitted given by equation (1).

Step 2: Input the degree of the approximating spline and the control parameters of the GA.

Step 3: Create an initial population by using random numbers.

Step 4: For each individual, compute the data fitting with knots made by real-number genes, and obtain the fitness value e. Save the best individual, i.e., that giving the smallest fitness value.

Step 5: Do selection by using the fitness values and generate candidates for the individuals of the next generation.

Step 6: For each individual, compute the data fitting with knots made by real-number genes, and obtain the fitness value e. Select the best fitness e_{min} and obtain the average fitness $\bar{e}$.

Step 7: Do crossover and generate candidates for the individuals of the next generation.

Step 8: For each individual, compute the data fitting that uses real number genes as knots, and obtain the fitness value e. Select the best fitness e_{min} and obtain the average fitness $\bar{e}$. Save the best individual, i.e., that giving the smallest fitness value.

Step 9: Do mutation and generate a set of individuals of the next generation.

Step 10: For each individual, compute the data fitting with knots made by real-number genes, and obtain the fitness value e. Select the best fitness e_{min} and obtain the average fitness $\bar{e}$. Save the best individual, i.e., that giving the smallest fitness value.

Step 11: Select the better individual between the best individual of Step 8 and the best individual of Step 10. Save the better individual between the selected individual and the best individual of the last generation.

Step 12: Have we arrived at the final generation? If yes, output the best individual obtained so far and the result of the data fitting for the individual, and then stop the computation. Otherwise, go to step 5.

Here, the recommendable values of the control parameters mentioned in Step 2 are as follows. The order of splines: 4 to 6 (third to fifth degree splines), the number of individuals K: an even number between 50 and 150, the number of final generations: 100 to 300, the number of trials: 20 to 40, the minimum of the mutation rate: 0.01.

The algorithm generates random numbers distributed uniformly in the interval of data fitting (a,b) by L times. By permuting them in ascending order, it creates an initial individual. The initial population consists of a set of K initial individuals, where K is the size of the population.

As our fitness function, we use Bayes Information Criterion (BIC) [24]. For our data-fitting problem, this criterion is written as follows.

$$BIC = N\log_e Q + (\log_e N)(2n + m). \tag{5}$$

In this equation, N is the number of data, Q is given by equation (4), and $(2n + m)$ is the number of parameters in the model function (3). In the second term on the right-hand side of equation (5), $n + m$ and n stand for the number of B-spline coefficients and the number of interior knots, respectively. In equation (5), the first term means the fidelity of the model function to the data to be fitted, and the second term means the simpleness of the model function. Therefore, BIC can balance the fidelity with the simpleness. By using BIC, we can choose the best model among the candidate models automatically. The value of BIC is called *fitness*.

As our method of crossover, we use a two-point crossover approach. That is, we cut the interval of fitting $[a, b]$ at two points determined randomly, and exchange the genes between the cut points for two individuals. Here, the numbers of the genes exchanged are not equal in general, but we can perform a crossover without any problem. For mutation, we use a method that inserts or deletes a gene (knot) in the interval of fitting $[a, b]$ by some probability. Here, the insertion and deletion probabilities are the same. The crossover and mutation probabilities are determined adaptively by the method of Srinivas et al. [25].

By using the above method, we can perform data fitting and segmentation of measurement data from a clay model of a car.

3.2.3 Computation of the Values of an Approximating Function and its Curvature

In geometric design, we frequently need both an approximating function $S(x)$ and its curvature. Let the curvature be denoted by k. Then, the curvature of $S(x)$ is given by

$$k = S''(x)/|1 + S''(x)^2|^{3/2}. \tag{6}$$

Since B-splines have a local support property, we can compute the value of $S(x)$ by

$$S(\bar{x}) = \sum_{j=i}^{i+k-1} c_j N_{k,j}(\bar{x})(\xi_{i-1} \le \bar{x} < \xi_i). \tag{7}$$

Moreover, the m-th derivative of $S(x)$ can be computed by

$$S^{(m)}(\bar{x}) = \sum_{j=i}^{i+k-1-m} c_j^{(m)} N_{k-m,j}(\bar{x}) \quad (\xi_{i-1} \le \bar{x} < \xi_i). \tag{8}$$

Here,

$$c_i^{(0)} \equiv c_i \quad (i = 1,2,...,n+k) \tag{9}$$

$$c_i^{(j)} \equiv (c_{i+1}^{(j-1)} - c_i^{(j-1)})/(\xi_i - \xi_{i-k+j})/(k-j)$$
$$(i = 1,2,...,n+k-j; j = 1,2,...,m). \tag{10}$$

Using equation (8), we can compute the first and second derivatives of $S(x)$. By substituting the values into equation (6), we obtain the curvature k of $S(x)$.

If we have the plane curve mentioned in section 3.2.1, we can obtain its curvature by

$$k = \{x'(t)y''(t) - y'(t)x''(t)\}/\sqrt{(x'(t)^2 + y'(t)^2)^3}. \tag{11}$$

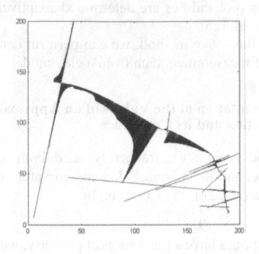

Fig. 12. *Result of data fitting by GA spline.*

3.2.4 Result of GA Spline Fitting

In Figure 12, a result of GA spline fitting for the measurement data of the key-line in Figure 11 is shown. The key-line was obtained from a clay model of a car. The curvature profile of the approximating curve is also given in Figure 12. From the curvature profile, we can obtain curvature monotonic curves for use in the analysis in chapter 4.

4. Simulation

4.1 Application to the key-line

In this section, a clay model 1/4 the size of a car is actually measured, the key-line in the 0X section (the central line of the car) is extracted, and an experiment using following five steps is performed.

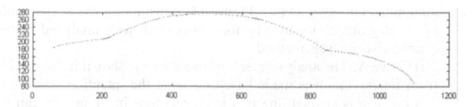

Fig. 13. *A set of points forming a key-line (side view of a clay model).*

1) Measurement of a clay model

The 1/4 size clay model created by a certain automaker was actually measured by the "lattice arrangement measurement" using Inspec 3D-M mentioned in section 3.1. Here, since the clay model was of a size exceeding the frame of the camera, it was measured after being divided into three parts: a bonnet part, a windshield part, and a roof part. Figure 13 is the set of point data of the key-line obtained by the actual measurement.

2) GA spline fitting

GA spline fitting was performed for the set of point data. Consequently, the curve shown in Figure 14 was obtained.

3) Division to curvature monotonic curves

First, the curvature of a curve obtained by GA spline fitting was computed. Next, a search was made for the inflection points and feature points of this curve on the basis of the curvature (■ mark in Figure 14). Here, the feature points of the curve are the points for which the curvature changes from a reduction to an increase. Between these calculated inflections points or feature points are curvature monotonic curves.

4) Analysis of the "characteristics" in each curvature monotonic curve

The following results were obtained when the main curvature

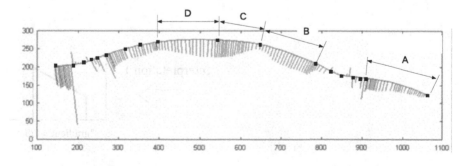

Fig. 14. *Curve obtained by GA spline fitting.*

319

monotonic curves A-D (see Figure 14) constituting the key-line 0X of the clay model obtained by this experiment were analyzed by the quantitative analysis method.

i) Curve A: The analysis result of this curve is shown in Figure 15. If an interpretation is made based on what the intention was when this curve was created, the two interpretations in Figure 15 can be considered. Interpretation 1 is an often seen logarithmic distribution diagram of curvature, when a parabola, etc., is analyzed, and it is obtained by the bonnet of a common car. Interpretation 2 is the result, when some single R tends to be connected and it proceeds to approximate a curve like a parabola. Here, interpretation 1 is chosen. Of course, the selection of the interpretation 2 is also possible. When a curve is generated using interpretation 1, it comes (see Figure 8) to have more of a convergent impression than interpretation 2.

ii) Curve B: The analysis result of this curve is shown in Figure 16. This curve is clearly made from a single R. The key-line of the windshield part is actually approximated in many cases by a single R to facilitate the modeling work. Generally speaking, moreover, a visual language with a certain "gradient" of plus is used for the key-line of the windshield part in many cases. Therefore, we temporarily

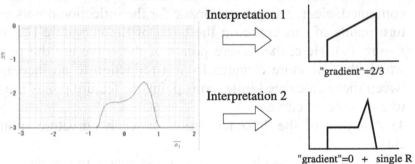

Fig. 15. *Analysis result of curve A.*

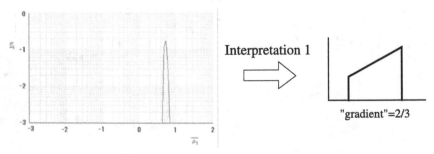

Fig. 16. *Analysis result of curve B.*

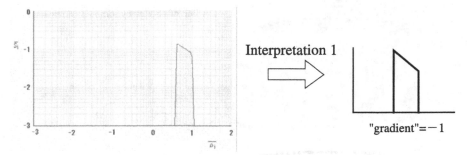

Fig. 17. *Analysis result of curve C.*

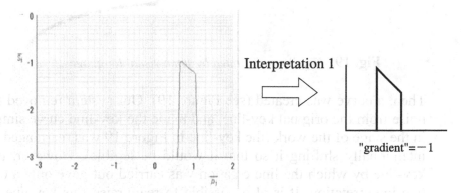

Fig. 18. *Analysis result of curve D.*

choose a visual language whose "gradient" is 2/3 in the meaning of correcting what was produced in the facilitation. Of course, we may choose to use the single R as it is.

iii) Curve C: The analysis result of this curve is shown in Figure 17. It is supposed that this curve is made by a curve whose "gradient" is -1. It is believed that a forceful and sharp impression of the roof results from this curve (see Figure 8).

iv) Curve D: The analysis result of this curve is shown in Figure 18. It is supposed that this curve is made by a curve whose "gradient" is −1, like curve C. That is, it is believed that a "sharp" and "forceful" impression of the roof also arises from this curve (see Figure 8).

5) Substitution to a visual language and rearrangement of each curvature monotonic curve

Using a curve creating system developed by writers on the basis of the above-mentioned interpretation results, the following items were input (into the system): "gradient" (selection of a visual language) in each curvature monotonic curve, volume, and position.

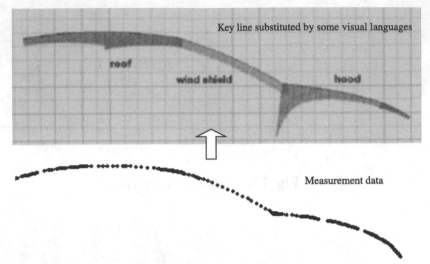

Fig. 19. *Key-line substituted by some visual languages.*

Then, a curve was created (see Figure 19). Our system removed the noise from the original key-line, and made the key-line curve simple in the stage of the work (the key-line in Figure 19 was rearranged by intentionally shifting it so that it would be legible). However, this key-line by which the line creation was carried out gave only a certain interpretation. It is also possible to reappraise this key-line, to reinterpret "characteristics" of the curve, by replacing it with new visual languages, and then carrying out the re-creation of the curve.

4.2   Application to the curved surface

In this section, a clay model 1/4 the size of a car is measured actually, key-lines in the section of roof's curved surface are extract, and an experiment is performed using substitution and rearrangement with visual languages. Furthermore, on the basis of key-lines rearranged, we generate a curved surface, and compare it with the curved surface made from a cluster of points measured directly.
1) Generation the curved surface by automatic curve fairing system
We extracted 7 key-lines in the section of roof's curved surface, and analyzed them (Figure 20). On the basis of this result, these key-lines were substituted and rearranged with visual languages. And the curved surface was made from the key-lines rearranged using "skin" command of Alias (Made by Wavefront Company, Canada).

322

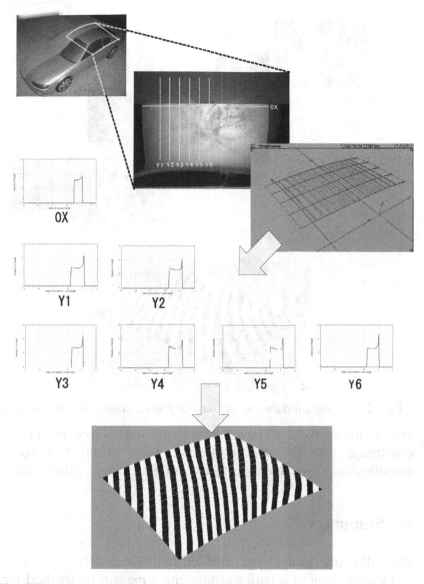

Fig. 20. *Generation the curved surface by automatic curve fairing system.*

2) Generation the curved surface from a cluster of points measured
We extracted a cluster of points of the roof's curved surface. And the curved surface was generated from the cluster of the points using "corners" command of Alias directly (Figure 21).

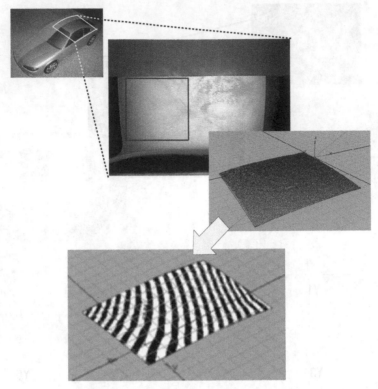

Fig. 21. *Generation the curved surface from a cluster of points measured.*

We compared two curved surfaces mentioned above. As a result, we confirmed that the curved surface with visual languages was smoother than the curved surface from the cluster of the points.

5. Summary

The following results were obtained in this research.
1) Development of a lattice arrangement measuring method for key-line extraction by a 3D-digitizer. A set of point data tracing a key-line can now be obtained easily.
2) Development of an algorithm for GA spline fitting for extracted key-lines. While obtaining a curve without ogee parts, i.e., parts not intended by the designer, this algorithm divides the curve into curvature monotonic curves easily.
3) Simulation of the automatic fairing of a key-line using a clay model, created by an automaker, 1/4 the size of a regular car. Consequently, the key-line was found to fit the GA spline and the curve

could be divided into a curvature monotonic curve. To analyze the "characteristics" of each curvature monotonic curve, visual languages were substituted. It was found that the key-line rearranged by the curvature monotonic curves replaced by the visual languages, was practical enough to use. Therefore, by using this automatic fairing system, it is possible to obtain curves designers intend in a short period of time compared with conventional fairing systems. In addition, we believe that this system can also be used to create beautiful curved surfaces from key-lines replaced by visual languages.

The following subjects remain as future work.

1) This time, a key-line was extracted using Inspeck3D-M. However, Inspeck3D-M was originally designed for measuring curved surface forms and extracting them as sets of point data. If it were possible to only measure a key-line, a 3D-digitizer could measure every constitutional point of a curve with a sufficient accuracy. In the future, the room for improvement category will include a form measuring method along with investigations on the measurement devices, etc.

2) Despite the many kinds of interpretations available on dividing a curve into a curvature monotonic curves, and analyzing the "characteristics" of each curvature monotonic curve after visual language substitution, it will be necessary to further acquire the know-how on which interpretation to choose.

3) In the case of rearrangement involving curvature monotonic curves replaced by visual languages, only the curve is arranged at present. However, if future utilization is considered, it will be necessary to inquire further about the connections between curves.

Acknowledgements

The authors would like to thank Mr. Kurihara and Mr. Nakashima for their collaboration with the simulation. This work was partly supported by a Grant-in-Aid for Scientific Research of the Japan Society for the Promotion of Science.

References

1. de Boor, C. (1978) A Practical Guide to Splines, Springer-Verlag, New York.
2. Dierckx, P. (1993) Curve and Surface Fitting with Splines, Oxford University Press, Oxford.

3. Farouki, R. T. (1997) Pythagorean-hodograph quintic transition curves of monotone curvature, Computer-Aided Design, 29, 601-606.
4. Goldberg, D.E (1989) Genetic Algorithms in Search, Optimization, and Machine Learning, Addison-Wesley.
5. Harada, T., Yoshimoto, F., Moriyama, M.(1999) An aesthetic curve in the field of industrial design, Proc. of IEEE Symposium on Visual Language, 38-47.
6. Harada, T., Mori, N., Sugiyama, K. (1994) Algorithm for creating a curve by controlling its characteristics, Bulletin of Japanese Society for the Science of Design, 41, 1-8 (In Japanese).
7. Kepes, G. (1994) Language of Vision, Paul Theobald and Co., Chicago.
8. Nutbourne, A. W., Martin, R. (1988) Differential Geometry Applied to Curve and Surface Design Volume 1:Foundations, Ellis Horwood Limited, 9-141.
9. Pal, T. K., Nutbourne, A. (1977) Two-dimensional curve synthesis using linear curvature elements, Computer-Aided Design, 9, 121-134.
10. Sarfraz, M, (2003), Weighted Nu Splines with Local Support Basis Functions, Advances in Geometric Modeling, Ed.: M. Sarfraz, John Wiley, 101 - 118.
11. Sarfraz, M. (2003), Optimal Curve Fitting to Digital Data, International Journal of WSCG, Vol 11(1).
12. Sarfraz, M. (2003), Curve Fitting for Large Data using Rational Cubic Splines, International Journal of Computers and Their Applications, Vol 10(3).
13. Sarfraz, M., and Razzak, M. F. A., (2003), A Web Based System to Capture Outlines of Arabic Fonts, International Journal of Information Sciences, Elsevier Science Inc., Vol. 150(3-4), 177-193.
14. Sarfraz, M, and Raza, A, (2002), Visualization of Data using Genetic Algorithm, *Soft Computing and Industry: Recent Applications*, Eds.: R. Roy, M. Koppen, S. Ovaska, T. Furuhashi, and F. Hoffmann, ISBN: 1-85233-539-4, Springer, 535 - 544.
15. Sarfraz, M, and Raza, A, (2002), Towards Automatic Recognition of Fonts using Genetic Approach, *Recent Advances in Computers, Computing, and Communications*, Eds.: N. Mastorakis and V. Mladenov, ISBN: 960-8052-62-9, WSEAS Press, 290-295.
16. Sarfraz, M., and Razzak, M. F. A., (2002), An Algorithm for Automatic Capturing of Font Outlines, International Journal of Computers & Graphics, Elsevier Science, Vol. 26(5), 795-804.
17. Sarfraz M. (2002) Fitting Curves to Planar Digital Data, Proceedings of IEEE International Conference on Information Visualization IV'02-UK: IEEE Computer Society Press, USA, 633-638.
18. Sarfraz, M. (1995) Curves and Surfaces for CAD using C^2 Rational Cubic Splines, Engineering with Computers, 11(2), 94-102.
19. Sarfraz, M. (1994) Cubic Spline Curves with Shape Control, Computers and Graphics 18(4), 707-713.
20. Sarfraz, M. (1993) Designing of Curves and Surfaces using Rational Cubics. Computers and Graphics 17(5), 529-538.

21. Sarfraz, M. (1992) A C^2 Rational Cubic Alternative to the NURBS, Computers and Graphics 16(1), 69-77.
22. Sarfraz, M. (1992) Interpolatory Rational Cubic Spline with Biased, Point and Interval Tension, Computers and Graphics 16(4), 427-430.
23. Schwarz G. (1978) Estimating the dimension of a model, The Annals of Statistics, 6, 461-464.
24. Srinivas M., Patnaik L.M.(1994) Adaptive probabilities of crossover and mutation in genetic algorithms, IEEE Trans. Systems, Man and Cybernetics, 24, 656-667.
25. Yoshimoto, F., Moriyama, M., Harada, T. (1999) Automatic knot placement by a genetic algorithm for data fitting with a spline, Shape Modeling International'99, IEEE Computer Society Press, 162-169
26. Yoshimoto, F., Harada, T., Moriyama, M. Yoshimoto, Y. (2000) Data fitting using a genetic algorithm with real number genes, Trans. of the Information Processing Society of Japan, 41, 162-169 (In Japanese).

Chapter 17

Hair Design Based on the Hierarchical Cluster Hair Model

Hair Design Based on the Hierarchical Cluster Hair Model

Tao Wang and Xue Dong Yang
Department of Computer Science, University of Regina, Regina, Saskatchewan
S4S 0A2 Canada

Modeling and rendering human hair is an important and challenging issue in computer graphics. The main difficulty comes from the large number of hairs, the fine geometry of individual hairs, and the complex illumination interactions among hairs and surrounding objects. In this chapter, an enhanced interactive hair styling system -- V-HairStudio, based on our previous work, is developed to design complex hair styles represented by a hierarchical hair model. This design tool provides a rich set of functionalities with multiple levels of abstraction. Due to the hierarchical representation, complex hair styles can be handled with reduced design efforts.

1 Introduction

Hair modeling and rendering is an important and rewarding problem, from both the practical and the theoretical points of view. It relates to many aspects of computer graphics technologies, such as object representation, shape design, rendering, and dynamic simulation. This makes the topic of hair modeling very attractive. The difficulties of hair modeling mainly come from the large number of hairs to be modeled, the finely detailed shapes of individual hairs, and the complex illumination interactions among hairs and surrounding objects. Existing hair models can be broadly classified into two categories: (1) explicit geometric models and (2) volume density models. The hierarchical model [14] falls into the second category.

1.1 Review of previous work

A number of geometric models have been used to model animal fur or human hair, for examples, the early pyramid mode formed from triangles proposed by Miller [8], the curved cylinders, approximated as a sequence of straight cylindrical segments used by LeBlanc et al. [7], and the trigonal prisms presented by Watanabe and Suenaga [12]. The main advantage of using explicit hair models is that the rendering power for geometric surfaces in today's advanced graphics workstations can be fully utilized. A main drawback of this type of technique is the severe aliasing problem. Another drawback is the enormous number of hair primitives to be handled, which not only consumes very large amount of storage space, but also has the associated high rendering costs. Furthermore, the problem of hair styling with explicit hair models is regarded as a formidable topic for the same reason [7].

Perlin and Hoffert [9] used a volume density model, called hypertexture, to create the soft fur-like object with strikingly natural looking. Kajiya and Kay [6] introduced the texel model, which is a generalization of 3D texture and has generated impressive Teddy bear images. While the texel is suitable for short straight hairs, the hypertexture can produce waving fur through turbulence functions.

A few efforts have also been made to construct hair styling systems, most of which are based on explicit models. Since volume models are primarily used for fur or short hair, there is no urgent need for sophisticated designing tools.

Watanabe and Suenaga [12] developed a hair modeling tool, based on their trigonal prism model. The tool contains two modes: wisp-modeling and hair-drawing. In the first mode, a set of parameters such as thickness, length, randomness, and color can be specified interactively. The hairs can then be created on a human head model in the second mode. However, the design can only be performed globally, implying that adjusting the parameters will affect all hairs. The system lacks the flexibility for editing individual wisps, which is required for detail refinement.

An interactive technique for modeling fur, but not for long and styled human hair, was presented by Gelder and Wilhelms [4]. In this system, the user selects a variety of parameters (for example, fur length, density, stiffness, and color properties) to achieve the desired fur appearance for a particular animal. The undercoat and overcoat may have separate specifications, and the randomness may be adjusted for a natural effect. The system provides four methods of drawing: single line, polyline, NURBS curve, and NURBS surface. A furry monkey example is demonstrated using this technique.

Chen em et al. [1] proposed a hair style editing tool (HET) to design hair styles based on the wisp model. This tool consists of five windows. The command window includes most of the command buttons. The view window is used to observe the hair in the 3D space. In the other three windows, the user can edit the 3D position and orientation of the wisps with a mouse. These wisps in the scene can be viewed from different viewing angles.

Daldegan et al. [2] and Daldegan and Thalmann [3] proposed a complete methodology and integrated system for hair modeling called HairStyler. It provides various interactive functions. A scalp surface is represented by a triangular patch and a single hair is represented by a 3D curved cylinder. Initially, all of the hairs placed on the triangular mesh have the same length, orientation, and symmetrical position. To make hair styles look more natural, these three parameters can be edited interactively. Random length, orientation, and position can be assigned to each set of hairs grown on the same triangular patch.

The commercial software Shag: Hair, developed by Digimation company, is a plug-in system for 3D Max and is used to add hair and fur to any 3D object [5]. Shag: Hair allows the user to draw out a number of individual "control" strands that will define the basic look of the hair or fur. Then it interpolates between the strands and the selected surface to create a complete style. In addition, Shag: Hair also has its own built-in dynamics engine for hair animation.

1.2 Motivation and main contribution of this research

In the field of hair modeling and rendering, a suitable hair model must first represent the necessary information of hairs (such as the geometry, distribution, position, and orientation of hairs). Ideally, a general purpose hair model should have the following properties:
- a compact and efficient representation;
- the ability to support interactive manipulation and allow easy design of complex hair styles;
- the flexibility to render photo-realistic or near photo-realistic hair images, which can be generated at multiple resolutions; and
- an efficient support of hair animation.

In an attempt to meet the above requirements, the hair model in our early work [15] was constructed based on the consideration that the model should have more than one level of abstraction. In this early model, hair shapes can be globally specified at the cluster level; the detailed geometry and distribution of individual hairs within a cluster is then defined at a lower level by an implicitly defined density function. The main idea is that users can control the styling of hair model easily at the cluster level. Though it is a significant improvement to some explicit models that deal with individual hairs separately, we realized that this model is effective only if the number of clusters is not too large. When the number of clusters is in the order of 100 (a sophisticated hair style often needs more than 70 clusters), manipulating the clusters individually can be tedious, and sometimes difficult. To overcome the above problem, a hierarchical concept is introduced in this chapter, which is an extension of our early work [14].

The concept of multi-level abstraction [15] is systematically generalized into a hierarchical structure: from the top to the bottom, the multiple levels are the hierarchical group levels, the cluster level, and the individual hair level. A large group is composed of several small groups, which can be recursively divided into a set of finer groups. Depending on the complexity of a hair style, the hierarchy levels can be adaptively chosen. At the lowest group level, a group consists of a set of clusters. The group at this level which directly contains clusters is called *base group*, and the level is called *base group level* correspondingly. However, at all other higher group levels, a group does not contain clusters directly. Instead, it only contains groups that are at the next lower level. By using this model, the structure of a complex hair style is hierarchically organized. As a result, the efficiency of styling is greatly improved; this is particularly noticeable when designing and modeling complex hair styles.

The hierarchical concept is also consistent with our daily observation. Although there are many individual hairs on a head, they are often grouped into large or small clusters, especially for permed and styled hair. Hairs within a cluster usually have similar shapes, but with small and random differences. Furthermore, some clusters may be close to each other and possess similar shapes (for example, as a result of combing). These clusters may compose a base group. Within a base group, the shapes and shading of clusters, though similar, may have variations. The groups close to each other may further form a larger group (corresponding to a higher group level in the hierarchical model). Usually, our visual perception is mainly based on the collective regional appearance of the clusters and groups, rather than the shape or placement of individual hairs. The design philosophy of the hierarchical model is that hair can be modeled at multiple levels of abstraction. In the modeling process, the shapes of hair styles can be manipulated at higher levels (the group levels and the cluster level) but mainly at the top group level. The detailed geometry and arrangement of individual hairs within a cluster is specified at the lowest level - the individual hair level - by an implicitly defined density function.

The hierarchical model possesses the capability to represent very complex hair styles. In order to fully utilize the advantage of the new model, a hair styling tool should be provided accordingly. Xu and Yang [13] developed a designing system for the cluster hair model. Based on their previous work, the system is extended to the hierarchical model [11].

This enhanced design tool provides a set of rich functionalities for interactive manipulation in designing complex hair styles at multiple levels of abstraction. Various head models can be imported and then modified. The light property may also be adjusted to achieve a satisfactory visual effect. With this system, the hierarchical model can be defined at multiple levels, and the complexity of hair styles subsequently increases with reduced design effort.

This chapter describes the enhanced hair modeling system based on the hierarchical model. A brief overview of the system is presented in section 2. Section 3 introduces the basic manipulations. Section 4 describes the hair group design. Section 5 focuses on the cluster axis curve, vertex and group editing. Section 6 describes the cross-section shape modification. Section 7 presents the head model editing. Light property editing is introduced in section 8. Section 9 gives the preview of hair and head models. The implementation issues along with the experimental results are discussed in section 10. Section 11 summarizes the contributions of this chapter and discusses topics for future work.

2 Overview of the V-HairStudio System

The whole system mainly contains two parts: hair modeling and hair rendering. V-HairStudio focuses on the hair modeling part in which hair groups and clusters can be created, edited, and arranged interactively on a head model. The head model consists of several components such as face, eyes, brows, and lips. Each part is represented by triangular meshes and can be edited individually. The head image with Z-buffer information is rendered using OpenGL functions.

In the hair rendering part, the projection-based sweeping technique is used to generate the final hair image. For the purpose of producing the natural hair shadow effect, the fuzzy shadow buffers generated for each light source are applied in the hair shading computation.

Figure 1 illustrates the structure of the hair modeling and rendering system.

The hierarchical model uses a set of large groups to represent a complex hair style, and each large group is further composed of small groups. At the base group level, several clusters with similar shapes constitute a base group. The user can define the group levels recursively according to the complexity of hair styles. Creating a complete representation of hair groups at higher levels and hair clusters at lower level is the central task. In V-HairStudio, the shape of a hair cluster is defined by a generalized cylinder, which contains two components: the axis curve and the cross-section contour shape. Within the generalized cylinder, a base density map defines the distribution of individual hairs on cross-sections along the axis curve. In addition, hair color and light properties have to be specified for the hair display. So, a hair cluster is completely represented by the following data sets: axis curve, cross-section contour shape, base density map, and hair color and light properties.

For the purpose of convenient styling, nine modules are provided in V-HairStudio:

- Light position, direction, and shading editing;
- Head loading, editing, and shading specification;
- Group creating and editing;
- Cluster axis curve drawing and editing;
- Cross-section shape editing;
- Base density map editing;
- Hair color and light property;
- Hair arrangement;
- Hair style preview.

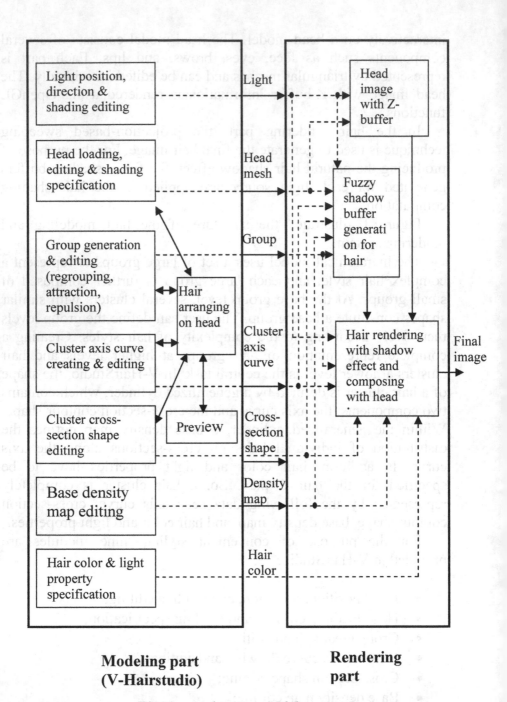

Modeling part (V-Hairstudio)   **Rendering part**

Fig. 1. *Architecture of the system*

The first two modules can render the image of a head model with Z-buffer values using GL functions. Modules three to seven are used to produce the data structures for hair groups and clusters.

Combined with the light and head mesh information provided by the first two modules, these data are used in fuzzy shadow buffer generation and hair image rendering. With the hair arrangement module, the groups and clusters can be fitted onto a head model interactively. To observe the global styling result during a design process, the hair style preview module can display the whole hair style (including both the hair and head models) in surface mode. The designer can then modify the unsatisfactory parts interactively by using the corresponding modules.

Figure 2 shows the main menu of the V-HairStudio.

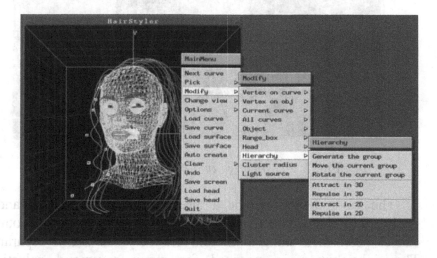

Fig. 2. *V-HairStudio main menu*

3 Basic Manipulations in 3D Space

Some primitive manipulations in 3D space are provided in V-HairStudio, for example, selection, insertion, and translation of control points of axis curves; translation, rotation, scaling, duplication, and deletion of clusters and groups; generation, regrouping, attraction, and repulsion of groups; and translation, rotation, scaling, and centralization of the head model.

The system supports two viewing modes in 3D space: one is the perspective view, and the others are the orthogonal projection views from x-, y- and z-axes, respectively. These two modes can be used simultaneously. Any operation performed in either the perspective view window or one of the orthogonal view windows will update the display in all other windows at the same time (Fig. 3).

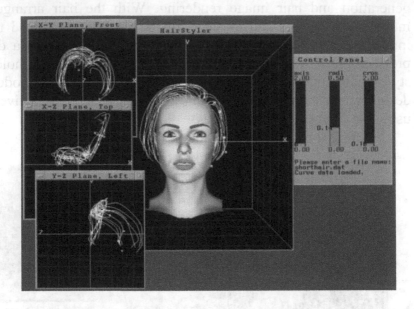

Fig. 3. *Perspective and orthogonal windows*

The input device can be a three-button mouse or a standard keyboard. In the perspective view window, the translation or rotation can be performed along or around the x-, y-, or z-axis separately. The right button is reserved for menu command selections. Combination of the buttons is used to distinguish operations along or around the x-, y-, or z-axis. The input operations by the keyboard keys will be discussed later.

A cubic reference frame is displayed in the perspective view. In order to provide an intuitive perception and accurate measurement of cursor position in the 3D space, three cross lines project the cursor onto the side walls of the cubic frame. The viewpoint can be rotated to different view directions, and be moved near or far so that objects can be zoomed in or out.

It is intuitive and easy to perform directly in the 3D perspective view. For example, a designer can quickly move the cursor to a place around a desired location. However, sometimes it is difficult to accurately position the cursor to the desired location, owing to the restriction of the perspective viewing angle itself. In this situation, the three orthogonal projections which are top, front, and side view windows, can be very useful. Accurate manipulations can be performed much more effectively in orthogonal views.

4 Hair Group Design

In the hierarchical model, hair is manipulated at multiple levels: the recursively-defined hierarchical group levels, the cluster level, and the individual hair level. The number of the hierarchical levels is determined by the complexity of a hair style. Even at different locations of a hair style, the level numbers do not have to be the same. That is, hairs with complex shapes and fine details may need more levels to be represented (such as very dense, curly, and randomly distributed hairs), while hairs with relatively regular shapes may require fewer levels. Our experiments show that in most cases, three or four levels are reasonably sufficient for the manipulation of relatively complex hair styles. However, in order to model very complex hair styles, the number of the hierarchical levels should be higher.

Because the implicitly defined density function supplies the geometric properties of individual hairs, explicit geometric representation of hair is avoided, thus making the hierarchical hair model very compact. The nature of the multiple levels of abstraction provides the capability of interactive design for sophisticated hair styles.

Figure 4 illustrates the concept of the hierarchical model, with the cluster groups and human head model shaded. The hair style has three hierarchical levels. It consists of a total of 71 clusters, divided into 10 base groups. These groups are located at different parts of the head model. During the designing process, 10 curves are created first, which are used as base curves for each group. Around each base curve, several curves are subsequently generated. The base curve and the generated surrounding curves form a group, and they serve as the axes of individual clusters of the group.

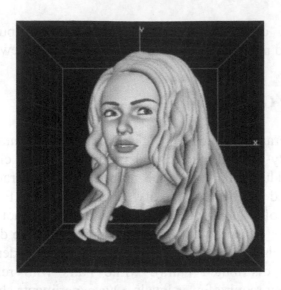

Fig. 4. *Illustration of the hierarchical model*

4.1 Generation of the Group

In the hierarchical model, a base group is composed of a set of similar clusters. Using V-HairStudio to generate a group of clusters, first one cluster is created as the base cluster by the user. Then, in the control panel, the number of clusters to be generated around the base curve (the axis curve of the base cluster) is specified. As a result, the base cluster and its surrounding clusters form a group. Two to 10 clusters usually constitute one base group. The diameters along the axis curve of the base cluster provide the basic shape of the generated group. If fine details are needed at that hair part, group refinement should be used. The generated clusters should be approximately within the original global boundary of the base cluster. Obviously, after an initial base group (called group *A*) is produced, the basic global shape of the original base cluster is still kept, but the generated group provides much richer details of hairs. Similarly, if more meticulous details are required for a cluster within group *A*, then that cluster should be selected as a new base cluster and a new base group (called group *B*) is generated accordingly. Within group *A*, the clusters other than those within group *B* form a new group (called group *C*). Therefore, in this case, group *A* is no longer a base group. Instead, it includes two new base groups *B* and *C* and is raised to a higher level of abstraction. Note that the group generation can be applied recursively, according to the complexity of hair styles.

When a group is generated, the control points of the base curve (Fig. 6), are used as references for the generated curves. The control points of the generated curves are distributed around the corresponding control points of the base curve. For example, all of the third vertices of the generated curves are within a cube, having its center in the third vertex of the base curve. In other words, this cube determines the distribution range of the vertices of the surrounding curves within the same group. The side length of the cube is determined by the original thickness of the base cluster. We note that the distribution of the vertices can not be completely random within the cube; otherwise, the curves generated will be either tangled together, or too far away from each other, thus giving the group a very unnatural appearance (Fig. 7).

To regulate the distribution of vertices of the surrounding curves in a group, once l, the side length of the cube, and n, the number of curves to be generated, are specified, the distance unit u (roughly the average distance between the generated curves) is defined as $l / (n - 1)$. To distribute the vertices more naturally within the cube, the coordinates of the jth vertex on the ith generated curve can be specified as follows:

$$x_{ij} = x_{base} + sign_i * coef_{i,j,x} * u + perturbation_x$$
$$y_{ij} = y_{base} + sign_i * coef_{i,j,y} * u + perturbation_y \quad (1)$$
$$z_{ij} = z_{base} + sign_i * coef_{i,j,z} * u + perturbation_z$$

The second term in each formula, that is, $sign_i * coef_{i,j,x/y/z} * u$, is used to make the distribution more regular. $sign_i$ is either 1 or -1, and it is used to balance the number of vertices of the generated curves around the corresponding vertices of the base curve. $coef_{i,j,x/y/z}$ is positive and is used to manipulate the distance between vertices at different positions along the curve. It is related to the distance between the scalp and the vertices. Assuming there are m control points for a curve, j varies from 0 to $m-1$, corresponding to the root and tip of a hair cluster. By varying $coef_{i,j,x/y/z}$, the group size may expand or shrink along the hair length. For example, increasing $coef_{i,j,x/y/z}$ will create a spreading hair group. For this kind of hair style, the smaller the values of $coef_{i,j,x/y/z}$ are, the closer the vertices are to the hair root and therefore the closer the vertices are to each other. This observation also conforms to our real life experience. On the other hand, a pony tail group will require increasing the values of $coef_{i,j,x/y/z}$ in the first half and decreasing its values in the second half. The choice of $coef_{i,j,x/y/z}$ is determined by the hair styles to be created.

The third term $perturbation_{x/y/z}$ in the above formulas, which is a random value within a suitable range, represents the perturbation of the vertices of the surrounding curves. This term is used to prevent the neighboring clusters from being uniformly parallel to each other. In other words, it creates some random variations between neighboring clusters, giving the group of clusters a natural appearance.

The above idea of the group generation can be viewed as a jittering technique as well. First, a 3D regular grid is formed around the control point of the base axis. Each grid point represents the corresponding control point of the generated curves, as is expressed in the second term of equation (1). Then, all of the grid points are jittered by three uncorrelated random values, for each direction along the x-, y-, or z-axis respectively, as shown in the third term of equation (1).

At the same time, cross-section shapes along the axis of the generated clusters can be similarly specified according to the base cluster. Minor perturbations are added to the parameters for the generated clusters. This results in shape variations among the clusters within a group and makes the group look more natural. During the process of group generation, the problem of penetration among different clusters has to be considered. Strictly speaking, different clusters should not penetrate into each other. However, precise penetration detection and avoidance can be very complicated and computationally intensive. On the other hand, it may not be necessary or practical to detect and remove all the penetrations. Based on daily observation, it is found that sometimes the boundary between different clusters is not very clear and may indeed have a little overlapping. In practice, hair clusters are fuzzy objects. Therefore, a small amount of overlapping between two clusters is often visually non-detectable from the rendered images. Our experiments confirm this observation as well. In the hierarchical model, minor penetration is permitted, but severe, deep penetration must be avoided.

The penetration detection approach is applied to detect and then prevent deep penetration among different clusters and groups. For a hair cluster, the cross-section shape is a closed planar contour. This contour is expressed by a closed B-Spline. The shape of the cross-section can be easily manipulated by moving the control points of the B-Spline. The contour of the cross-section can vary along the space curve. Several cross-sections can be assigned at various positions along the curve. Each can take a different shape, then the shape between any two neighboring cross-sections is achieved by interpolation. However, in most cases, the contour shapes are very similar along the space curve. Therefore, a simplified model is employed. Instead of specifying a set of contours, only one contour is defined but this contour can be scaled arbitrarily along the space curve. Meanwhile, a cubic B-spline curve is provided to represent the scaling factors along the space curve. This simplified definition has been used to create many complex hair clusters.

The existence of penetration between the surfaces of the generalized cylinders should be checked. Since a generalized cylinder is a curved surface, penetration computation directly based on the curved surface is very complex and expensive. Polyhedra are introduced to approximate the curved surface. Suppose for each cluster, the axis curve is a piecewise 3D quadric B-Spline curve. If there are n control points, there are $n-2$ segments. Each segment can be further divided into m sections; this results in a total of $(n-2)*m$ sections.

An accurate computation of penetration using a general intersection algorithm between polyhedral objects may still be expensive. A simple and effective approximation algorithm is developed in our system. For the purpose of penetration detection, a sufficient number of sample points on the contour can be selected to test against other clusters or scene objects (such as the head). At the axis center of each cross-section, the central angle 2π can be evenly divided by t, and accordingly there are t points distributed roughly evenly on the contour of the cross-section. In total, there are $(n-2)*m*t$ points for penetration detection.

It is relatively easy to determine if a point p is in another hair cluster c. Suppose the axis curve of c is already approximated as a polyline, and can be divided into $(n-2)*m$ line segments. It is necessary to calculate the distance between the point p and each line segment. If the distance between point p and the jth segment is less than the radius of the contour shape on the corresponding cross-section, then the point p penetrates into the cluster c. If this is not true for any segments, then point p does not penetrate into the hair cluster c.

If penetration exists, the penetration depth d between two clusters is known. If d is shorter than $k*r$, where k is a predefined constant and r is the radius of the thinner cluster, penetration is negligible and thus does not need to be further handled. Otherwise, a "repulsion force" can be applied to both clusters in the opposite directions. The force can be fragmented into two parts: one is used to change the cross-section shape by modifying the position of the control points on the contour of cross-section, and the other is transferred to the cluster axis to move the control point of the axis curve. This step iterates until all the deep penetrations are removed.

4.2 Operations of the Group

After a group is generated, it can be refined using many functions provided by our hair styling tool.

4.2.1 Regrouping

A curve can be added to a base group by first choosing the desired curve as the *current* curve, and then the desired group as the *current* base group. After choosing "Add cluster to a base group" from the submenu, the *current* curve is included in the *current* base group. If the *current* curve belongs to another base group previously, it can not join the *current* base group directly. Instead, the user has to choose "Detach cluster from a base group" to separate the required curve from its previous base group before adding this curve to the desired base group. This can prevent the user from accidentally changing the subordinate cluster within a group. Note that once a curve is regrouped, all the group information relevant to this curve has to be updated simultaneously.

4.2.2 Attraction and Repulsion

Attraction and repulsion functions are also offered to change the shape of the clusters at multiple group levels. It is observed that after a group is generated, the shape and mutual relationship of the clusters in the group may not be very satisfactory and may need to be further refined. Consequently, attraction and repulsion between the clusters are provided in V-HairStudio to manipulate the curves. After the level of abstraction is specified (the default is the base group level), the functions will be performed at this level. When "Attract in 3D" is chosen from the submenu, the mouse cursor is regarded as the center of attraction and all the vertices of the curves in the group are attracted to the cursor position. The closer the vertex is to the cursor, the greater the attraction force it receives. As a result, the vertices closer to the cursor are attracted more rapidly to the cursor. At the same time, the attraction to the vertices is gradually reduced when the vertices are farther away from the cursor. Repulsion works in a similar manner but in the opposite direction. It is seen that the force of either attraction or repulsion is inversely proportional to the distance. Therefore, an exponential function is chosen to approximate the attraction and repulsion forces:

$$F = k_1 e^{-k_2 d} \quad (k_1, k_2 > 0) \tag{2}$$

The function is illustrated in Fig. 5. The forces applied to a cluster act on each vertex and change the positions of the vertices, as well as the cross-section shapes. The horizontal axis (corresponding to d in equation (2)) represents the distance from the cursor to a vertex of a cluster, while the vertical axis (corresponding to F in equation (2)) is the corresponding attraction or repulsion force.

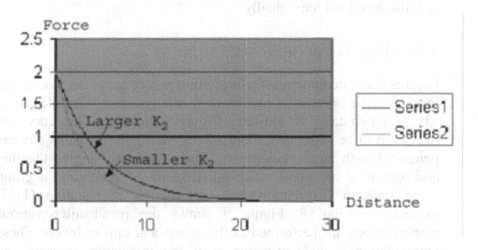

Fig. 5. *Attraction/repulsion force variation related to distance*

The coefficient k_1 represents the maximum force. That is, when the cursor and one of the vertices overlap, the force acting on the vertex is the largest. In this case, the force direction is usually first achieved by linear interpolation of the two adjacent vertices' normalized force vectors and then normalized. Particularly, if the vertex is the first or last one of a curve, the force direction is then directly specified as the normalized force vector of the adjacent vertex. k_2 is used to control the force attenuation with the increase of d. A larger k_2 means faster attenuation. k_2 can be adjusted to apply to different groups. The attraction force directions are from the vertices to the cursor, and the repulsion forces point from the cursor to individual vertices. The attraction and repulsion forces are dynamically changed according to the variation of distances from the vertices to the cursor. Once the attraction/repulsion operation is acted on a specified group, the positions of the vertices on curves within this group are all changed. The function calculation is always based on the current positions of the vertices and the mouse cursor.

4.2.3 Group Translation and Rotation

Translation and rotation operations are also provided at multiple group levels. The group can be treated as a whole, and can be moved or rotated along or around the x-, y-, or z-axis individually. The group can also be duplicated or deleted. For more precise manipulation of a group, the vertices and curves within the group can also be edited individually.

4.2.4 Illustration of Group

Figures 6 to 9 demonstrate the designing procedure at the base group level. For the purpose of clarity, only one group is illustrated here. The group contains six clusters. Figure 6 shows the base curve for the group to be generated. Figure 7 shows a group with its clusters generated with large randomness. The curves are tangled together and thus look unnatural. The initial result of generating a group using uniform distribution and perturbation in equation (1) is presented in Fig. 8. Figure 9 shows the result after various manipulations are performed at the group and cluster levels. These manipulations include attraction, repulsion, translation, rotation, and so on. Figure 10 shows the picture of the original base cluster corresponding to Fig. 6, while Fig. 11 shows the generated base group corresponding to Fig. 9. Neither of them considers the head effects. It is clearly seen that Fig. 11 provides much finer detail than Fig. 10.

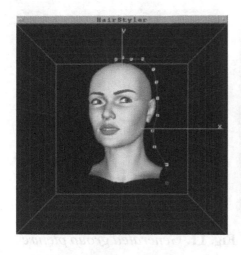

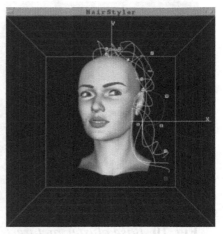

Fig. 6. *Base cluster*

Fig. 7. *Group of randomly generated cluster*

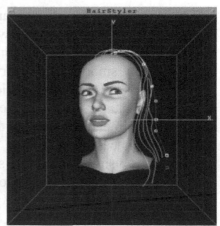

Fig. 8. *Group of naturally generated clusters*

Fig. 9. *Group after modification*

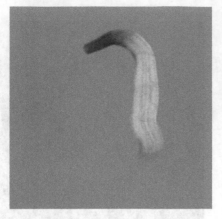 

Fig. 10. *Base cluster picture*   **Fig. 11.** *Generated group picture*

5 Cluster Axis Curve, Vertex and Group Editing

5.1 Curve Creating

The axis curve of a cluster can be represented in many forms. The modeling system provides two representations of the curve: the cubic B-spline curve and the quadric B-spline curve. When the curve is relatively smooth or the curvature is not very large, the cubic B-spline may be a suitable choice. However, when modeling a very curly hair cluster, it is relatively difficult to draw the axis curve in cubic B-spline form, due to the large curvature of the very curly hair. In this case, it would be more flexible to control the shape of the curve using quadric B-spline, although the smoothness of quadric B-spline is slightly poorer than the cubic B-spline.

To create a curve, a sequence of control points has to be specified. On the main menu, after the menu item "Next curve" is selected, the system enters the *drawing* mode. The designer can then move the mouse in the 3D space to specify the position of each control point of a 3D B-spline in sequential order.

5.2 Curve, Vertex and Group Selection

Three major modes for object (such as curve, vertex, or group) selection are provided in the system: "sequential selection", "index input", and "direct selection". The designer can use one of the modes to access the desired object at multiple levels. After a set of groups is created, any object can be selected as the *current object* for further modification. In the following, the curve selection will be discussed. Vertex and group selection can be performed similarly.

One selection mode is "sequential selection". When the number of curves is not too large, say around 20, the identity number (or index number) of a curve can be used to select the desired curve. In the system, each vertex is assigned an identity number in the vertex set of a curve. Similarly, each curve or group has an identity number in the corresponding curve or group set, respectively. The identity number of an object is an index for quick selection. In the "sequential selection" mode, whenever the right mouse button is clicked, the curve with the next index number is highlighted and selected as the current curve. If the last curve is reached, then the next selection will move back to the first curve. The designer can also click the middle mouse button for the selection in reverse order. Each time a curve is selected, its index will be displayed in the control panel.

If the index of the desired curve is known, a more direct mode "index input" can be used to select the curve. The designer can directly input the identity number of the curve in the control panel to highlight that curve.

The drawback of the above two selection modes is their inflexibility. When many curves exist, it is obviously inconvenient to specify a desired curve. As a result, the mode "direct selection" is provided to make curve selection convenient. With the reference of the cross lines at the cursor position in the three orthogonal windows, the designer can move the mouse cursor in 3D space to an approximate location around the desired curve. In the system, the distance between the cursor and a curve is evaluated by an approximation method. A set of points is sampled along the curve, and the distances between the cursor and all sample points on the curve are calculated. The shortest distance is regarded as the distance between the cursor and the curve. Among all curves, the closest curve to the cursor is selected as the current curve. When the designer moves the mouse cursor in the perspective window, the current curve changes accordingly.

However, in some situations, it is not intuitive to judge the 3D distance between the cursor and a curve in the main perspective window. For example, the designer may find a desired curve appears to be nearer to the cursor than other curves are in the perspective view, but the highlighted curve is not the desired curve. To solve this problem, a 2D selection method in "direct selection" mode is also provided for the perspective view. Using the perspective projection, the pixel positions of the cursor and sample points of curves on the 2D window plane can be evaluated. The 2D distance between the cursor and a point is defined as the distance between their corresponding projected pixel positions. The current curve then should be the curve with the shortest 2D distance to the cursor on the window plane. By so doing, it is more intuitive and convenient to select the desired curve with the mouse cursor only according to their 2D positions.

The vertex and group selections are similar to the curve selection. When the "direct selection" mode is used in the vertex selection, the closest vertex on a curve to the cursor is chosen as the current vertex, and the curve on which the current vertex lies becomes the current curve consequently. In the system, after the group level is specified (the default is the base group level), various groups are displayed with different colors. The designer can first move the mouse cursor to specify the current group, and then click the middle or right mouse button to pick the vertex on the current curve. It should be pointed out that the current curve does not have to be within the current group, while the current vertex must always lie on the current curve. This means that the group and curve may be manipulated separately, but the vertex editing should be applied on the current curve.

5.3 Curve and Group Modification

After a set of curves has been created, modification can be made on the currently selected curve. The shape of a curve may vary by manipulating its control points. Its position and orientation can be adjusted through translation and rotation. The position of a vertex can be translated along the x-, y-, or z-axis respectively, using the combinations of mouse buttons. By pressing the left mouse button, the mouse movement leads to the translation along the x-axis; the left and middle buttons together, the y-axis; and all three buttons together, the z-axis.

This operating status can be accessed by the following simple menu selections. In the main and sub menus, after an item sequence "Modify → Vertex on curve → Move by mouse" is selected, the system enters the mouse manipulation mode to move the current vertex on the current curve (Fig. 12). The sub-menu under the item "Vertex on curve" contains five options to edit the selected vertex: move by mouse; move by keyboard keys; delete; insert before; and insert after. To make the curve shape simpler or smoother, the current vertex may be deleted, and then the next vertex will become the current vertex. Inversely, to increase the complexity of the curve shape, a vertex may be inserted before or after the current vertex. The position of the added vertex is determined by the cursor movement.

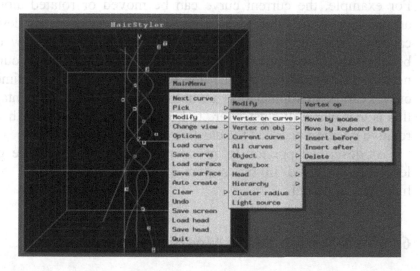

Fig. 12. *Curve vertex modification*

The option "move by mouse" is suitable for editing a curve vertex in many situations. However, when a hair style is very complex, the interactive design process is repetitive and takes a relatively long time before a satisfactory visual effect is obtained. It is very inconvenient and tedious to press the mouse button and drag the cursor simultaneously and continuously during the design process. To overcome this shortcoming, a quicker operating option, "move by keyboard keys", is provided. With this option, the default moving direction of the vertex is along the x-axis. The designer can switch the moving direction to be along any of the three axes by pressing key "X", "Y" or "Z". A set of keys is used to control the movement of the current vertex with different moving step sizes. For example, when the vertex is translated along the positive direction of a specified axis, keys "A", "U", "I", "O" and "P" stand for five step sizes, in increasing order. Similarly, keys "C", "V", "B", "N" and "M" can also perform the vertex translation, except that the moving direction is along the negative axis. Using keys with larger step sizes moves the vertex to a rough location quickly, changing the curve shape greatly. Conversely, the designer may use keys with smaller step sizes to precisely move the vertex to a desired position, achieving the desired result.

To edit a curve as a whole, V-HairStudio provides some manipulations such as translation, rotation, duplication, and deletion. For example, the current curve can be moved or rotated along or around any of the three axes in the 3D space. As well, the designer can copy the current curve. After the duplication, the new curve becomes the current curve for further modification. When a curve is not needed, it can be selected and removed. In the mean time, all curves can be manipulated as a whole object. The main advantage of this is that when all clusters are to be moved or rotated together, the relative position between the clusters remains the same.

As described previously, group modification at multiple group levels is provided in the system, based on the concept of the hierarchical model.

6   Cross-section Shape Modification

The cross-section shape of a hair cluster, $R(s,a)$, is a closed planar contour, represented by a closed B-Spline. The shape of the cross-section can be manipulated easily by adjusting the control points of the B-Spline. The contour of the cross-section can vary along the space curve. In the current system, along the space curve, all cross-sections have the same shape but can be scaled by different factors. In Fig. 13, the bottom-left window, in which contour editing is performed, displays the cross-section contour. The top-left window shows a curve representing the scaling factors along the space curve. The current simplified definition is used to create the examples in this chapter.

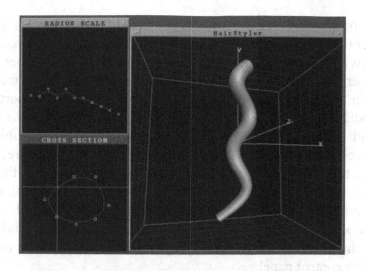

Fig. 13. *Modification on cross-section shape*

7 Head Model Editing and Manipulation

The hierarchical hair model is placed onto a head model, such as in Fig. 14. In V-HairStudio, the head model is imported from another system, called 3DCAFE. Though creating a 3D head object is not a task of our modeling system, several head editing and manipulation functions are provided.

Fig. 14. *A head model with hair*

In the system, a head model consists of eighteen parts: face, right eyebrow, left eyebrow, right pupil, left pupil, right eyewhite, left eyewhite, right iris, left iris, right lower lash, right upper lash, left lower lash, left upper lash, lower lip, upper lip, lower teeth, upper teeth, and, finally, tongue. V-HairStudio supports some modifications of the head model (Fig. 15). As a whole object, the head model can be translated, rotated, scaled, or automatically moved to the center of the coordinate system. Furthermore, each part on a head model can be edited separately. For example, the position and orientation of each eye (including eyewhite, pupil and iris) can be adjusted by translation and rotation, so that limited expression can be achieved. The color of each part (such as face and lips) can be changed as well, by specifying the RGB color values of that part in the control panel.

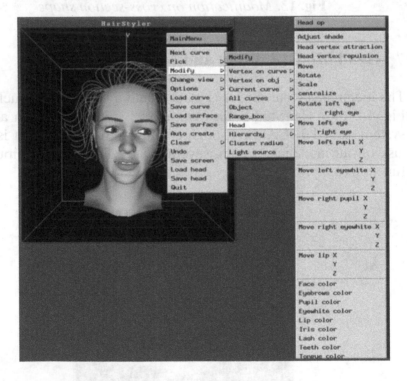

Fig. 15. *Menu for editing the head*

Since sometimes the head shape may need to be changed, each vertex of the head mesh can be selected and then moved or deleted. However, when the shape of the head model needs a relatively large change, it is obviously inconvenient to edit each vertex individually, owing to the large number of the head vertices (usually greater than 10,000). To solve this problem, the system provides two functions: "Head vertex attraction" and "Head vertex repulsion", which are similar to the idea of the hair group attraction and repulsion operations. These two functions can make the modification of the head shape more efficient and easy.

8 Light Property Editing

The light source provided in V-HairStudio can be either parallel or point. If the light source is parallel, the vector of the light direction is specified by three values through a control panel. If a point light source is used, these three values define its position. Multiple light sources can be used to display the hair and head models, and the colors of each light source are specified in RGB mode in the control panel. Various visual effects can be obtained by using light sources with different positions, orientations, and colors. In order to achieve desired shading effects, the designer can edit the light source properties interactively.

Figures 14 and 16 show the same hair style with different light sources, which are from the front and front-top-right, respectively. A single parallel light source is used to display the scene objects in both pictures.

Fig. 16. *A head model with front-top-right light source*

9 Preview of Hair and Head Models

The shape preview mode can be turned on at any time during the design process. The hair clusters, together with the head model, are displayed in surface mode. The designer can clearly observe the spatial relationship between hair clusters, and the fitting between the hair clusters and the head model. Rotation, translation, and zoom in or out help the designer to inspect the hair styles at different distances and from arbitrary view angles.

Further modification can be applied repeatedly to any part of the hair and head models until a satisfactory result is achieved. Figure 17 shows the axis curves of hair clusters on a mesh formatted head model and Fig. 18 illustrates the surface preview of this hair style.

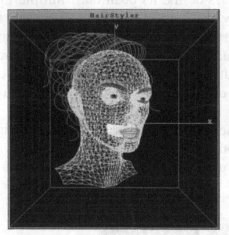

Fig. 17. *Preview of models in curve and mesh mode*　　**Fig. 18.** *Preview of models in surface mode*

When a user becomes reasonably familiar with the system, the design of a hair style with similar complexity to the above examples (such as Fig. 4 takes approximately one hour, which is much shorter than the design time using the earlier cluster hair model (usually two to three hours). Therefore, due to the hierarchical representation, complex hair styles can be handled with reduced design efforts.

10 Implementation and Results

The experiments were conducted on a Silicon Graphics System. The modeling part is developed using C and GL. In the hair designing process, hardware display and rendering functions are fully utilized for highly interactive manipulation. The hair rendering part is implemented in C without using any graphics hardware acceleration. The rendering technique is a projected-based sweeping algorithm based on the hierarchical hair model. Parallel techniques can be applied to accelerate the process of the hair rendering.

Figure 19 and Fig. 20 are two examples of rendered images. V-Hairstudio is used to create the hair styles.

Fig. 19. *A wavy hair style*   **Fig. 20.** *A hair style*

11 Summary

Based on Xu and Yang's previous work [13], an enhanced hair design tool, called V-HairStudio, is developed for the hierarchical model. This model allows a complex hair style to be manipulated efficiently at multiple levels of abstraction: the group levels, the cluster level, and the individual hair level, from top to bottom. In V-HairStudio, a set of rich functionalities is provided for creating and manipulating the hair groups and clusters in an interactive manner. For example, the axis curve of a cluster is first drawn and modified in the 3D space. It may then be selected as the base curve for the generation of a base group. The designer can further use functions such as group attraction and repulsion to change the cluster shapes within the group. The main advantage of this tool is that the manipulation can be performed at multiple levels, significantly improving the efficiency of hair modeling. The cross-section shape contour and the hair distribution map can also be edited easily. For a fine visual effect, light properties such as positions, directions, and colors may be adjusted. With the preview functions, the designing process is repeated until a satisfactory hair style is achieved.

As for future improvements of the styling tool, the interface may be refined to provide simpler and easier selection and manipulation of the group or cluster. A pre-created collection of different hair styles should also be considered. Instead of creating a new hair style every time, a pre-created style can be selected and instantiated as an initial shape. Further modification and refinement can be performed based on the initial shape. This may reduce the design period significantly. Each pre-created style may also have a number of parameters that allows variations to be controlled by designers during the instantiation.

The hierarchical hair model provides a novel basis for efficient dynamic hair simulation. Instead of considering dynamics of individual hairs (for example, [10]), a group of hair clusters may be considered as an elastically deformable object. This will allow hair dynamics to be handled at a very abstract level. It can potentially reduce the complexity and computational cost of dynamic simulation very significantly. At the same time, satisfactory visual effects can be obtained for graphical animations. A dynamic simulation module is also under development by our research group.

References

1. L. Chen, S. Saeyor, H. Dohi, and M. Ishizuka. (1999) A System of 3D Hair Style Synthesis Based on the Wisp Model. The Visual Computer, 15(4), 159-170.
2. A. Daldegan, T. Kurihara, N. Thalmann, and D. Thalmann. (1993) An Integrated System for Modeling, Animating and Rendering Hair. Proc. Eurographics 93, Computer Graphics Forum, 12(3), 211-221.
3. A. Daldegan and N. Thalmann. (1993) Creating Virtual Fur and Hair Styles for Synthetic Actors. Communicating with Virtual Worlds, Springer-Verlag, Tokyo, 358-370.

4. A. Gelder and J. Wilhelms. (1997) An Interactive Fur Modeling Technique. Proc. of Graphics Interface 97, 181-188.
5. http://www.digimation.com/asp/product.asp?product_id=86.
6. J. Kajiya and T. Kay. (1989) Rendering Fur with Three Dimensional Textures. Computer Graphics, 23(3), 271-280.
7. A. LeBlanc, R. Turner, and D. Thalmann. (1991) Rendering Hair Using Pixel Blending and Shadow Buffers. Journal of Visualization and Computer Animation, 2(3), 92-97.
8. G. Miller. (1988) From Wire-frame to Furry Animals. Proc. Graphics Interface, 138-146.
9. K. Perlin and E. Hoffert. (1989) Hypertexture. Computer Graphics, 23(3), 253-262.
10. R. Rosenblum, W. Carlson, and E. Tripp. (1991) Simulating the structure and Dynamics of Human Hair: Modeling, Rendering and Animation. Journal of Visualization and Computer Animation, 2(4), 141-148.
11. T. Wang and X. D. Yang. (2001) A Design Tool for the Hierarchical Hair Model. CAGD Symposium, International Conference on Information Visualization IV'2001, IEEE Press, 186-191.
12. Y. Watanabe and Y. Suenaga. (1991) Drawing Human Hair Using the Wisp Model. The Visual Computer, Vol. 7, 97-103.
13. Z. Xu and X. D. Yang. (2001) V-HairStudio: An Interactive Tool for Hair Design. IEEE Computer Graphics and Applications, May/June, 36-43.
14. X. D. Yang, T. Wang, Z. Xu, and Q. Hu. (2000) Hierarchical Cluster Hair Model. Proc. IASTED, Int. Conf. Computer Graphics and Imaging, Las Vegas, November, 75-81.
15. X. D. Yang, Z. Xu, J. Yang, and T. Wang. (2000) The Cluster Hair Model. Graphics Models and Image Processing, 62(2), 85-103.

Chapter 18

Shape Modeling for 3D Archaeological Vessels

Shape Modeling for 3D Archaeological Vessels

Anshuman Razdan, Dezhi Liu, Myungsoo Bae and Mary Zhu
Partnership for Research in Stereo Modeling (PRISM), Arizona State University,
Tempe, AZ 85287-5906, USA

Arleyn Simon
Department of Anthropology, Arizona State University, Tempe, AZ 86287, USA

Gerald Farin
Department of Computer Science and Engineering, Arizona State University,
Tempe, AZ 86287, USA

Mark Henderson
Department of Industrial Engineering, Arizona State University, Tempe, AZ
85287-5906, USA

This chapter presents a method for archiving and searching three-dimensional Native American ceramic vessels using geometric modeling techniques. Archaeological vessels are scanned and defined as a set of 3D triangulated meshes composed of points, edges and triangles. Our work includes modeling the data with parametric surfaces, extracting features to raise the level of abstraction of data, and organizing vessel data based on XML schema. A visual query interface on the web was developed that permits users to sketch or select sample vessel shapes to augment text and metric search criteria to retrieve original and modeled data, and interactive 2D and 3D models.

1 Introduction

The understanding of 3D structures is essential to many scientific endeavors. Recent theoretical and technological breakthroughs in mathematical modeling of 3D data and data-capturing techniques present the opportunity to advance 3D knowledge into new realms of cross-disciplinary research. 3D knowledge plays an important role in archaeology too. Archaeologists study the 3D form of Native American pottery to characterize the development of cultures. Quantitative methods of reasoning about the shape of a vessel are becoming far more powerful than was possible when vessel shape was first given a mathematical treatment by G. Birkhoff[2].Conventionally, vessel classification is done by an expert and is subjective and prone to inaccuracies. The measurements are crude and in some case eyeballing is the method of choice.

This chapter describes geometric modeling techniques used in our research to analyze the 3D archaeological vessels from the Classic Period (A. D. 1250 – 1450) of the prehistoric Hohokam culture area of the Southwest (Salt/Gila River Valleys) near present-day Phoenix, Arizona. Our research involves obtaining shape information from the scanned three-dimensional data of archaeological vessels, using 2D and 3D geometric models to represent scanned vessels, extracting features from geometric models and storing the feature information in database for Web-based retrieval. This chapter is structured as fellows. Part two describes archaeological vessel features from the point of view of archaeologists. Part three introduces geometric modeling for vessels. Some feature recognition methods used to extract archaeological vessel features are presented in part four. The Visual Query Interface (VQI) for archiving and searching vessels on the Web are described in part five. Conclusions and further research direction can be found in part six.

2 Features of Archaeological Vessels

Features have different definitions in different application domains. In this chapter, features of archaeological vessels mean form features. Traditionally, much attention has been paid to extract form features from mechanical parts. One hypothesis of this feature extraction problem is that all form features are face oriented. A form feature is defined as a set of faces with distinct topological and geometrical characteristics [6]. Three kinds of traditional methods to extract form features: (i) rule-based method, (ii) graph-based method, and (iii) neural net method also mainly deal with "face features", regions of interest in a part model. Form features in our project have an extensive application domain. Further more, form features are not limited to faces. Feature information can be divided into four categories: Points, Curves, Regions and Volumes. Following is our feature classification (Fig. 1).

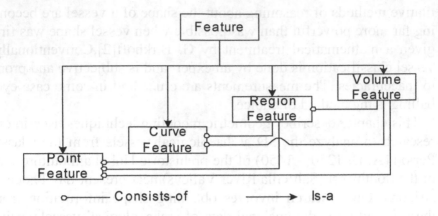

Fig. 1. *Feature Classification*

2.1 Features of vessel profile curves

Mostly archaeological vessels are (approximately) surfaces of revolution, and studying contour shape will suffice to gather shape information about the whole object. According to archaeological definition there are four kinds of feature points on profile curves to calculate dimensions and proportions of vessels. They are End Points (EPs), Points of Vertical Tangency (VTs), Inflection Points (IPs) and Corner Points (CPs) found on the vertical profile curve of a vessel:

- *End Points* - points at the rim (lip) or at the base (i.e. top and bottom of vessels).
- *Points of Vertical Tangency* - points at the place where is the maximum diameter on spheroidal form or minimum diameter on hyperbolic form.
- *Inflection Poinst* - points of change from concave to convex, or vice versa.
- *Corner Points* - points of sharp change on a profile curve. See Fig. 2.

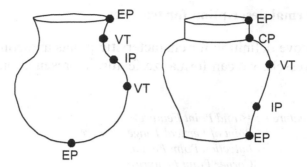

Fig.2. *Feature points of profile curves.*

2.2 Features common to all vessels

Next four features are common to all vessels:

- *Orifice* - the opening of the vessel, or the minimum diameter of the opening, may be the same as the rim, or below the rim.
- *Rim* - the finished edge of the top or opening of the vessel. It may or may not be the same as the orifice. It may have a larger diameter.
- *Body* - the form of the vessel below the orifice and above the base.
- *Base* - the bottom of the vessel, portion upon which it rests, or sits on a surface. The base may be convex, flat, or concave, or a combination of these. See Fig. 3.

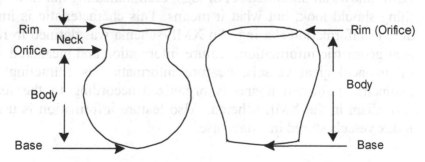

Fig.3. *Common features of vessels*

2.3   Formal description for features

From above definition for characteristic points and common features for all vessels, we can formalize feature representation of vessels as below.

<Point Feature> :=<End Point Feature>
             |< Point of Vertical Tangency Feature>
             |< Inflection Point Feature>
             | <Corner Point Feature>;
<Curve Feature> := <Rim Curve Feature>
             |< Orifice Curve Feature>
             |< Base Curve Feature>;
<Rim Curve Feature> :=<End Point Feature>< End Point Feature>;
<Orifice Curve Feature> :=<Corner Point Feature> <Corner Point Feature>;
<Base Curve Feature> := <End Point Feature>
             |<End Point Feature>< End Point Feature>
<Region Feature> :=< Neck Region Feature>
             |< Body Region Feature>
             |< Base Region Feature>;
<Neck Region Feature> := <Rim Curve Feature><Orifice Curve Feature>;
<Body Region Feature> := <Orifice Curve Feature>< Base Curve Feature>;
<Base Region Feature> := <Base Curve Feature>;
<Volume Feature> :=< Unrestricted Volume Feature>
             |< Restricted Volume Feature>.

Extensible Markup Language (XML)[3] is used to represent information of vessels. XML is the standard format for structured document/data interchange on the Web. Like HTML, an XML document holds text annotated by tags. However, unlike HTML, XML allows an unlimited set of tags, each indicating not how something should look, but what it means. This characteristic is invaluable to information sharing. An XML schema was defined to represent geometric information, feature information and measured value of archaeological vessels. Feature information is extracting from geometric information and is organized according to the feature formalism in the XML schema. Also feature information is used to index vessels stored in a database.

3   Geometric Modeling for Vessels

3.1   3D geometric models

(1) Polygonal Meshes

   After scanning an archaeological vessel via a 3D laser scanner (Cyberware 3030), we can get a polygonal mesh that is constituted of faces, edges and vertices. The polygonal mesh is used as a raw data for further analysis. Polygon Meshes M are 3-tuples, i.e. M = (V, E, F) where V is vertex set, E is edge set, and F is face set.

(2) Surface Models

   One of representing or modeling surfaces is via parametric surfaces such as B-Spline or NURBS. Surface models are generated by fitting points of polygonal meshes with least squares approximation. Parametric representation enables us to rebuild models, analyze properties such as curvatures, and make quantitative measurements as well "repair" incomplete models. A NURBS surface can be represented as

$$\vec{P}(u,v) = \frac{\sum\limits_{i=0}^{m}\sum\limits_{j=0}^{n} w_{i,j}\vec{d}_{i,j} N_{i,k}(u) N_{j,l}(v)}{\sum\limits_{i=0}^{m}\sum\limits_{j=0}^{n} w_{i,j} N_{i,k}(u) N_{j,l}(v)} \tag{1}$$

where $\vec{d}_{i,j}$, $i = 0, 1, \ldots, m$; $j = 0, 1, \ldots, n$, are control points, $w_{i,j}$ are weights, $N_{i,k}(u)$ and $N_{j,l}(v)$ are B-Spline basis functions. When weights equal 1.0, it reduces to a non-uniform B-Spline surface.

   NURB surfaces are generated by fitting points of raw data with least squares approximation [1, 4]. The combination of cylindrical projection and spherical projection is used to parameterize data points for surface approximation because vessels are round and symmetrical, and closed at bottom. Fig. 4 is a sample of surface modeling.

3.2   2D geometric models

Contour shape information plays an important role in analysis of archaeological vessels. We use two kinds of models, chain codes and NURB curves to represent profile curves of archaeological vessels.

Using 2D geometric models can make problem simple, and reduce 3D problem to 2D problem.

(1) Chain codes

In order to get a 2D profile curve from a vessel, archaeologists use a cutting plane to intersect the vessel (polygonal mesh) and can get intersection points, then connect all the points according to some order, and get the chain code.

(2) NURB Curves

NURB curves are generated by fitting points of chain codes with least squares approximation. Since curvature has useful information such as convexity, smoothness, and inflection points of the curve needed by vessel analysis, cubic NURB curves are sufficient to approximate profile curves of vessels

$$\vec{P}(u) = \frac{\sum\limits_{i=0}^{n} w_i \vec{d}_i N_{i,k}(u)}{\sum\limits_{i=0}^{n} w_i N_{i,k}(u)} \qquad (2)$$

where $\vec{d}_i$, $i = 0, 1, \ldots, n$, are control points, w_i are weights, $N_{i,k}(u)$ are B-Spline basis functions. See Fig.5.

3.3 Signed curvatures for profile curves

One of the important characteristics of a curve is the curvature. The curvature of a curve is the magnitude of the rate of change of the tangent vector with respect to arc length. It is very useful for analysis and classification of vessel shape. In 3D space the curvature of curves is unsigned. However, for planar curves in 3D space, positive curvatures can be converted into signed curvatures[1,4], see Fig. 6.

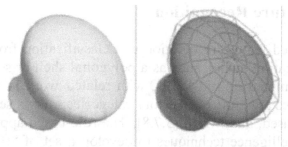

Fig. 4. *(L to R) An archaeological vessel and its surface model*

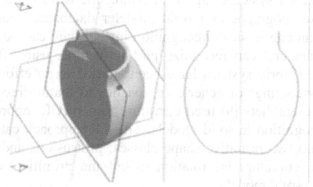

Fig. 5. *(L to R) An archaeological vessel and its profile curve*

Fig.6. *A vessel, its profile curve and signed curvature plot of the curve.*

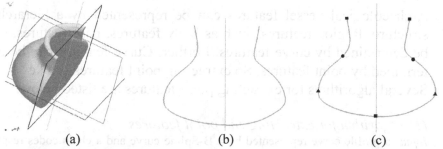

Fig.7. *Profile curve generating and feature point extracting.*

4 Feature Recognition

Automated feature recognition and classification from a boundary-representation model, such as a polygonal shell or solid model, was first attempted in mid-1970s, with related work in the 1960s. The techniques are grouped into four categories: rule-based, graph-based, neural-based, and hybrid [5,7,8]. The rule-based approach uses artificial intelligence techniques to develop a set of "If...Then...Else" feature rules. The work includes pattern-matching techniques. Graph-based systems require matching the feature graph to the appropriate subgraphs in a solid modeler database. Neural-nets have the advantage in shape recognition in that they can be taught using exemplars and can recognize partial features using floating point outputs. Hybrid systems have been studied in an effort to supplant the shortcomings of generic approaches to feature recognition. Recently researchers [9] used curvature region (CR) approach for feature recognition in solid models. The CR approach categorizes features into two primitive shape classes: protrusions and depressions, and use curvature information to get the primitive shape feature from the solid model.

Curvatures are characteristic in this chapter too. Mostly pots are rotation volumes, and contour shape can represent characteristic for pots. One of geometric features of a contour is its curvature. Feature points can be extracted from the contour by analyzing its curvatures. For example IPs can be located where the curvature changes from negative to positive or vice versa. CPs and EPs can also be researched from analyzing curvature plot of contour curves.

4.1 Extracting point features

Archaeological vessel features can be represented by a hierarchic structure. Region features, such as body features, base features, can be determined by curve features. Further, Curve features can be determined by point features. So extracting point features is a key here. Several algorithms for extracting point features are listed below.

(1) Algorithm for extracting end point features
Input: a profile curve represented by a B-Spline curve and a chain codes respectively.
Output: end point features

1.   end point 1 ← start point of the chain code; end point 2 ← end point of the chain code;
2.   center point ← center point of the chain code;
3.   find the base section around center
4.   *if* base section is flat or concave *then*

> total end point number ←4; end point 3 ← left terminate of base section; end point 4 ← right terminate of base section;

    else { base is convex}

> total end point number ←3; end point 3 ← center;

5. calculate feature information for each end points, include space coordinates, parameter value, position on the chain code, and so on;

(2) Algorithm for extracting corner point features
Input: a profile curve represented by a B-Spline curve and a chain code respectively.
Output: corner point features
1.   calculate curvature value for each points on the chain code;
2.   find points with local maximum (minimum) curvature value as candidates for corner points;
3.   for each candidate *do*

> *if* angle at the candidate point < a predefined value *then*
>> the candidate point is a corner point.

4. calculate feature information for each corner points, include space coordinates, parameter value, position on the chain code, and so on;

As for inflection features and point of vertical tangency features, they are easy to find because by analyzing curvature value and tangent lines.

When computing the angle between points (x_1, y_1), (x_0, y_0) and (x_r, y_r) in algorithm (2), the value of angles is sensitive to sample errors. In order to reduce errors due to sampling, instead of taking (x_1, y_1) and (x_r, y_r) as points of the curve, the coordinates of these points are calculated by averaging the coordinates of a group of neighbors to perform a less noise prone resampling.

Let us consider the mid point (x_0, y_0) of n contiguous points in a chain code of a curve, where n is an odd number, and let $p = n/2 + 1$ be the point (x_0, y_0). Thus, the initial point of the angle (x_1, y_1) is calculated from the n/2 + 1 previous point as

$$x_l = \frac{\sum_{i=1}^{p} x_i}{n/2+1}, \; y_l = \frac{\sum_{i=1}^{p} y_i}{n/2+1} \tag{3}$$

and similarly for the end point of the angle (x_r, y_r)

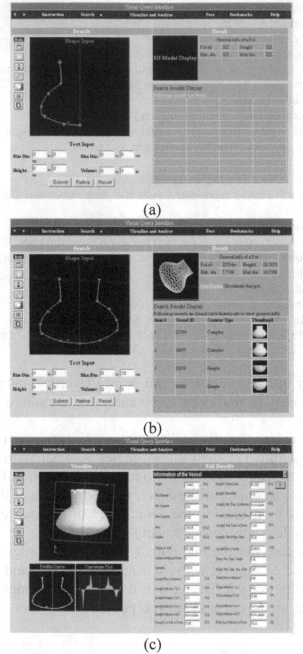

Fig.8. *Query interface screen with sketch–based, numeric, and text-based input fields. Initial response screen with resulting thumbnail images and summary data(a), and wire frame of first matching vessel(b). Detail individual display screen with 2D, 3D, and descriptive vessel data(c).*

$$x_r = \frac{\sum\limits_{i=p}^{n} x_i}{n/2+1}, \; y_r = \frac{\sum\limits_{i=p}^{n} y_i}{n/2+1} \tag{4}$$

4.2 Some results for point feature extracting

Figure 7 (b) shows a profile curve that is generated by intersecting a vessel with a user defined cutting plane (Fig. 7(a)). Fig. 7 (c) shows the result of finding end points (rectangular points), inflection points (circular points) and points of vertical tangency (triangular points).

5 Visual Query Interface (VQI)

After getting point features we continue finding curve features and region features based on feature hierarchical definition. Then we use XML to represent the result. The purpose of using XML to represent information is that we can develop a distributed and web based visual query interface for archiving and searching 3D archeological vessels. Embedding data in XML adds structure and web accessibility to the inherent information of archeological vessels. Fig. 8 describes the flow chart of the Web-based VQI.

The query process in VQI combines a sketch-based interface and searches by traditional text and metric data. Representative vessel shapes can be selected from the supplied palette and modified, or a freeform profile sketch can be created in the interface window. Text and numeric fields support parallel query of descriptive and derived data within the databases. Query results from database are stored in XML format, and are visualized via a pre-designed Extensive Stylesheet Language (XSL) file.

6 Summary

We present a method for archiving and searching 3D objects, Native American ceramic vessels using geometric modeling techniques. We have (i) modeled raw data of 3D archaeological vessels with parametric curves and surfaces, (ii) extracted features to raised the level of abstraction of data, (iii) organized vessel data based on XML and

(iv) developed a visual query interface on the web for sharing information.

During our research we have found some problems that need solving in the future. First we observe that vessels are handmade, and some vessels are of irregular shape. It is impossible to use profile curves to describe the shape of these vessels. We often get broken vessels because they are "historical" objects. Completely refitting broken vessels is difficult task. All of these need our further work.

References

1. Bae, M. (1999) Curvature and Analysis of Archaeological Shapes. MS Thesis, Arizona State University.
2. Birkhoff, G. (1933) Aesthetic Measure. Cambridge University Press, Cambridge, Massachusetts.
3. Extensible Markup Language (XML) 1.0, http://www.w3.org/TR/REC-xml
4. Farin, G. E. (1996) Curve and Surface for Computer Aided Geometric Design. Academic Press, Boston, fourth edition.
5. Henderson, M. (1994) Boundary Representation based Feature Identification. In Advances in Feature Based Manufacturing, J.J. Shah, M. Mantyla, and D.S. Nau (editors), Elsevier Science.
6. Nalluri, S. (1994) Form Feature Generation Model for Feature Technology, PhD Thesis, Dept. of Mechanical Engineering, IISc, India.
7. Razdan, A. and Henderson, M. (1989) Feature Based Neighborhood Isolation Techniques for Automated Finite Element Meshing, Geometric Modeling for Product Engineering, Editors Josh Turner, Ken Press, North Holland.
8. Razdan, A., Henderson, M., Chavez, P. and Erickson, P.E. (1990) Feature-Based Object Decomposition for Finite Element Meshing, The Visual Computer, pp. 291-303.
9. Sonthi, R., Kunjur, G. and Gadh, R. (1997) Shape Feature Determination using the Curvature Region Representation. Solid Modeling'97, Atlanta GA USA, 285 – 296.

Chapter 19

Archimedean Kaleidoscope: A Cognitive Tool to Support Thinking and Reasoning about Geometric Solids

Archimedean Kaleidoscope: A Cognitive Tool to Support Thinking and Reasoning about Geometric Solids

Jim Morey
Department of Computer Science, Cognitive Engineering Laboratory, The University of Western Ontario, London, Canada

Kamran Sedig
Department of Computer Science, Faculty of Information and Media Studies, Cognitive Engineering Laboratory, The University of Western Ontario, London, Canada

This chapter presents an interactive visualization tool, Archimedean Kaleidoscope (AK). AK is a cognitive-scaffold tool aimed at supporting users' cognitive activities while exploring and making sense of polyhedra visualizations. AK uses a three-dimensional kaleidoscopic metaphor to generate the visualizations. To aid users with their cognitive activities, AK produces the three-dimensional visualizations dynamically and provides a high level of interactivity with them. Additionally, the three-dimensional visuals are made metamorphic so as to help users investigate transitions and relationships among the different polyhedra.

1   Introduction

Perceiving, thinking and reasoning about three-dimensional structures, such as polyhedra figures, is not always straightforward. One often needs to think about these structures in terms of their external form, constituent grammar, characteristics, and inherent relationships [17]. Norman [14] states that "Without external aids, memory, thought, and reasoning are all constrained. ... The real powers come from devising external aids to enhance cognitive abilities". External aids can be in the form of interactive computer-based tools can support thinking and reasoning process. These tools are referred to as *cognitive tools* or *mind tools* – tools that augment, enhance and support people's cognitive activities and reduce their cognitive load during thinking, reasoning, and problem solving [11]. A group of these tools are referred to as *interpretation tools* [4, 8, 9]. These tools allow users to view, make sense of, and develop mental models of available information through interaction with them.

Often times, visual, rather than algebraic, representations of mathematical concepts can promote development of insight into the underlying general principles [16, 21]. Mathematical visualization tools can act as interpretation support tools [2, 15]. These tools can help users interpret and make sense of structural and causal relationships among the constituent elements of visualized mathematical concepts [8, 15], particularly geometry [6].

Visual representations by themselves, however, are not sufficient. Often times, users need to understand the relationship between one family of mathematical objects to another. Such relationships can be explored by observing transitions between these associated objects [15]. This suggests that visualizing such transitional processes between these objects can support cognitive activity and sense making [15, 17]. However, transition-based visualizations can have different degrees of interactivity—from highly passive observation of animations, to highly active control of the pace, rate, and sequencing of the transitional visuals [20]. While interacting with mathematical structures, a high degree of continuous visual feedback and control is important as it allows users to adapt their actions and to reflect on their goal-action-feedback cycle [12].

This chapter presents the Archimedean Kaleidoscope (AK), an interactive computer-based cognitive tool to augment and support users' thinking and reasoning about polyhedra figures, their visual structures, their characteristics, and their relationships. The next section provides conceptual and terminological background for the further discussion of AK.

2 Background

A polyhedron (plural: polyhedra) is a geometric object that has many polygonal faces [3, 7, 10]. These faces are bound together by edges, and these edges meet at corners called vertices. Certain polyhedra are called *regular solids,* or *Platonic solids*. These solids have special properties: their faces are identical regular polygons, and the same number of polygons meets at each vertex. Fig. 1 depicts the five Platonic solids—namely, tetrahedron, cube, octahedron, dodecahedron, and icosahedron. The dodecahedron in Fig. 1, for instance, has twelve faces (all regular pentagons) and twenty identical vertices.

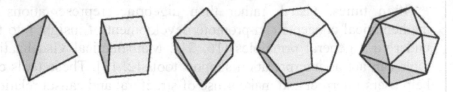

Fig. 1. *Platonic solids—L to R: tetrahedron, cube, octahedron, dodecahedron, icosahedron.*

Another group of polyhedra are *Archimedean solids*. What distinguishes the Archimedean from the Platonic solids is the number of different types of polygons in these solids. The Archimedean solids have at least two different types of polygons (see Fig. 2), unlike the Platonic solids that have only one type of polygon. The cuboctahedron in Fig. 2, for instance, has two types of polygons: equilateral triangles and squares.

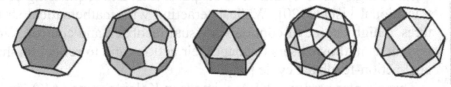

Fig. 2. *Archimedean solids—L to R: truncated octahedron, truncated icosahedron, cuboctahedron, rhombicosadodecahedron, rhombicuboctahedron.*

The Platonic and the Archimedean solids are closely related. Their relationships are multifaceted. One example of these relationships between polyhedra has to do with planes of symmetry. For instance, the planes of symmetry of the cube are the same as the planes of symmetry of the octahedron. Mathematicians understand and describe these relationships though abstract algebra, in particular through group theory [3, 13]. However, these relationships can also be understood through observing correspondences in the structures of the solids (i.e., their faces, edges, and vertices). For instance, there is a correspondence between the octagonal faces of a truncated cube and the square faces of a cube. Fig. 3 depicts this correspondence by superimposing the cube faces on the truncated cube. Another instance of correspondence occurs between the vertices of the octahedron and the faces of the cube. One can observe a correspondence between the centers of each of the faces of the cube and the vertices of the octahedron (see Fig. 4).

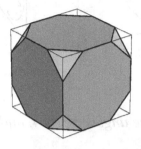
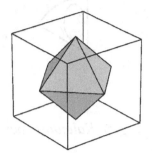

Fig. 3. *A truncated cube obtained by cutting the vertices of the cube.*

Fig. 4. *An octahedron inside a cube.*

The relationships described above can either be explicitly communicated through verbal explanations, or they can be implicitly experienced through interaction with the visual solids themselves [18, 21]. Allowing users to experience such relationships through interactive exploration of the environment before explicit verbal explanations is conducive to better construction of mental models of the concepts [21]. This can be done by allowing the user to interact with the visuals and to observe the effects of such interaction. The users can observe how controlling parameters of a visual object (in this case, polyhedral solids) can change the object from one form to another passing through intermediate stages. This process of a shape gradually transforming into another shape is referred to as metamorphosing (or in short form, morphing) [5]. Metamorphosis can play an important role in understanding difficult-to-visualize mathematical processes [15].

Platonic and Archimedean solids are symmetric structures. Kaleidoscopes can aid in the visual exploration of symmetric structures [10]. A kaleidoscope is an instrument made of two mirrors used to produce a collage of images [1]. The produced collage consists of an original wedge, the area bounded by the mirrors, and many reflected copies of the original wedge. The collage's symmetries are determined by the angle between the mirrors. To produce a simple collage, one with an exact number of reflected copies, the angle must be of the form π/n. Fig. 5 depicts a number of simple two-dimensional kaleidoscopes with a smiling face contained in the original wedge. The asymmetry in the face is intended to highlight the orientation of the image.

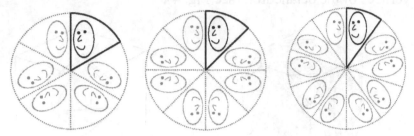

Fig. 5. *Kaleidoscopes based on the angles π/3, π/4, and π/6.*

A three-dimensional kaleidoscope is constructed using three mirrors. Whereas the construction of simple two-dimensional kaleidoscopes is only dependent on the value of the angle between the adjacent mirrors, the construction of simple three-dimensional kaleidoscopes is dependent on the value of the angles among all three mirrors. There are only three sets of angles that produce simple three-dimensional kaleidoscopes [1]. Fig. 6 depicts these possible kaleidoscopes as spheres covered with the wedges. Each wedge is a spherical triangle determined by three great circles. Ball refers to these mathematical constructions as polyhedral kaleidoscopes [1].

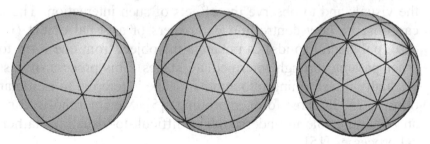

Fig. 6. *The polyhedral kaleidoscopes.*

3   Archimedean Kaleidoscope

Archimedean Kaleidoscope (AK) is an interactive tool designed to provide cognitive support for users to explore, visualize, think about, and reason about complex polyhedra solids and their relationships.

The design of AK is inspired by the kaleidoscope metaphor. However, rather than focusing on the placement of the mirrors, AK is concerned with the framework within which the mirrors are placed. The framework for a two-dimensional kaleidoscope with angle π/n can be a regular polygon with n sides, cut into 2n triangles. The polygon acts as a framework for the reflected wedges. Fig. 7 shows three kaleidoscopes that use the angles $\pi/3$, $\pi/4$, and $\pi/6$ as subdivided regular polygons. Familiarity with the symmetries of these shapes aid in the understanding of the symmetries of corresponding kaleidoscopes.

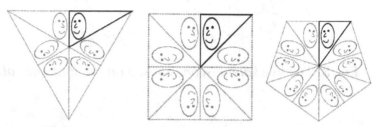

Fig. 7. *Kaleidoscopes based on a triangle, a square, and a pentagon.*

The kaleidoscope in AK produces 3-dimensional collages. The symmetries of these collages come from one of the five Platonic solids rather than that of a regular polygon. This means that the wedges of the AK's kaleidoscope cover all the faces of the solid. Fig. 8 depicts screen captures of the five kaleidoscopes in AK that are based on the five Platonic solids. These kaleidoscopes are similar to their respective 2-dimensional counter parts. However, there are many more wedges in the 3-dimensional versions. For instance, the cube has 48 wedges and the dodecahedron has 120.

As the 2-dimensional polygon-based kaleidoscopes produce 2-dimensional images, AK's 3-dimensional Platonic-based kaleidoscopes produce 3-dimensional images; that is, each wedge contains a 3-dimensional image. These 3-dimensional images seamlessly fit together to create the collage—i.e., the produced polyhedron. Fig. 9 shows a 3-dimensional wedge in AK, and Fig. 10 shows the resulting polyhedron produced by AK. The highlighted wedge (upper right corner) contains a 3-dimensional image represented by three different colors.

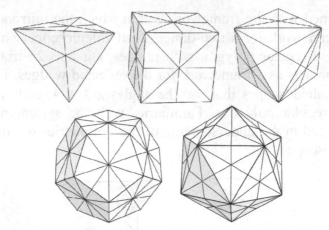

Fig. 8. *The five kaleidoscopes based on the Platonic solids.*

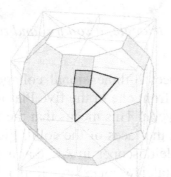

Fig. 9. *Image in one wedge.*

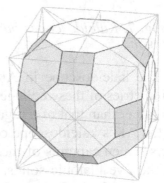

Fig. 10. *Archimedean solid produced as a collage of wedge images.*

This construction differs from the description of the polyhedral kaleidoscope by Ball [1]. AK's construction, based on the Platonic solids, has the advantage that all its constituent wedges are made of the familiar 2-dimensional polygons rather than the less familiar spherical wedges used in Ball's kaleidoscope.

Fig. 11 shows a screen capture of AK. There are two main panels in AK: 1) Kaleidoscope Menu, where the user selects one of five base platonic kaleidoscopes, and 2) Interactive Visualization Area, where the user can interact with a selected kaleidoscope to visualize and explore different polyhedra solids. In the figure, the selected base kaleidoscope is the circled icosahedron on the right, and the visualized solid is the truncated icosahedron (or the soccer ball).

Fig. 11. *Screen capture of Archimedean Kaleidoscope.*

The following subsections discuss AK's main components and features and their roles in supporting the users' thinking and reasoning about these geometric solids.

3.1 Kaleidoscopic Vertex

Vertices play an important role in the structure of polyhedra. AK uses one single vertex in the kaleidoscope to control the shape of the produced polyhedra. We refer to this vertex as the Kaleidoscopic Vertex (KV). KV is represented as a black dot (see the dot near the center of the visualized solid of Fig. 11). The location of KV determines the size and shape of the 3-dimensional image in each wedge (see Fig. 8 for an example). AK uses this location and the mirroring quality of the kaleidoscope to replicate the wedge and generate 3-dimensional images of the polyhedra solids. For example, in the cube-based kaleidoscope, a single point within the original wedge will generate 47 other points on the kaleidoscope. Fig. 12 shows how the location of KV, in the original triangular wedge, is used to generate a polyhedron. The left image shows KV as a sole vertex within the kaleidoscope. The next image results from KV's replication in the cube-based kaleidoscope, generating 48 vertices in total, where each of the vertices lies near a corner of the cube, slightly off its diagonals. The 48 vertices define the polyhedron shown as the third image from the left in Fig. 12, where the polyhedron is the convex hull of the vertices. The last image in Fig. 12 shows the final visualization generated by AK where the replications of the original wedge can also be seen as a wire frame. The wire frame provides a visual context to support the users' thinking about the generated polyhedron and its symmetric nature.

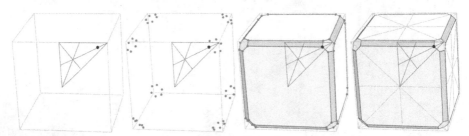

Fig. 12. *L to R: How KV is replicated in the cube-based kaleidoscope to generate a polyhedron.*

3.2 Interaction

AK provides the user with two levels of interaction: formative and perspective.

Formative: This level of interaction is related to KV and allows the user to manipulate the formation of different polyhedra. At this level, KV becomes an interactive control that the user can manipulate in a continuous fashion to dynamically generate new polyhedra images—i.e., to simultaneously change all generated kaleidoscopic images. This control allows the user to adjust the placement of KV and experiment with different formations of polyhedra through a manipulation-observation-modification feedback loop. This interaction takes place within the bounded area of the original wedge. For instance, aiming to find out where KV should be placed to form a rhombi-truncated cuboctahedron, the user can manipulate KV and search the available visualization space until the desired solid in Fig. 13 is generated. Although interaction at the formative level takes place within a bounded area, yet the number of different polyhedra that can be explored and observed is countless.

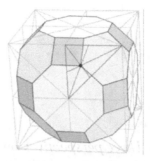

Fig. 13. *A rhombi-truncated cuboctahedron.*

As stated before, every kaleidoscope has a wired Platonic frame associated with it. As the user interacts with KV and forms new polyhedra, the original frame stays the same. This allows the user to make mental connections between the base Platonic solid and the newly generated polyhedra.

Perspective: This level of interaction is related to the entire kaleidoscopically-generated visual structure. Once a structure is generated, the user can manipulate it and change its orientation in a continuous manner to observe it from different perspectives. Fig. 14 shows three snapshots of a visual structure (a truncated octahedron) during the process of viewing it from different perspectives. As can be seen in the figure, it is not easy to perceive that the three static images represent the same structure. Continuous interaction and viewing of the whole structure from different perspectives supports the user's cognitive processes to make this observation.

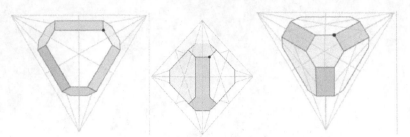

Fig. 14. *Three perspectives on a polyhedron.*

3.3 Metamorphosis

As stated before, visualizing metamorphic processes can assist people in understanding and reasoning about mathematical concepts. This is especially true about geometric concepts and structures. Observing a continuous transition from one polyhedron to another can help the user to see some of their existing properties and relationships. In AK the user can experience metamorphic processes in two ways: by formative interaction with KV and by observing AK's simulator.

Fig. 15 shows five snapshots of the intermediary stages of a continuous metamorphic transition between two polyhedra: the cube and the cuboctahedron, where one of the stages results in the creation of the truncated cube (the third image in the figure).

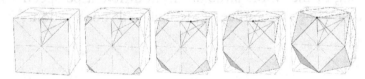

Fig. 15. *L to R: Metamorphic transition from cube to cuboctahedron.*

Fig. 16 shows another metamorphic transition and some of the resulting intermediary stages. This time the transition is from a dodecahedron to a truncated icosahedron. As a result of the movement of KV from the upper-right corner of the original wedge to the middle of the left side of the wedge, the vertices and edges of the dodecahedron get truncated. During metamorphic transitions many structural changes take place in the visualized solid. All constituent polygons can change in size, shape, and orientation, as seen in Fig. 16.

An example of a change in polygons' sizes can be seen in the regular hexagons present in all but the top-left image. An example of a change in the polygons' shapes can be seen in the rectangles metamorphosizing to widthless lines where the hexagons meet. Finally, an example of a change in orientation of polygons can be seen in the pentagons in the top-left and the bottom-right images. As the metamorphosis process is continuous and the visualizations are rendered in real time, the user can observe countless numbers of polyhedra while interacting with KV at the formative level.

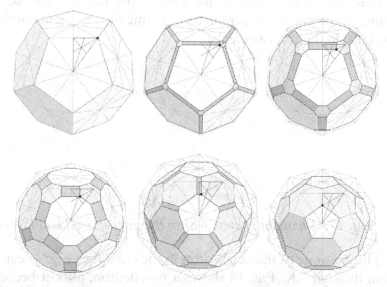

Fig. 16. *TL to BR: Metamorphic transition from dodecahedron to truncated icosahedron.*

Another method by which the user can observe transitions among polyhedra is through the simulator. The simulator in AK automates the metamorphosis process. When the simulator is active, it walks through and displays all the transitional polyhedra generated along pre-specified paths. The path that KV walks is associated with these transitions and is intended to aid the user to think and reason about the common properties of the polyhedra encountered along the way.

3.4 Relationships

As mentioned before, many relationships among polyhedra can be seen as correspondences between faces, edges, and vertices. AK supports these correspondences by color-coding the faces of the visualized polyhedra according to their relationship to the Platonic frame. The white faces of a constructed polyhedron correspond to the faces of the base solid. The blue faces of a constructed polyhedron correspond to the edges of the base solid. And, the yellow faces of a constructed polyhedron correspond to the vertices of the base solid. For example, the relationship between the truncated cube and the cube can be seen by the correspondences of their structures (see Fig. 17). The yellow triangles in the truncated cube correspond to the vertices of the cube. The white octagons in the truncated cube correspond to the faces of the cube. This relation can be viewed statically or can be experienced dynamically by metamorphosizing the cube to the truncated cube.

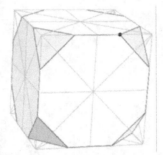

Fig. 17. *A truncated cube within the cube-based kaleidoscope.*

The example of the octahedron in the cube (see Fig. 4) can also be seen through AK. Fig. 18 shows a polyhedron, almost becoming an octahedron. Its triangular faces are yellow, its thin rectangular faces (on the verge of becoming edges) are blue, and the small octagonal faces (on the verge of becoming vertices) are white. This suggests the following correspondences: the octahedron's faces correspond to the cube's vertices; the octahedron's edges correspond to the cube's edges; and the octahedron's vertices correspond to the cube's faces. This intermediary polyhedron is more suggestive than viewing an exactly-formed octahedron. Even though this intermediary polyhedron is not a Platonic solid (or an Archimedean solid), it suggests a relationship between two Platonic solids. The correspondence between the cube's edges and the octahedron's edges is often difficult for people to understand. The exploration of intermediary polyhedra through AK can make these subtle relationships more apparent to users.

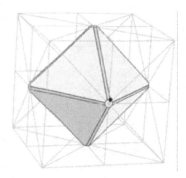

Fig. 18. *An intermediate polyhedron.*

3.5 Extension of the Metaphor

One of AK's goals is to allow users to interact with all Archimedean solids. Not all of these solids have a symmetry of reflection. Some have rotational symmetries only. Fig. 19 shows rotational symmetries based on the triangle, the square, and the pentagon. To include rotational symmetries, AK extends its kaleidoscopic metaphor. The extended metaphor generates collages consisting of either the original wedge and many reflected copies or the original wedge and many rotated copies.

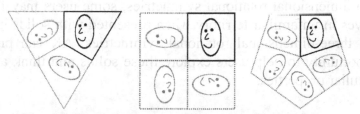

Fig. 19. *Two-dimensional rotational symmetries.*

Although the same image is replicated in the wedges of Figs. 19 and 5, the number of replicated images in Fig. 19 is half as many as those in Fig. 5. This is due to the relationship between reflections and rotations. This same principle applies to the Platonic solids when using rotational symmetries. Fig. 20 shows the rotational symmetries, depicted by wedges, of all five Platonic solids.

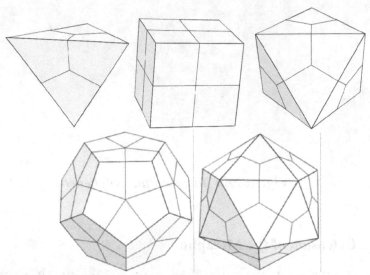

Fig. 20. *Three-dimensional rotational symmetries based on the five Platonic-based kaleidoscopes.*

Two of the Archimedean solids with rotational symmetries are the snub cube (see Fig. 21) and the snub dodecahedron (see Fig. 22). Many of the polyhedra with the rotational symmetries appear to have twists in them. For example, in Fig. 21 the squares in the polyhedron appear to be vertically twisted. Despite being familiar with two-dimensional rotational symmetries, some users may find themselves in unfamiliar territory when presented with solids incorporating three-dimensional rotational symmetries. AK can provide the opportunity to help users explore these solids and think about their peculiar properties.

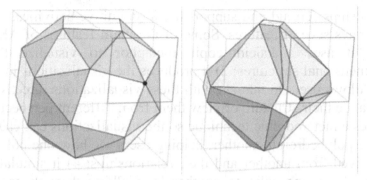

Fig. 21. *A snub cube and an intermediate polyhedron with rotational symmetries.*

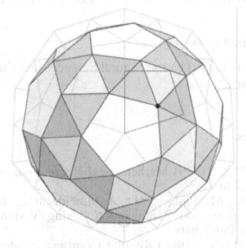

Fig. 22. *A snub dodecahedron.*

4 Summary

People do not always find it easy to visualize and reason about polyhedra figures. Computer-based tools, in the form of cognitive or mind tools, can enhance, support and scaffold people's mental activities while dealing with such three-dimensional structures.

This chapter has presented Archimedean Kaleidoscope (AK), a cognitive tool designed to support users' mental processes to conceptualize and understand Platonic and Archimedean solids. AK is intended to help users visualize, make sense of, and reason about these structures, their properties and characteristics, and their relationships. To aid and support users' cognitive activities, AK incorporates several features. Some of the main features of AK include: 1) it uses a kaleidoscopic metaphor to visualize the three-dimensional structures; 2) it produces the visualizations in real time; it allows users to interact with these visualizations directly to either change their structure or view them from different perspectives; 3) it uses a metamorphic technique so the visualizations can change from one polyhedron to another, helping users see how one polyhedron is derived from another and their relationships; 4) it simulates transition from one solid to another by walking through pre-specified paths to assist users in visualizing intermediary structures; and 5) it extends the kaleidoscopic metaphor to include rotational symmetries so as to allow users to visualize all possible Archimedean solids.

Acknowledgements

This research is funded by the Natural Sciences and Engineering Research Council of Canada.

References

1. Ball, W.W.R. (1939) Mathematical Recreations and Essays. Macmillan & Co. Ltd.
2. Card, S. K., Mackinlay, J. D., & Shneiderman, B. (Eds., 1999) Readings in Information Visualization: Using Vision to Think. Morgan Kaufmann Publishers.
3. Coxeter, H.S.M. (1991) Regular Complex Polytopes (2nd ed.). Cambridge University Press.
4. Gordin, D.N., Edelson, D.C., & Gomez, L. (1996) Scientific Visualization as an Interpretive and Expressive Medium. In D. Edelson & E. Domeshek (Eds.), Proceedings of the Second International Conference on the Learning Sciences, 409-414.
5. Gregory, A., State, A., Lin, M.C., Manocha, D., & Livingston, M.A. (1999) Interactive Surface Decomposition for Polyhedral Morphing. The Visual Computer, 15, 453-470.
6. Hanson, A., Munzner, T., & Francis, G. (1994) Interactive Methods for Visualizable Geometry. IEEE Computer, 27, 78-83.
7. Holden, A. (1971) Shape, space, and symmetry. Columbia University Press.
8. Jonassen, D. H., & Carr, C. S. (2000) Mindtools: Affording Multiple Knowledge Representations for Learning. In S. P. Lajoie (Ed.), Computers as Cognitive Tools, Lawrence Erlbaum Assoc., NJ.
9. Jonassen, D.H., Peck, K.L., & Wilson, B.G. (1999) Learning with Technology: A Constructivist Perspective. Prentice-Hall, Inc.

10. Kinsey, K., & Moore, T. E. (2002) Symmetry, Shape, and Space. Key College Publishing, PA.
11. Lajoie, S. (Ed., 2000) Computers as Cognitive Tools. Hillsdale, NJ: Lawrence Erlbaum Assoc.
12. Laurillard, D. (1993) Rethinking University Teaching: A Framework for the Effective Use of Educational Ttechnology. Routledge.
13. Möbius, August Ferdinand, Gesammelte Werke (1967) hrsg. auf Veranlassung der Königlichen Sächsischen Gesellschaft der Wissenschaften. Wiesbaden: M. Sändig, [Neudruck der Ausg. 1885-87].
14. Norman, D. A. (1993) Things That Make Us Smart: Defining Human Attributes in the Age of the Machine. Addison-Wesley Publishing.
15. Palais, R.S. (1999) The Visualization of Mathematics: Towards a Mathematical Exploratorium. Notices of the American Mathematical Society, 46(6), 647-658.
16. Peterson, D. (ed., 1996) Forms of Representation. Exeter, UK: Intellect Books.
17. Rieber, L. P. (1999) Animation in Computer-Based Instruction. Educational Technology Research & Development, 38(1), 77-86.
18. Sedig, K., Klawe, M., & Westrom, M. (2001) Role of Interface Manipulation Style and Scaffolding on Cognition and Concept Learning in Learnware. ACM Transactions on Computer-Human Interaction, 1(8), 34-59.
19. Senechel, M., & Fleck, G. (Eds., 1988) Shaping Space: A Polyhedral Approach. Birkauser.
20. Shedroff, N. (1999) Information Interaction Design: A Unified Field Theory of Design. In R. Jacobson, Information Design, 267-292.The MIT Press.
21. Skemp, R. R. (1986) The Psychology of Learning Mathematics (2nd ed.). Middlesex, UK: Penguin Books.

Chapter 20

Towards a Web Based CAD System

Towards a Web Based CAD System

Frédéric Danesi, Laurent Denis and Yvon Gardan
CMCAO Team / LERI, IFTS, 7, boulevard Jean Delautre, 08000 Charleville-Mézières, France

Yann Lanuel and Estelle Perrin
CMCAO Team, Metz University - UFR MIM - Ile du Saulcy, 57045 METZ Cedex 01, France

Actual CAD systems are still monolithic and are based on complex geometrical models. Our aim is to implement a new approach taking advantage of web possibilities and providing a novice end-user with an easy-to-use system. In order to get the same performances as a local system, it is necessary to provide our system with adequate data representation and architecture.

A study realized in our team has shown that classical models are too less-level information for our specifications and that a functional model is needed.

This chapter deals with our first results concerning the architecture which will be the basis of different implementations and with a first operational prototype. The first section introduces the tests we processed in order to understand advantages and drawbacks of different alternatives and describes the final choices for the architecture. The second section describes a prototype, proving the feasibility of our approach. It is based on new intuitive interaction techniques even if we only outline this point in this chapter.

1   Introduction

Actual CAD systems are still monolithic and are based on complex geometrical models. Our aim is to implement a new approach taking advantage of web possibilities and providing a novice end-user with an easy-to-use system. This implies that such a system has no knowledge about the remote computer (operating system, memory, processor and storage capabilities, etc.) and its user. Consequently, the whole CAD system cannot be uploaded on the client and the model has to be stored on the server. During a work-session, numerous specific CAD transmissions between the server and the client will be added to classical ones. Moreover, CAD systems use specific representation models and by the way carry huge data volume. Therefore, it is necessary to provide our system with adequate data representation and architecture in order to get the same performances as a local system.

A study has been realized in our team [1] that concerns the transmission of the three main models used in CAD (CSG, octree and BRep). It has shown that such classical models are too less-level information for our specifications and that a functional model is needed.

This chapter deals with our first results concerning the architecture which will be the basis of different implementations and with a first operational prototype. The first section introduces the tests we processed in order to understand advantages and drawbacks of different alternatives and describes the final choices for the architecture. The second section describes a prototype, proving the feasibility of our approach. It is based on new intuitive interaction techniques even if we only outline this point in this chapter.

2 Architecture of a Distributed CAD System

We recall that in order to have a CAD system available on Internet, the architecture has to allow the client to use CAD data (or representation model) stored on the server. This can be done by two ways:

- the client has to invoke remote methods on the server. The section 2.1. studies different available tools for integrating distributed objects into programming languages: CORBA, DCOM and Java RMI and shows that all these solutions are not adequate;
- the server provides the client with a replica of the data. Then, the client is able to work by its own and this is explained in section 2.2.

2.1 Remote methods invocation

In order to let the programmer call remote methods as he does with local methods, we could use tools like CORBA (Common Object Broker Architecture) [2], DCOM (Distributed Component Object

Model) [3], or JAVA RMI (Java Remote method invocation) [4]. All these tools provide a solution to manipulate objects from a distant machine. They use the notion of client stub and server stub. When the program on the client machine calls a method on a remote object, a request is sent to the client stub which dialogs with the server stub, by transferring the call and the arguments. The server stub uses them to call the real method before returning the result by the reverse way. CORBA and DCOM have drawbacks which lead us to give them up: CORBA requires installing a specific software on the client, by an experienced user, and DCOM is only available on MICROSOFT^{TM} operating systems.

That is the reason why we have only studied Java RMI in order to measure the time consumed by calling remote methods. These tests have been made on INTEL^{TM} Celeron and INTEL^{TM} Pentium processors and consist in invoking methods that return one of the coordinates of a 3D point by three ways:

- the method is locally called;
- the method is remote, but the server is the same computer as the client;
- the method is remote and the client and the server are relied by a 10 Mbits network.

Methods invocation times are summarized in Table 1.

Table 1. *Comparison between spent time on a call to a local method, a call to a remote method on a unique computer, a call to a remote method on two computers on a local network.*

	INTEL^{TM} Celeron 533 MHz	INTEL^{TM} Pentium 166 MHz
Local method	0.0001 ms	0.0005 ms
RMI on one machine	0.9096 ms	1.5086 ms
RMI on a 10Mbits network	1.1800 ms	1.9357 ms

Results show that a call to a remote method on a unique computer spends almost the same time as the one on two computers on a local network, but the spending time on the call of a local method is very lower than the two previous ones. This is explained by the fact that the use of RMI involves numerous extra treatments which imply very important overhead time.

As the system spends time on numerous extra treatments for each invoked method, we can decrease it by invoking methods on higher granularity objects. New tests have been realized on an outline of 3D points. The outline used representation is either N distinct 3D points, or an array of N 3D points. In the first case, we invoke N times the remote method that returns one of the coordinates of a 3D point. In the second case, we invoke the remote method that returns an array of the coordinates of the outline 3D points. Results of this test with N equals to 500 are shown in table 2.

Table 2 *Comparison between spent time on 500 calls to a remote method and a call to a remote method on an array.*

500 3D points	590 ms
Array of 500 3D points	10.9 ms

As expected, a higher granularity provides better time results, but in our case, such a solution often transfers too many useless data. For example, if the client program needs only 20 points of the outline, it is obvious that it is better to get the entire array, but it will have to store 500 points instead of 20 and to extract these 20 points among the 500 ones.

In conclusion, with a low-level granularity, Java RMI is too slow to satisfy our specifications, and with a high-level one, we loose control over transferred data and tasks sharing out between the server and the client. Finally, the different studied available tools are inadequate for our system. We study in the next section data replication to provide client with local treatments.

2.2 Replication

Replication process consists in copying on the client the data that the remote program needs among those stored on the server, and in ensuring the consistency at each modification realized on a data or on one of its replicas. Only few works have been published on this subject, and only one is under our context [5]. This work is a general-purpose object-oriented library for developing distributed 3D graphics applications called Repo-3D. Repo-3D uses three kinds of distributed object semantics:

- simple objects with no special distributed semantics;

- remote objects with a client-server distribution semantics. All method invocations are forwarded to the original object;
- replicated objects with a replicated distribution semantics. If any replica is changed, the change is reflected in all replicas.

As the time delay of synchronous remote method invocations is unsuitable for rapidly changing graphical applications, in Repo-3D the data needed are updated asynchronously with replicated objects. In order to update the replicas, all the operations are performed in the same order on all copies.

Repo-3D shows that it is possible to use replicas in order to obtain an application with satisfying access-time. For our system, it will allow us to realize more or less complex computations on the client. Nevertheless, the server will still have to take into account specific tasks like the high-functional level ones. If the tasks natures between the server and the client are different, it seems that the client and the server may have different data representations. As replicas are realized in Repo-3D, they do not allow us to use different representations.

We present in the next section our own architecture, based on replica.

2.3 System Architecture

Because usual commercial tools are inadequate to help us to exchange data, we use a scene description language. In this language, a component part may be thus described by the way it has been built rather than by listing the elements which composes it. For example a box is described by two points (the point more in bottom on the left and the point more in top on the right when the sides of the box are parallel to the axes) and a matrix of transformation to position it in space. Although this technique requires additional treatments to reconstitute the part on the distant machine, it makes it possible to decrease the transfer times significantly.

Our main model is based on a construction history using the scene description language mentioned above. Such a model is relatively abstract (we explain it in section 2.4), it is thus difficult to construct the whole treatments necessary for design process directly from its data. Then the main model is translated into specialized models adapted to each family of tasks. For example, if the user needs to visualize and interact with its component part, a boundary representation model is generated. Or if user's needs relate to simulation, the system then creates a model based on finite-elements. During the translation, the specialized model interprets each instruction saved in the main model. Consider an instruction that creates a face: in a boundary representation model, the translation will produce a face composed by one or more loop(s) whereas in a model for simulation, it will produce a mesh.

Several specialized models can exist at the same time in the same machine, but they all depend on the main model. Obviously, if a modification is made in one of the specialized models, information is transmitted to the main model which redistributes information on the whole of the specialized models.

As we mentioned it above, the language is also used to communicate between the server machine and the client machine. This language ensures the independence of the models used on the client and the server. At the beginning of a session, the server machine transfers data to the client machine. Then the system based on the client machine begins to work and creates new data following user's interactions. So the model stored in the client machine differs from the one stored in the server machine. We explain in section 2.5 that we have developed a special naming mechanism to deal with this problem.

2.4 Multi-level language

The DIJA Project aims to propose tools adapted to the trade of the user. As it is relatively difficult to directly develop a whole of high level tools, we proceed step by step, increasing the semantic. We start from a level of relatively low abstraction (near to the physical model) then we build various layers having an increasingly high level of abstraction. This approach is also used in [6]. The instructions used in the main model are gathered in three distinct levels. Each level being expressed using the preceding level.

- **level 1:** is a basic language. It is used to communicate with specialized models. All specialized models must implement the whole of the instructions present at this level. The vocabulary is based on the name of the primitives used to access to the information present in specialized models. At this level, instructions like "creating a face" can be found;

- **level 2:** is a language based on dialog elements. All actions that's the user can make on the system are at this level. Level two instructions can be translated in level one instructions. All the actions made by the end user could thus be stored in the specialized models. Instruction like "selecting line" can be found at this level;
- **level 3:** is a trade directed language. At this level all the instructions are associated with a particular trade (although some can be common to several trade associations). The translation towards level two uses a knowledge model which brings additional information. According to the profession, the results will be thus different. For example, "to hollow out a part" gives different results according to the context used. In the case of plastic processing, "to hollow out" means to withdraw material and create a network of ribs in order to increase the rigidity of the part, whereas in the case of machining, it mean only withdraw material.

On one hand, the more the language is abstracted, the more the treatments for the specialized models are significant. On another hand, the more the language is abstract, the less the quantity of data to be exchanged via the network is significant. According to the machines used and the flow of the network, we use the most adapted level of abstraction.

2.5 Data exchanges

The client is self-sufficient when it is not connected to the server. So the client's model may evolve without the server having knowledge of it. At a given time, it is possible that none of the two models entirely knows all the data (those existing at the same time on the server and the client). However the server or the client may create new elements in its respective model. To solve the problem of persistence, any element has a single name usually called identifier. In order not to have two entities having the same name, it is necessary to develop a naming mechanism which takes into account the characteristics of our architecture.

Data exchanges between the server and the client are based on the interpretation of a certain number of instructions of more or less high level. A three (respectively two) level instruction is interpreted into a two (respectively one) level instruction. Then each level one instruction, when necessary, generates a whole of specialized models data (segments, points, ...) which will be then used for specific treatments (visualization, handling or simulation ...). The naming mechanism must take into account the localized relations between the client and the server. In general, the server will create new elements at the client request. This is why we name the data relatively to the number of the request on which they depend. Thus the 3rd element created by the 4th request will bear the name (4,3).

In the following sections, we validate our naming mechanism by confronting it with the following cases:
- when the client loads data from the server;
- when the client updates data on the server;
- when the client requests the server to do some treatments.

Loading data from the server

When the client begins to work, in addition to model data, the server sends it the first instruction number it can use. Thus, the client has the possibility to start working during the loading of the model. As the instructions are sent, the client interprets them to build its specialized model. The only instructions sent are those useful for the user. Determining the parts useful for the user is made by the interface (in using the point of view, knowledge about the behaviors of the user...) [1].

Updating data on the server

To update server model, the client has to send all the instructions corresponding to the modifications realized by the user since the last update. The server then interprets each instruction and consequently modifies its model. The resulting elements are named according to the usual naming mechanism.

Requesting the server to do some treatments

When the client requests some treatments to the server, these treatments are supposed to be applied on an up to date model. Thus, the server has to update its model before realizing the task. The client can continue to propose a certain number of services while waiting for the server communicates the results of the treatment. In order not to have conflict, the client is only authorized to create new elements. As the client and the server can both create new elements at the same time, it is necessary to prevent that those elements have the same name. For that, all the elements created by the client are numbered in a temporary way from 1*. When the server finishes its treatment, it returns to the client a whole of instructions that will have to be interpreted. The elements created by the client are then renamed (we recall that there names were temporary) by adding to their instruction number the higher instruction number in its model. We give hereafter an example of creation of entities integrating a request to the server.

- **stage 1:** the client creates elements (1,1), (1,2), (1,3), (2,1), (2,2);
- **stage 2:** the client asks the server to carry out a treatment. The server updates its model and thus knows the number of the last instruction used by the client. As from this moment, the client cannot create new elements;
- **stage 3:** the client creates new elements. They are named temporarily in the following way: (1*,1), (1*,2), (2*,1), (2*,2);
- **stage 4:** the server returns two instructions corresponding to the treatment carried out: instructions 3 and 4;
- **stage 5:** the client interprets these instructions and creates the elements (3,1), (3,2), (3,3), (4,1), (4,2);
- **stage 6:** the elements created by the client at stage 3 are renamed. The naming mechanism adds 4 (the highest number of instruction present in the model) to the number of instruction in order to obtain the elements (5,1), (5,2), (6,1), (6,2);

At the end of the process, the specialized model contains the following elements: (1,1), (1,2), (1,3), (2,1), (2,2), (3,1), (3,2), (3,3), (4,1), (4,2), (5,1), (5,2), (6,1) and (6,2). The result is the same one performed by a local work. If the client modifies one of the elements created by instruction 1 to 6, this is done by the way of a new instruction. The elements coming from this modification are thus prefixed by the instruction number which generated them.

2.6 Conclusion

We have presented an architecture for web-based CAD system. This architecture responds to problems of reactivity by duplicating a part of the model of the server on the client computer using a language history. When a modification occurs on one of the two models, a description of the modifications, represented by a series of language instructions, is sent. As a language describes the modifications, two different models can be used for client and server. To ensure the consistency between the two models an adequate naming mechanism for entities used in instructions has been described. In this way, we can improve the reactivity of the system by downloading data in an appropriate format according to the use made by the client.

3 Prototype Implementation

A prototype has been implemented, proving the feasibility of our approach. In this section, we describe its features.

3.1 Specifications

As our goal is to provide a web-based CAD system, the later has to be adapted to each possible user: from CAD system expert to neophyte. For somebody that has never used a CAD system, creating geometry is a very difficult task. We are so interested in providing a more intuitive interface than a classical one.

By intuitive interface, we mean interface that does not require the user to think about it. For example, if the user will create a cube, he has not to ask himself where the adequate menu, button or command is. If we want the interface to be intuitive, he has to know naturally what to do. Sketching is a natural method for everyone to show roughly shape specifications. Since sketch modeling has been largely studied (see related work in the section 3.2.), we present in the section 3.3. new sketch modeling tools based on deformation of primitive shapes.

3.2   3D Sketch Modeling

Knowing that 3D design is very important but often not very obvious to be realized by an operator, we are interested in 3D sketch modeling. In this section, we present a brief state-of-the-art in the context of CAD, on the 3D sketch modeling. We show then how to extend this concept of sketching in order to improve CAD systems.

We present here a study based on the complete work in [7]. This study is based on numerous operational systems. All proposed 3D modeling systems are based on sketch in various forms (from interaction techniques like navigation or selection, to input methods or shape approximation). Most of the studied systems are CAD systems which can be designated by lax; namely, CAD systems which let more freedom to the operator during the design stage than in traditional ones. It is the case for example, of 3-Draw [8] which makes possible to draw directly in 3D free forms by capturing specific points and drawing curves. In the same way, [9] defines Holosketch like a 3D drawing (and animation) system, providing an easy to use environment. It is also the case of the following systems: 3D Palette [10], COVIRDS [11], JDCAD [12] and SKETCH [13] which aim to approximate modeling, within the meaning of the size, of an object or a 3D scene by aggregation of primitive shapes. However, this modeling method limits the possibilities of the systems. To compensate this lack, COVIRDS ensures the creation of surfaces, just as the systems DO-IT [14] and Virtual Clay [15] which are based on NURBS deformation using specific input devices.

In a CAD system, the user must usually precisely know his goal as well as the step to catch it (primitive shapes, CSG operator and so on). Some systems propose an opposite approach: the user roughly draws (sketch) his goal and the system proposes a solution. Let us cite the example of 3DSketch [16], QuickSketch [17], Digital Clay [18], TEDDY [19] and SKETCH13. These systems use a gesture-based engine of pattern recognition which proposes a corresponding model (Figure 1).

3.3   Interface Software Tool

In our prototype, we use a shape recognition method. The user draws in two dimensions a sketch of the shape he wants to model. Then he validates (always by sketching, but without noise) his drawing

(Figure 2). The stroke is then sampled as a sequence of points from which the program interprets the type of the shape (Figure 3a).

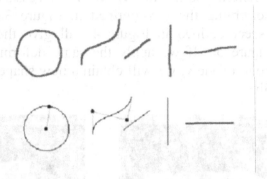

Fig. 1. *Sketch and recognised pattern in the Quicksketch system.*

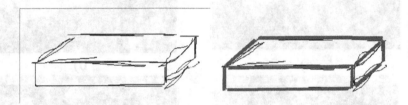

Fig. 2. *Left : the user free sketch, with noise. Right : the user validation sketch.*

With this method, we are limited to a set of primitive shapes (box, sphere, cylinder…). But the user, especially novice CAD system user, wishes to create free form 3D shapes (an expert too, but he can do it by using Boolean operators on primitive shapes), so we cannot limit his creativity with a countable set of forms. We provide him with intuitive deformation modeling tools based on sketch input. Once he has created a shape, the user can deform it, by stretching one of the shape surfaces (surface deformation), or by deforming the shape globally (shape deformation):

- the surface deformation tool lets the user deform a selected surface by redefining the global aspect. For example, the user can introduce notion like "to arch" (Figure 3d: deformation applied on the box defined in Figure 3a) or "to dig" (Figure 3b), or precise himself the future global aspect of the selected surface (Figure 3c);

- the shape deformation tool lets the user deform the entire shape by redefining the global aspect (in several times, from different point of view, if necessary). For example, deforming the box defined in Figure 3a using the global aspect defined in Figure 4 will give the shape viewed in Figure 5a. If we apply the same deformation on a second point of view, we will obtain a new shape, viewed in Figure 5b.

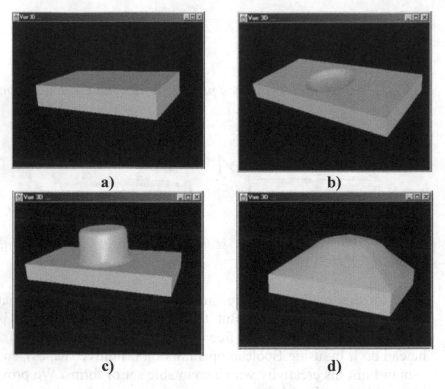

a)

b)

c)

d)

Fig. 3. *a) the recognized box, b) the recognized box after the "to dig" action, c) the recognized box after a global aspect deformation, d) the recognized box after the "to arch" action.*

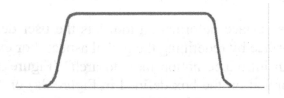

Fig. 4. *Global aspect for shape deformation.*

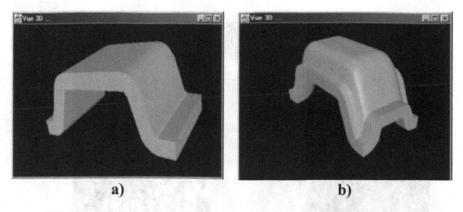

a)   b)

Fig. 5. *Shape deformation with Figure 4 global aspect.*

Many surface deformation methods exist. Some are intuitive, but often require specific or dedicated 3D input devices. 3DSketch uses a repulsive / attractive virtual source, attached to a 3D arm input, to deform shape surfaces. COVIRDS uses data gloves to locate and interpret user's hand position and movement. DO-IT uses as input tool a dedicated deformable object that has the same shape as the virtual object. Our system will be used over the World Wide Web. By the way, the interface software tool has to be adapted to each kind of input and output devices which user could have. Our surface deformation method is adequate to every kind of input device. These two methods bring new and intuitive way of 3D shape modeling.

3.4 Implementation

The acceptance of Java applets is very high among Internet users. A Java applet can be interpreted by any modern web browser. Furthermore, Java provides developers with a powerful 3D-rendering tool named Java3D. Therefore we choose Java as programming tool, for realizing the two-dimensional interface and the client-server communication process, and Java3D for realizing the 3D rendering.

According to these choices, we develop a first prototype, implementing and validating the proposed architecture (see section 2.). This prototype provides the feasibility of our approach. It is based on new dialog techniques outline in the section 3. Some results of complex shape design using this system are shown on figure 6.

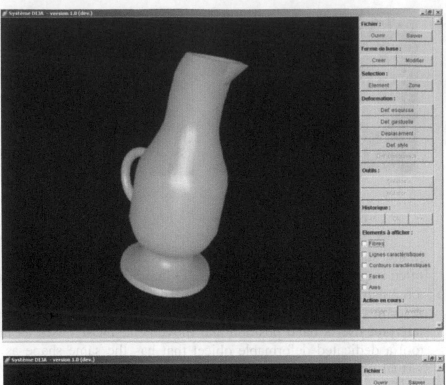

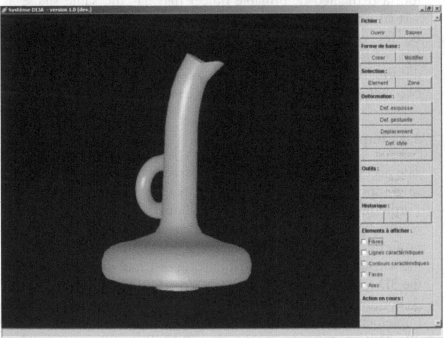

Fig. 6. *3D shapes created with the DIJA Prototype.*

4 Summary

We have presented in this chapter an architecture to implement Web-based CAD software, and some ideas on new approaches for man-machine dialog. The implementation of a prototype based on the proposed architecture and using our interface software tools has shown the feasibility of our approach. Our future work will consist in using the proposed architecture to develop a complete CAD system, including some previous results about function-to-shape translation [20]. Moreover, the ideas about man-machine dialog will be extended in this complete CAD system.

References

1 F. Danesi, C. Dartigues, Y. Gardan and E.Perrin, "World Wide Web Adapted Geometric Model in the Context of Functional Design", 4th International Conference on Design of Information Infrastructure Systems for Manufacturing 2000, 15–17 November 2000, Melbourne (Australia), pp.432-440.

2 Object Management Group, "The Common Object Request Broker: Architecture and Specification", February 2001, http://www.omg.org.

3 Microsoft, "DCOM Technical Overview", January 2000, http://www.microsoft.com.

4 Sun Microsystems Inc., "Java Remote Method Invocation - Distributed Computing For Java", November 1999, http://java.sun.com.

5 B. MacIntyre and S. Feiner, "Distributed 3D graphics library", ACM SIGGRAPH'98, 18-24 July 1998, Orlando, pp. 361-370.

6 R. Gadh and R. Sonthi, "Geometric shape abstractions for Internet-based virtual prototyping", Computer Aided Design vol 30(6), pp 473-486

7 F. Danesi, Y. Gardan, B. Martin and I. Pecci., "La conception 3D par esquisses", Revue internationale de CFAO et d'informatique graphique, vol. 15(1), june 2000, pp. 71-86.

8 E. Sachs, A. Roberts and D. Stoops, "3-Draw: A tool for designing 3D shapes", IEEE Computer Graphics & Applications, vol. 11(11), 1991, pp. 18-26.

9 M. F. Deering, "Holosketch : A Virtual Reality Sketching / Animation Tool", ACM Transaction on Computer-Human Interaction, vol. 2(3), 1995, pp. 220-238.

10 M. Billinghurst, S. Baldis, L. Matheson and M. Philips, "D Palette : A Virtual Reality Content Creation Tool", ACM Virtual Reality Software Technology VRST'97, Lausanne, Switzerland, 1997, pp. 155-156

11 C. Chu, T.H. Dani and R. Gadh, "Multi-sensory user interface for a virtual-reality-based computer-aided design system", Computer-Aided Design, vol. 29(10), 1997, pp 709-725.

12 J. Liang and M. Green, "JDCAD : A Highly Interactive 3D Modeling System", Computer and Graphics, vol. 18(4), 1994, pp. 499-506.

13 R. C. Zeleznik, K.P. Herndon and J.F. Hughes, "Sketch: An interface for Sketching 3D Scenes", Computer Graphics, vol. 30(4), 1996, pp. 163-170.

14 T. Murakami and N. Nakajima, "DO-IT: deformable object as input tool for 3-D geometric operation", Computer-Aided Design, vol. 32(1), 2000, pp. 5-16.

15 K. Kameyama, "Virtual Clay Modeling System", ACM Virtual Reality Software and Technology VRST'97, Lausanne, Switzerland, 1997, pp. 197-200.

16 S. Han and G. Medioni, "3DSketch: Modeling by Digitizing with a Smart 3D Pen", ACM Multimedia'97, 1997, pp. 41-49.

17 L. Eggli, C. Hsu, B. Bruderlin and G. Elber, "Inferring 3D models from freehand sketches and constraints", Computer-Aided Design, vol. 29(2), 1997, pp. 101-112.

18 M. D. Gross and E. Y.-L. Do, "Drawing on the Back of an Envelope: a framework for interacting with application programs by freehand drawing", Computer & Graphics, vol. 24, 2000, pp. 835-849.

19 T. Igarashi, S. Matsuoka and H. Tanaka, "Teddy: A Sketching Interface for 3D Freeform Design", ACM SIGGRAPH'99, 8-13 August 1999, Los Angeles, pp. 409-416.

20 Y. Gardan, C. Minich and D. Pallez, "On shape to specifications adequacy", Proceedings of the IEEE International Conference on Information Visualisation IV'99, 14-16 July 1999, London, England, pp. 315-320.

Chapter 21

Interactive Handling of a Construction Plan in CAD

Interactive Handling of a Construction Plan in CAD

Caroline Essert-Villard and Pascal Mathis
LSIIT, UMR CNRS 7005, Université Louis Pasteur de Strasbourg, Boulevard Sebastien Brant, 67400 Illkirch, France

Solving geometric metric constraints is a topical issue in CAD. An original way to solve a constraint system is to use geometric methods, providing a symbolic construction plan. Then, this plan can be numerically interpreted to generate the required figure. If multiple solutions are produced, most solvers propose to scan the entire space of the solutions found, that is generally tedious. This chapter shows how the inner properties of a symbolic solver allow to deal more efficiently with this case. After briefly recalling our sketch-based selection method, that enables to easily eliminate most of the solutions and to keep the only, or at the worst the few solutions that have the best likeness with the original drawing, we introduce a new step by step interpretation mechanism implemented as a debugger-like tool, that allows to browse the remaining solutions tree in order to help the user choosing the required solution.

1 Introduction

In Computer-aided design (CAD), a geometric object can be precisely described by constraints. They concern distances between points, angles between lines, tangency of circles and lines, etc. Generally, constraints are declaratively placed by a user on a sketch. If we wish to carry out calculations, simulations or manufacturing, the object must really respect the constraints. Thus, a CAD system must be able to solve them and give the possible solution figures. This kind of approach was initiated by I.E. Sutherland [10] with Sketchpad and was then studied by many authors.

Whatever the approach, a constraint system does not usually define a single figure. In the case of a well-constrained system, the exploration of the solutions space is not as easy as it seems. In most cases, CAD users only want one solution figure when they design an object. That is why an important matter of geometric solvers is identifying the solution that is most consistent with the user's expectations, as we can see in [3] and [8]. The most common response to this problem is the use of heuristics to filter the results. When using a numerical method, the constrained figure is compared with each of the numerical solutions. This is generally characterized by slow runtimes, and there is often more than one solution left. Our symbolic approach allows us to take advantage of the construction plan to compare the sketch with a solution, and to define an easy-to-use debugger-like tool if the solution space has even so to be explored.

The rest of the chapter is structured as follows. Section 2 presents the constraint solving framework. Then, Section 3 explains a basic construction plan evaluation. Section 4 shows how the sketch can be used to find a good plan's evaluation and Section 5 how using interactive tools. Finally, Section 6 concludes.

2 Solving with *YAMS*

YAMS (Yet Another Meta Solver) is the prototype resulting from the merging of the 3D topology-based geometric modeller *TOPOFIL*, and a 2D geometric constraint solver. A precise description of the modeller and of this association can be found in [2], so we'll only present here the solver part that supplies the \emph{construction plans} on which we work.

The solver belongs to the family of symbolic solvers. The solving process acts in two steps: first, a symbolic phase that produces a construction plan according to the constraints; then a numerical phase interprets this construction plan. The symbolic stage is obviously the most costly.

2.2 Solving the Constraints

Constraints are predicative terms of the form $P(x_1, ..., x_n)$, where P is a predicative symbol, and x_i are typed identifiers of geometric elements. Then, denoting that the distance between a point *p1* and another point *p2* is a length *k1* can be written *distpp(p1, p2, k1)*. About thirteen different kinds of constraints exist in *YAMS*. Among them, we distinguish metric constraints (such as distances, angles) and Boolean constraints (such as incidence or tangency). Note that this way of writing the constraints is quite usual, and can be found, for instance, in [1,4].

During the symbolic solving, the numerical values of distances and angles are not taken into account, whereas they are given by the user with the rest of the constraints. They only appear in a symbolic way in the constraints under the form of typed identifiers (for instance *k1* in the example above, to represent a length). The numerical values are associated to these identifiers by functional terms, in *definitions* of the form: $x := f(x_1, ..., x_n)$, where x is the defined identifier, f a functional symbol, and x_i the parameters that can be either other identifiers or numerical values. For example, if the user imposes a length to be 100 units from point *p1* to point *p2*, we express it by a constraint *distpp(p1, p2, k1)* and a definition *k1 := initl(100)*, where *initl* initializes *k1* to the value *100*.

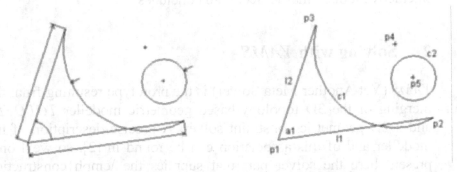

Fig. 1. *A sketch with constraints (left) and identifiers association (right).*

When capturing the data of a problem, these definitions are the first lines of the construction plan (that is a list of definitions), that will be supplemented during the symbolic solving with other definitions. Let's take an example: Fig.1 shows the placing of the constraints on a sketch, and the identifiers association. The symbolic transcription of the constraints and the definitions for this problem are the following:

Constraints	
egal_p(p5, p4)	onl(p1, l1)
centre(c2, p5)	distpp(p1, p2, k2)
centre(c1, p4)	distpp(p1, p3, k1)
radius(c2, k4)	fixorgpl(p1, l1, p2)
radius(c1, k3)	onc(p3, c1)
onc(p2, c1)	onl(p3,l2)
onl(p2, l1)	onl(p1, l2)
angle(p1, p2, p1, p3, a1)	

Definitions	
k4 = initl(200)	k3 = initl(400)
a1 = inita(1.570796)	k2 = initl(300)
k1 = initl(200)	p1 = initp(0,0)
l1 = initd(p1,0)	l2 = lpla(p1,l1,a1)
c3 = mkcir(p1,k2)	p2 = interlc(l1,c3)
c4 = mkcir(p1,k1)	p3 = interlc(l2,c4)
c2 = mkcir(p5,k4)	p5 = centre_of(c1)
c1 = medradcir(p2,p3,k3)	

Our solver gives a geometric answer to this problem, that has the advantage of producing several solutions. The construction plan given above expresses the geometric construction yielded by the solver, and describes, in the right order, the objects to build and the operations to apply so as to obtain a figure.

The numerical interpretation forms the subject of the rest of this chapter. For more details on symbolic solving, see article [5] that explains this part more precisely, notably the original general mechanism of decomposition in subfigures and assembling that *YAMS* uses to solve large systems.

2.3 Construction Plan

In the construction plan, the list of definitions is presented in *triangular solved form*, i.e. an identifier used as parameter in a definition must have been defined earlier in the plan. Note that by switching two definitions in a construction plan, it is possible to obtain an equivalent one, as long as the result is still in triangular solved form.

In a general way, a set of definitions can be structured as a Direct Acyclic Graph (DAG), called *dependence graph*. Its vertices are the definitions, and its oriented edges makes a link from a definition $x = f(x_1,...,x_n)$ to a definition $y = g(y1, ...,x,...,ym)$. A *topological sort* of a DAG gives a list of vertices such that a vertex does not appear in the list before its successors. For a DAG, there generally are several possible topological sorts which, in our case, correspond to the different possible construction plans. Note that all these possible plans provide exactly the same solutions, after a numerical interpretation.

Therefore, even if the solver gives a particular construction plan, we can choose another order for the definitions, taking into account the dependencies, without affecting the solutions.

3 Interpretation

3.2 Tree of Solutions

In this stage, the data given by the user are exploited as parameters for the numerical interpretation of the construction plan.

Each functional symbol is associated with a numerical function. But interpretation of a functional term may provide multiple results. For example, the intersection between two circles, symbolized by *intercc*, generally produces two points, and *medradcir* that builds a circle through two known points, with a known radius, generally produces two different circles. So these are not simple functions, but what we call *multifunction*, i.e. functions that can give more than one result.

The existence of multifunctions in a construction plan introduces choices in the interpretation process. So, we can consider the interpretation as the building of a tree labeled with numerical values. The interpretation of a multifunction that can produce up to k results generates a branching of degree k. By giving a numbering to the various solutions produced by each multifunction, we number the branches of the tree. At the end, the tree represents the solution space, and one solution corresponds to the labels of one branch.

We have to distinguish two kinds of trees :

- the tree of the possible solutions, made by only taking into account the degrees of multifunctions, and whose number of branches is maximum. This one is called *tree of possibilities*
- the tree of the effective solutions, made by interpretation with real values parameters, and that may have less branches than the tree of possibilities. This one is called *tree of solutions*

The difference is caused by several kinds of events that may occur during the interpretation process. A multifunction may provide less results because of particular data (for example if two circles are tangent, the intersection has only one result), or even a "failure" (for example if those circles have no intersection). In this last case, the interpretation stops in the branch.

Note that practically, in our prototype, the tree is not really built but explored by a depth-first backtracking.

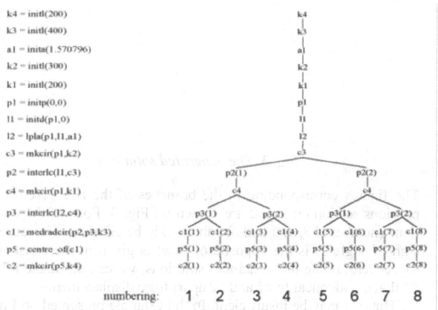

k4 - initl(200)
k3 - initl(400)
a1 - inita(1.570796)
k2 - initl(300)
k1 - initl(200)
p1 - initp(0,0)
l1 - initd(p1,0)
l2 - lpla(p1,l1,a1)
c3 - mkcir(p1,k2)
p2 - interlc(l1,c3)
c4 - mkcir(p1,k1)
p3 - interlc(l2,c4)
c1 - medradcir(p2,p3,k3)
p5 - centre_of(c1)
c2 - mkcir(p5,k4)

numbering: 1 2 3 4 5 6 7 8

Fig. 2. *Construction plan corresponding to Fig.1 and tree of solutions.*

3.3 Problems due to a high number of solutions

Even if the tree of solutions is lighter than the tree of possibilities, the number of solutions can be very important, and increases with the length of the construction plan (that depends on the number of geometric entities of the sketch). That is why, at first, we would like to minimize the size of the tree, in order to speed up the backtracking used to explore the tree.

A first pruning can be done by eliminating what we call the "false solutions". Actually, the computed construction plan enables to construct all the solutions as well as other figures which are not consistent with the constraints, because the geometric solver only uses necessary conditions to make the construction. This can be done with a simple test, by verifying if the constraints are satisfied.

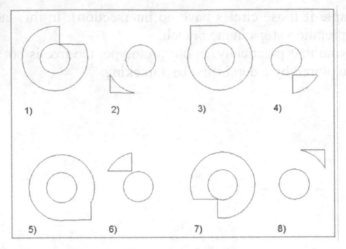

Fig. 3. *The generated solutions*

The figures corresponding to the branches of the tree given in the previous section on Fig. 2 are shown on Fig. 3. Four of these solutions (numbered *3, 4, 5* and *6*) can quickly be eliminated because the sign of angle a_1 is the opposite of what is given in the constraints. Moreover, among the remaining solutions, we can eliminate #7 and #8 that are identical to #1 and #2 apart from displacements.

But that may be insufficient. In the example presented on Fig. 5, there are 32768 different solutions for a geometric object made of 15 equilateral triangles figure, but the solution space can not be reduced because all of the figures are consistent with the constraints. Other heuristics are necessary to drastically prune the tree of solutions, eliminating the figures that does not look like the sketch.

4   Using the Sketch

4.2   Usual Criteria of Likeness

Likeness is generally defined as conformity in appearance between things. Two figures are usually said to look like each other if some geometric properties are similar, such as:
- orientation of points,
- relative placing of objects,
- angles acuteness,

- convexity of some parts of the figure.

This definition is used in most of the CAD frameworks to eliminate inappropriate solutions.

However, most of these criteria can be held in check by some simple examples. For instance, Fig. 4 illustrates that sometimes we are not able to decide between two solutions by only comparing the geometric properties: in this figure, all angles are acute and all points have the same relative placing and orientation.

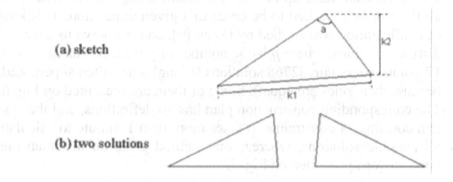

Fig. 4. *Lack of discrimination criterion.*

4.3 Freezing of a Branch

In order to eliminate a maximum number of solutions that do not look like the sketch, we proposed another definition of likeness (see [6]). This definition is based on the notions of geometric homotopy, continuous deformation of a constrained system, and continuous numbering of the solutions.

For each metric multifunction we use in our solver, we described a particular continuous numbering of its distinct results. This continuous numbering allows us to use an original method to find the figure that has the best likeness with the sketch.

We first make an interpretation of the construction plan, using as parameters the data measured on the sketch drawn by the user. This interpretation produces a tree of solutions, among which lies the branch corresponding to the sketch. We memorize the number of this branch. Then, it only remains to make another interpretation, using the user's data as parameters, and to follow the branch which number has been memorized. With the properties we explained before, we are sure that the figure we found has the same geometric characteristics than the sketch, and looks like it in the sense that we defined. We call this process *freezing of a branch*. The branch is selected and its number is kept for further interpretation, with new numerical values for the parameters. The other branches of the tree are not cut, so the other solutions are not lost and can be examined later.

This method gives very good results when all the multifunctions used in the construction plan are metric. As an example, Fig. 5 shows a sketch made up of fifteen adjacent triangles. The lengths of all their sides are asked to be equal to a given dimension. This kind of configuration was studied by Owen [9], and is known to have 2^{p-2} distinct solutions, where p is the number of points. In our case, with 17 points, we obtain 32768 solutions (triangles are often superposed, because their sides are equal). Some of them are presented on Fig. 6. The corresponding construction plan has 80 definitions, and the system contains 94 constraints. It takes more than 1 minute to calculate all possible solutions, whereas our method gives an instantaneous good answer, presented on Fig. 7.

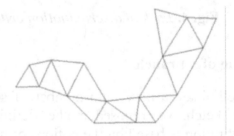

Fig. 5. *15 triangles configuration: the sketch*

However, our method is not appropriate when Boolean constraints (such as tangency or equality between objects) are present in the construction plan. Indeed, it is impossible to compare the solutions with the sketch when some information is missing in the sketch. Actually, unlike metric constraints that don't affect topology, these constraints are generally not respected on the sketch.

Because of these Boolean constraints, some systems have a tree of solutions that can not be reduced to a single branch. Its number of branches can be decreased down to a few branches, but there still remains a little subtree to be explored. It may also happen that the user is not satisfied with what the solver found, whether the sketch he drew was not precise enough, or he did not expect such a solution for the constraints he gave. For all these reasons, the user may want to interactively explore, either the subtree of solutions, or the rest of the entire tree of solutions, and to examine solutions that are close to the frozen branch.

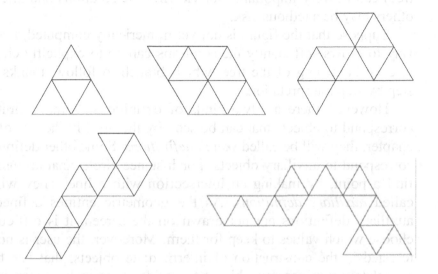

Fig. 6. *Five solutions among 32768 to "15 triangles".*

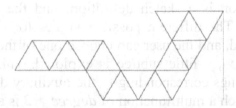

Fig. 7. *The required solution of the sketch given on Fig. 5.*

5   Interactive Solution Refining

All the above reasons led us to propose some functionalities to explore the solution space within *YAMS*. Remember this solution space is not simply a set of figures, but a structured space. The solutions tree and the construction plan structures we use offer us the possibility to define an exploration tool, inspired by *debug tools* provided by most of the development systems in software engineering.

5.2 A Step by Step Interpretation

First, remind that in the case where the user wants to explore the entire solution space, the number of solutions (i.e. of branches in the tree) can be very important. So, viewing the solutions one after the others may be a tedious task.

Suppose that the figure is not yet numerically computed. A good way to browse efficiently the solutions can be to explicitly choose, at each branching of the tree, which branch to follow, thanks to a step by step interpretation.

However, there are two kinds of definitions. Some definitions correspond to objects that can be seen by the user. In the rest of this chapter, they will be called *sketch definitions*. Some other definitions correspond to auxiliary objects. For instance circles that are used to find a point, by making an intersection with a line. They will be called *auxiliary definitions*. As the geometric entities defined by auxiliary definitions are not drawn on the screen, it is difficult to choose which values to keep for them. Moreover, the user is not interested in the construction of intermediate objects, that has to be completely transparent to him. So, an idea is to make a step in the interpretation only at the sketch definitions.

At each step, we work on a *layer* (see Fig. 8). In the layer, the last definition is a sketch definition, and the others are auxiliary definitions. The different possible values for the concerned object are proposed, and the user can choose one of them. It means that for this operation, a little subtree is explored. This subtree contains a few branchings corresponding to the auxiliary definitions within the layer to which a multifunction of *degree > 2* is assigned. So, a backtracking is done into this layer, but this backtracking is hidden from the user, in order to make it transparent. Then, when a interpretation is chosen for the current sketch definition, the corresponding branch is frozen in this layer. See Fig. 8, where the branch that has been frozen so far is in bold, the current studied layer is between dashed lines, and the visible objects in the sketch are framed.

The construction plan may not be provided by the solver in the best form for this operation. It can be necessary to perform a topological sort of the plan before the step by step interpretation.

Indeed, we need to have the following criterion on the construction plan: let *d1* and *d2* be two sketch definitions, *d1* being placed before *d2* in the construction plan, such that no other sketch definition exists between *d1* and *d2*. Then, all definitions between *d1* and *d2*, that are obviously auxiliary definitions, are the remaining definitions that are necessary to compute *d2* and that have not been required before *d1*.

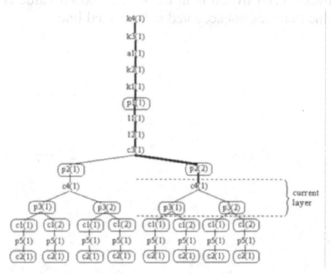

Fig. 8. *Backtracking on a little subtree, included in a layer of the solutions tree.*

In order to obtain such a form, we have to sort the construction plan. The topological sort is made by placing first the sketch definitions following the current order, and then interleaving the auxiliary definitions just before the first sketch definition that needs it (i.e. that contains it as an argument).

When a construction plan verifies the above criterion, the only backtracking to be done is located in the subtree between *d1* and *d2*, excluding *d1*. If the user is not satisfied with the numerical interpretations proposed for *d2*, and wishes to see other possible solutions, then we are sure that some of the sketch definitions have to be thrown back into question.

In such a case, we browse the sketch definitions that have been defined earlier, and on which *d2* depends. We suggest to the user to reconsider some of the values he had chosen for these previous sketch definitions. First, we propose him to review only a few of them, those that are placed closer in the tree. Then, progressively we put into consideration more definitions, including those that were defined a longer time ago.

On Fig. 9, we can see a step by step interpretation of the constrained sketch of Fig. 1. At each step ((a), (b), or (c)), the user chooses one of the two available results. The part of the figure that has already been frozen is in thick, the chosen value is in thin, and the value that was not accepted is in dashed line.

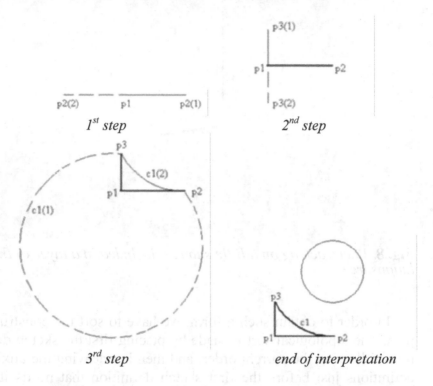

Fig. 9. *Interpretation in 3 steps, and final result*

5.3 Our Debugger-like Tool

The method exposed above is implemented as a module of *YAMS*, named *SAMY* and the user has the choice to use it or not, and to start it when he needs.

Practically talking, we draw the solution step by step as the interpretation goes along. For each new object drawn on the solution figure, the corresponding part of the sketch is highlighted. This way, the user can easily follow the construction process. At each step, *SAMY* proposes a set of possible choices for the current object to be drawn. When the user chooses one, it is constructed on the figure and *SAMY* goes on to the next step.

On Fig. 10, we can see a snapshot of our debbugger-like tool based on this step by step interpretation method, where a rail support is being constructed. The figure represents one of the intermediary steps of the construction. The figure represents one of the intermediary steps of the construction. The dialog box on top left of the figure allows the user to direct the interpretation by choosing a solution for each multifunction with arrows, and then to continue.

Figure 11 presents the final step of the interpretation, with the solution entirely built thanks to the process.

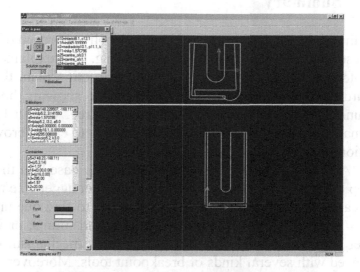

Fig. 10. *Step by step interpretation of a rail support*

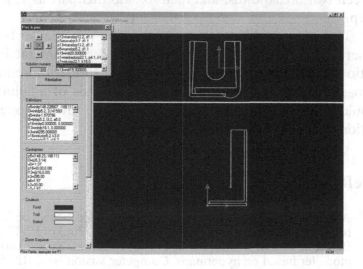

Fig. 11. *Solution of the rail support*

Note that, in order to be easier to use, the step by step process can be combined with the branch freezing. Then, every time a multi-function is reached, the first solution proposed to the user is the one that would have been automatically chosen by the freezing of a branch.

6 Summary

In this chapter, we first exposed our symbolic approach of geometric constructions for CAD constraints solving. We explained that our prototype *YAMS* provides a general construction plan, that is afterwards numerically interpreted. Then, after showing how we can prune the solution space represented by a tree, we put forward the remaining problems that led us to find a way to easily browse the solutions tree.

As a solution, we proposed a tool that is based on the idea of a step by step numerical interpretation. This debugger-like tool is used in case the pruning method did not manage to find one unique solution because of the presence of Boolean constraints, or in case the user is not satisfied with the solution. This mechanism can be enhanced with several kinds of breakpoint tools. Moreover, it is possible to offer the opportunity to freeze a part of a tree of solutions between two breakpoints, and then to skip this part as if it was a big step.

The debugger-like tool we presented in this chapter is the first of a series of exploration tools. Other tools, like the interactive manipulation of a yet computed solution, allow a further more intuitive approach of the selection problem. On the basis of a solution, a user could drag a misplaced element of the figure towards one of the positions allowed by the tree of solutions [7].

References

1. Aldefeld B. (1988) Variations of geometries based on a geometric-reasoning method. Computed-Aided Design. 20(3):117-126.
2. Bertrand Y. and Dufourd J.-F. (1994) Algebraic specification of a 3D-modeller based on hypermaps. Computer Vision – GMIP. 56(1):29-60.
3. Bouma W., Fudos I., Hoffmann C., Cai J. and Paige, R. (1995) Geometric constraint solver. Computer-Aided Design. 27(6): 487-501.

4. Brüderlin. (1988) Automatizing proofs and constructions. Proceedings of Computational Geometry'88. LNCS 333, Springer-Verlag. 232-252.
5. Dufourd J.-F., Mathis P. and Schreck P. (1997) Formal resolution of geometric constraint systems by assembling. Proceedings of the ACM-Siggraph Solid Modelling Conference. ACM Press. 271-284.
6. Essert-Villard C., Schreck P. and Dufourd J.-F. (2000) Sketch-based pruning of a solution space within a formal geometric constraint solver. Artificial Intelligence. Elsevier. 124:139-159.
7. Essert-Villard C. (2002) Helping the designer in solution selection: applications in CAD. Proceedings of ICCS 2002. LNCS 2330, Springer-Verlag. 2:151-160.
8. Lamure H. and Michelucci D. (1995) Solving constraints by homotopy. Proceedings of the ACM-Siggraph Solid Modelling Conference. ACM Press. 134-145.
9. Owen J. (1991) Algebraic solution for geometry from dimensional constraints. Proceedings of the 1st ACM Symposium of Solid Modelling and CAD/CAM Applications. ACM Press. 397-407.
10. Sutherland I.E. (1963) Sketchpad: A man-machine graphical communication system. Proceedings of the IFIP Spring Joint Computer Conference. 329-36.

Chapter 22

Computer Aided Chair Design System Using Form Features

Computer Aided Chair Design System Using Form Features

Mio Matsumoto and Hideki Aoyama
Keio University, Department of System Design Engineering, 3-14-1 Hiyoshi,
Kohoku-ku, Yokohama, Japan

This paper deals with "Computer Aided Chair Design System" based on form features of chair elements that constructs the 3-D product model of a chair from a sketch drawn by a designer. In design processes of a chair, a designer usually wants to express his/her idea of the form and structure of a chair by drawing sketches because sketching is one of the most desired manners for expressing designer's idea. Besides, CAD data of a chair are essential in its manufacturing processes. In this study, a system is proposed to construct the 3-D CAD data of a chair by arranging form features of chair elements with fitting them on a sketch. The system enables designers to easily make the 3-D chair models from their idea sketches and to evaluate the design-form by figures from the various visual angles.

1   Introduction

Sketching is the most convenient and easiest way to express designer's idea and thinking on the product shape because it can be executed by only a pen and paper and does not prevent a designer elaborating his/her idea. Sketches are then used to express the product shape in the initial stage of design processes. The same technique is applied for designing chairs [1]. After drawing sketches, a physical model such as a clay model, a wooden model, an expanded polystyrene model, and so on is created according to the sketches and the computer model is constructed from measurement data of scattered points on the physical model [2]. This process is called "Reverse Engineering" [3] as shown in Fig. 1(a), and the same process is also applied in designing and manufacturing processes of a chair.

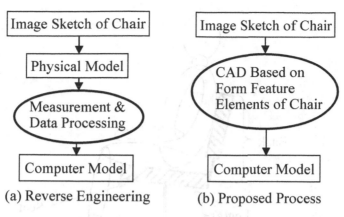

Fig. 1. *Chair design process*

Much time and a huge cost must be needed for creating a physical model in the process of "Reverse Engineering". However, in spite of investment of time and cost, the physical model falls into disuse after constructing the computer model. Therefore, a method to construct a computer model without creating a physical model is desired in order to avoid such a wasteful process in design.

In order to construct a 3-D model without a physical model, a chair design system [4] was developed. In the system, a chair model is constructed by combining components with very simple forms so that the form variations are restricted.

Therefore, a new process for chair design shown in Fig. 1(b) is proposed in this paper. In the proposed process, the computer model of a chair can be easily constructed from an idea sketch without creating the physical model by a chair CAD system developed in this study. The peculiarity of the CAD system is to directly make a chair model by combining the elements constructed by form features according to a sketch. The shapes of form features can be easily changed by easy operations using a computer mouse. The developed CAD system is described in the following.

2 Proposed Chair Design Process

As shown in Fig. 1(b), the proposed chair design processes start with drawing a sketch of a chair. The CAD model is constructed from a sketch in the following procedures.

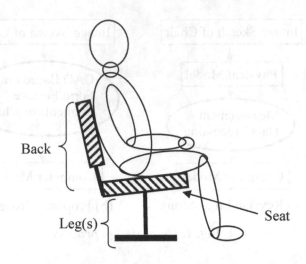

Fig. 2. *Office chair elements*

(a) A sketch is fed into the developed chair CAD system by a scanner and the sketch data is displayed on a TV screen.
(b) The visual point of the sketch is automatically identified from the sketch data.
(c) The most similar form feature is selected for a chair element drawn on the sketch and is arranged at the position accurately fitted to the sketch with adjusting the shape by the parameters defining the form feature.

When all chair elements drawn on the sketch are arranged on the TV screen, the 3-D model is completely constructed.

3 Form Features of Chair Elements

The objects of the chair CAD system developed in this study are office chairs. As shown in Fig. 2, an office chair basically has the following elements[5]:

- seat,
- back,
- leg(s), and
- armrest(s) (as optional element).

Figures 3, 4, and 5 show the parameters to define the forms of the seat feature, back feature, and leg feature. The parameters are easily changed by computer mouse operations so that the forms of the seat

434

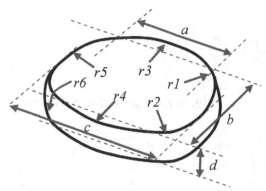

Fig. 3. *Parameters defining seat feature*

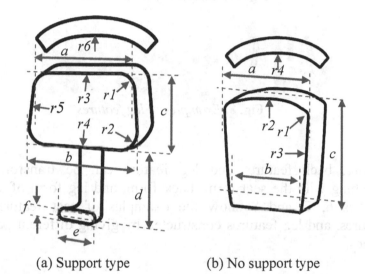

(a) Support type   (b) No support type

Fig. 4. *Parameters defining back feature*

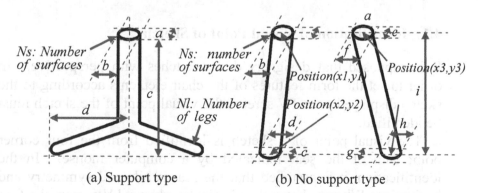

Ns: Number of surfaces

Ns: number of surfaces

Nl: Number of legs

Position(x1,y1)

Position(x2,y2)

Position(x3,y3)

(a) Support type   (b) No support type

Fig. 5. *Parameters defining leg feature*

Fig. 6. *Examples of seat features*

Fig. 7. *Examples of back features*

Fig. 8. *Examples of leg features*

feature, back feature, and leg feature can be transformed for matching with the seat form, back form, and leg form of a sketch. Figures 6, 7, and 8 show the examples of seat features, back features, and leg features constructed by giving different parameter values.

4 Procedure to Construct 3-D Chair Model

4.1 Identification of Visual Point of Sketch

It is supposed that designers draw sketches with perspective. In order to set the form features of the chair elements according to the sketch displayed on a TV screen, the visual point of the sketch must be identified.

The visual point of a sketch is identified from the four corner point data of the seat indicated by a computer mouse. In the identification, it is assumed that the seat is bilateral symmetry and horizontal. When a sketch data is displayed on a TV screen, the four corner points A, B, C, and D of the seat are indicated by mouse

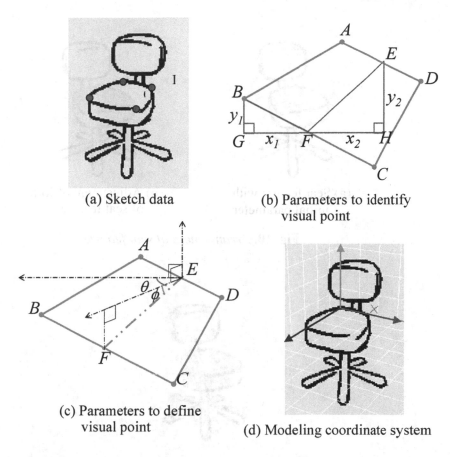

(a) Sketch data

(b) Parameters to identify visual point

(c) Parameters to define visual point

(d) Modeling coordinate system

Figure 9. Identification of sketch visual point

clipping operations as shown in Fig. 9(a). In Fig. 9(b), the points E and F are the center points between the points A and D and the points B and C, respectively. The line G-H is the horizontal line on the point F, and the angles B-G-F and E-H-F are 90 degrees. x_1 and x_2 are the length between the points F and G and the points F and H, respectively. y_1 and y_2 are the length between the points B and G and the points E and H, respectively. Consequently, the angle θ of the visual point for the front of a chair and the angle ϕ of the visual point for the horizontal are derived from equations (1), and (2).

$$\theta = \arctan\sqrt{\frac{x_1 \times y_2}{y_1 \times x_2}} \qquad (1)$$

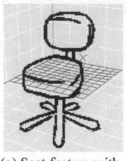
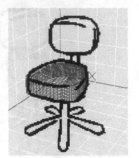

(a) Seat feature with
initial parameter

(b) Adjustment of form
of seat feature

Fig. 10. *Arrangement of seat feature*

Fig. 11. *Arrangement of features*

$$\phi = \arcsin\sqrt{\frac{y_1 \times y_2}{x_1 \times x_2}} \qquad (2)$$

Identifying the visual point: the angles θ and ϕ makes the coordinate system for chair modeling as shown in Fig. 9(c).

4.2 Arrangement of form features and adjustment of element forms

4.2.1. Arrangement of seat feature. When an operation of seat feature arrangement is executed, a seat defined with initial parameter values is arranged as shown in Fig. 10(a). Then the parameter values defining the seat feature are adjusted by easy mouse

438

(a) Wire frame model   (b) Solid model

Fig. 12. *Constructed 3-D model from sketch*

operations in comparing the form of the seat feature with the seat form of the sketch. Consequently, the 3-D seat element matching

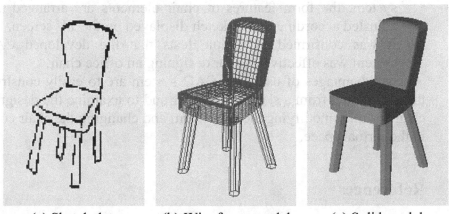

(a) Sketch data   (b) Wire frame model   (c) Solid model

Fig. 13. *3-D model construction from sketch*

the sketch seat is constructed as shown in Fig. 10(b).

4.2.2. Arrangement of back feature and leg feature. In the same manner as seat feature arrangement, the back feature and leg feature are arranged and their forms are adjusted to the sketch. Figure 11 shows the result of arrangements of the adjusted seat feature, back feature, and leg feature on the sketch.

4.2.3. Construction of 3-D chair model. Figure 12 (a) and (b) show the wire frame model and solid model of the final result of the 3-D model constructed from the sketch shown in Fig. 9(a).

Another test to construct the 3-D chair model from a sketch was executed. Figure 13(a) shows the sketch of the test. And Figure 13 (b) and (c) show the wire frame model and solid model, respectively, constructed from the sketch.

5 Summary

This study is summarized as follows:
(1) Design processes of an office chair starting from drawing a sketch was proposed.
(2) A CAD system using chair element features to easily construct the 3-D model of a chair from a sketch was developed.
(3) In the construction of the 3-D model using the developed CAD system, the form features of chair elements are arranged and adjusted according to the sketch displayed on the TV screen.
(4) It was confirmed by simple tests that the developed CAD system was effectiveness for designing an office chair.

The advantages of using the CAD system are to easily construct the 3-D model from a sketch of a chair and to examine the designed chair form by modifying the chair form and changing the chair color in the virtual space.

References

1. Charlotte, Peter Fiell, *1000 chairs*, TASCHEN, 1997, p. 175, 193, 226, 252, 256, 283, 325, 364, 421, 424, 555.
2. W. F. Gaughran, "Cognitive Modeling in Engineering Design", *10th International Conference on Flexible Automation and Intelligent Manufacturing*, 2 (2000), 1136.
3. T. Varady, R. R. Martin, J. Coxt, "Reverse engineering of geometric models –an introduction", *Computer-Aided Design*, Vol. 29, No. 4, 1997, 255.
4. P. Yunhe, G. Weidong, T. Xin, "An intelligent multi-black board CAD system", *Artificial Intelligent in Engineering*, 10, 1996, 351.
5. T. Jindo, K. Hirasago, M. Hagamachi, "Development a design support system for office chairs using 3-D graphics", *International Journal of Industrial Ergonomics*, 15(1995), 49.

Section on

Colored Pictures

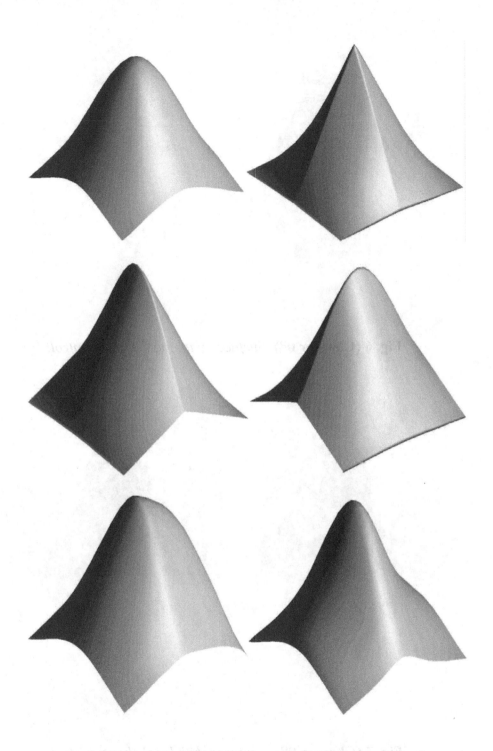

Fig. 5 (Chapter 08). *Surfaces with local shape control.*

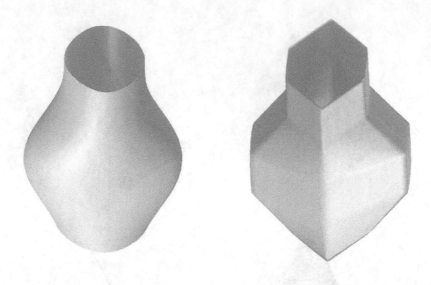

Fig. 6 (Chapter 08). *Surfaces with local shape control.*

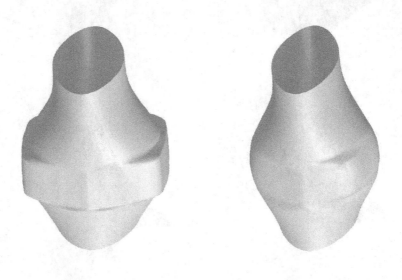

Fig. 7 (Chapter 08). *Surfaces with local shape control.*

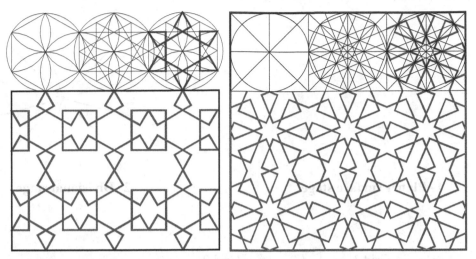

Fig. 3 (Chapter 13). *Hexagonal and octagonal patterns by Issam El-Said in "Islamic Art and Architecture, The System of Geometric Design" [9].*

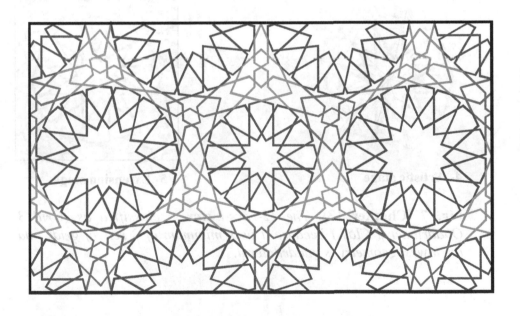

Fig. 6 (Chapter 13). *Pattern containing twelve (red) and fifteen (blue) rayed Star/Rosette by E. Hanbury, Mathematical Gazette "Some Difficult Saracenic Designs" [12].*

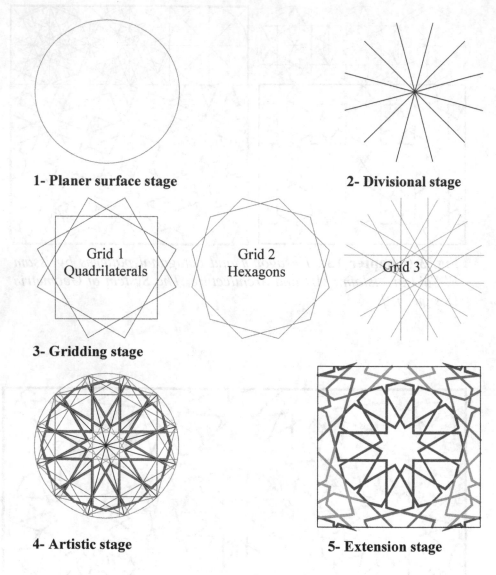

1- Planer surface stage

2- Divisional stage

Grid 1
Quadrilaterals

Grid 2
Hexagons

Grid 3

3- Gridding stage

4- Artistic stage

5- Extension stage

Fig. 7 (Chapter 13). *The 12-rayed star is classified as (Grid 3 Quadrilateral Class) because it uses minimum of 3 sets of grids and the lowest geometry is quadrilateral.*

446

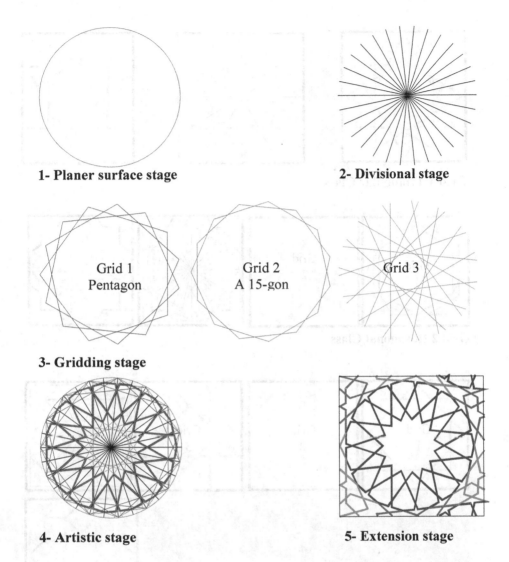

1- Planer surface stage

2- Divisional stage

Grid 1
Pentagon

Grid 2
A 15-gon

Grid 3

3- Gridding stage

4- Artistic stage

5- Extension stage

Fig. 8 (Chapter 13). *The 15-rayed star is classified as (Grid 3 Pentagonal Class) because it uses minimum of 3 sets of grids and the lowest geometry is a pentagon.*

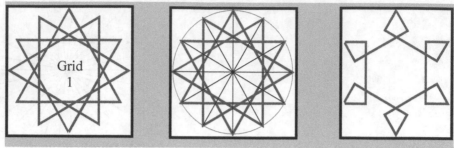

Grid 1 Triangular Class

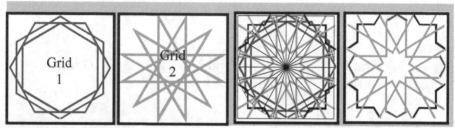

Grid 2 Hexagonal Class

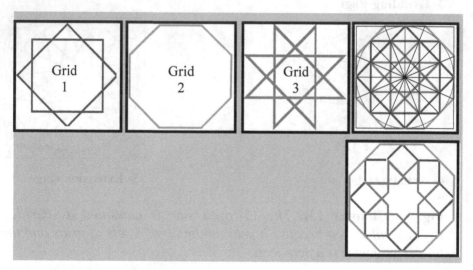

Grid 3 Quadrilateral Class

Fig. 9 (Chapter 13). *Showing the classification of some Islamic Geometric Star/Rosette.*

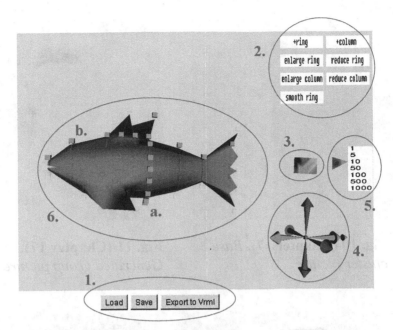

Fig. 1 (Chapter 14). *Annotated screenshot of the system.*

Fig. 3 (Chapter 14). *Example models.*

449

Fig. 10 (Chapter 17). *Base cluster picture*

Fig. 11 (Chapter 17). *Generated group picture*

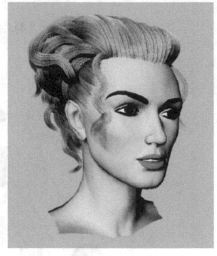

Fig. 19 (Chapter 17). *A wavy hair style*

Fig. 20 (Chapter 17). *A hair style*

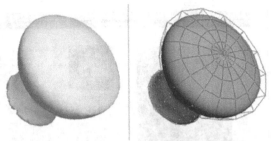

Fig. 4. (Chapter 18). *(L to R) An archaeological vessel and its surface model*

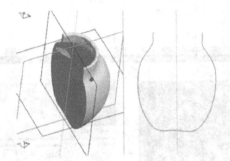

Fig. 5. (Chapter 18). *(L to R) An archaeological vessel and its profile curve*

Fig.6. (Chapter 18). *A vessel, its profile curve and signed curvature plot of the curve.*

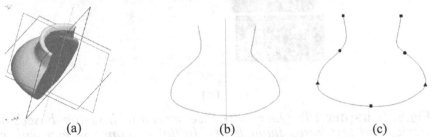

       (a)   (b)   (c)

Fig.7 (Chapter 18). *Profile curve generating and feature point extracting.*

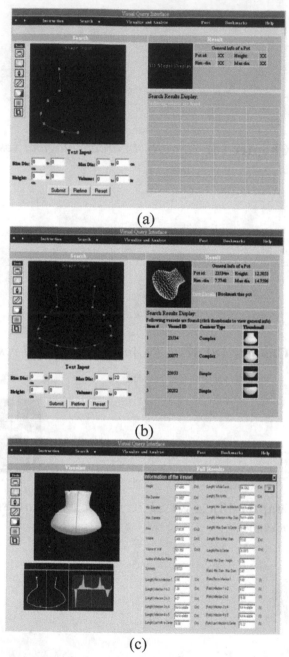

(a)

(b)

(c)

Fig.8. (Chapter 18). *Query interface screen with sketch–based, numeric, and text-based input fields. Initial response screen with resulting thumbnail images and summary data(a), and wire frame of first matching vessel(b). Detail individual display screen with 2D, 3D, and descriptive vessel data(c).*

Index of Authors